Control Systems in Textile Machines

Control Systems in Textile Machines

G. Nagarajan
Dr. G. Ramakrishnan

CRC Press is an imprint of the
Taylor & Francis Group, an **informa** business

WOODHEAD PUBLISHING INDIA PVT LTD

New Delhi

Published by Woodhead Publishing India Pvt. Ltd.
Woodhead Publishing India Pvt. Ltd.,
303, Vardaan House, 7/28, Ansari Road,
Daryaganj, New Delhi - 110002, India
www.woodheadpublishingindia.com

First published 2018, Woodhead Publishing India Pvt. Ltd.

Woodhead Publishing India Pvt. Ltd. ISBN: 978-93-85059-30-8
Woodhead Publishing India Pvt. Ltd. e-ISBN: 978-93-85059-80-3

Typeset by Allen Smalley, Chennai

Contents

Author Details

G NAGARAJAN

Ganapathy Nagarajan has around 23 years of experience in the spinning mills of repute in North India in production, maintenance, and in Quality Assurance department as General Manager (Technical) in Bangladesh. He has worked in The South India Textile Research Organization (SITRA) in the spinning division for 5 years. At present, he is associated with Aksum University Axum, Ethiopia as Asst. Professor in Textile Engineering Dept.

G RAMAKRISHNAN

Professor, Department of Fashion Technology
Coordinator-KCT-TIFAC CORE Kumaraguru College of Technology
Coimbatore-641049, Tamil Nadu
Email: ramakrishnan.g.core@kct.ac.in, g.ramki15@gmail.com
Mobile:9842313921

Presently working as a **Professor** in Department of Fashion Technology. Also Coordinator of KCT-TIFAC CORE, Research center of **Kumaraguru College of Technology, Coimbatore**. I have a total of 30 years of experience that **includes 12 years in industry and 18 years in academics.**
Recipient of **GOLD MEDAL** for securing **FIRST RANK** in Anna University PG DegreeExaminations in the year 2004. Completed my doctoral degree (Ph D in Textile Technology) from Anna University, Chennai during 2010.

Publications-Journals

Published **40 papers** in International peer reviewed journals like Journal of textile institute,Journal of Industrial Textiles, International Journal of thermo physics, Journal of IndustrialTextiles, Journal of thermal analysis and Calorimetry, International Journal of Fashion Designetc and **25 papers** in National Journals.

Conference Presentations

Presented more than **50 papers** in National/International conferences.

Books/Monographs

Co-authored an International monograph TEXTILE PROGRESS published by the TEXTILE INSTITUE, UK. Recently I have coauthored a book titled "CONTROL SYSTEM IN TEXTILE MACHINES" to be published by Wood head publishers.

Programmes Organized

Organized **2 International Textile conferences and 10 National conferences**. Organized more than 60 Seminars/FDP/Workshops

Centre of Excellence

Coordinated to set up the prestigious **376 Lakhs worth TIFAC CENTER OF RELEVANCE AND EXCELLENCE project** in Textile Technology and Machinery, the first of its kind in India, funded by Department of Science and Technology, Government of India in KCT campus

Funded Research Projects

- Principal investigator of DST-SEED project titled received approval for DST-SEED project "Enhancing Livelihood of Handloom Weavers through Technology up gradation and Training in Samathur Block, Pollachi in Tamilnadu" worth 15,65,200/
- Co-investigator of a DST-WTI sponsored research project titled on "Complete removal of recalcitrant synthetic dyes from textile wash liquor using organic solvents in multi stage centrifugal extractors" .The sanctioned fund for the project is Rs 24.96 lakhs and successfully completed.
- Co-investigator of research project titled "Design and Fabrication of Computerized Dynamic Knittability Tester for Hosiery Yarns" sponsored by Department of Science and Technology, Government of India for funding under Instrument Development Programme for a total cost of Rs.25.93 Lacs.
- Co-investigator of Management funded project titled "Greener approach of natural dyeing towards sustainable Development for handloom weaver's society" has been approved under Management Funded Projects Phase II for a sum of Rs 1.04 lakhs.
- Mentor for the student research project titled "SUSTAINABLE FASHION" and received Rs 69,800/ from Research Cell of KCT

PhD Scholars Guided

Presently approved supervisor to guide research scholars of Anna University, Chennai and so far **5 research** scholars have completed their research and **3 scholars** are pursuing Ph.D under my guidance.

Reviewer of International Journals

Reviewer of leading International journals for textile research such a Textile Research journal, Journal of Industrial Textiles, Indian Journal of Fibre and Textile research etc and reviewed more than **75 research papers.**

Industrial Projects/Consultancy

Carried out more than 100 project/consultancy work to various textile/apparel industries and earned revenue.

Awards Won

- Received OUTSTANDING FACULTY AWARD (in the category of Engineering / Textile) from Dr A Kalanidhi, Former Vice Chancellor AnnaUniversity,Chennai on 5th July 2015 at the Radha Regent Hotels, Chennai.
- Received Dr RADHAKRISHNAN AWARD for the year 2015-2016 on the eve of "TEACHERS DAY" celebrations held at Kumaraguru College of Technology on 03.09.2016.
- Received Recognition for paper publication in SCI journals and Research Grant recognition from Chairman, KCT on 22.10.2016
- Received appreciation from Institution of Engineers (India) for evaluating research papers for peer reviewed journals.

Foreword

Textile processes and operations are not only varied but also complex. The handling of delicate fibrous materials further add to the complexity of textile operations. In recent years, textile machines have undergone radical changes in terms of design, productivity and automation. The new generation textile machines have automatic controls which ensure better product quality and higher productivity. Though 'Control System' as a subject is taught in Electrical Engineering program, it is essential for textile engineers to have some fundamental understanding of machine control. This book provides the basic as well as advanced knowledge of "Control Systems for Textile Machines". Basic concepts underlying the electronic controls have been dealt with by the authors in a simple and comprehensive manner considering the requirements of undergraduate and postgraduate students. Besides the academic requirements, this book will also be valuable for budding textile engineers. The authors have meticulously planned the chapters and have covered every relevant topic in a very systematic way. Sufficient examples have been given wherever necessary with clear diagrammatic illustrations. Both the authors have worked in the textile industry and are presently teaching the textile and apparel students in leading engineering institution. The amalgamation of experiences of industry and academia has helped the authors to write the book in a well-defined manner.

I wish the book is well received by the Textile academia and Industry.

Dr. Abhijit Majumdar
Associate Professor
Department of Textile Technology
Indian Institute of Technology, Delhi

Foreword

The book "CONTROL SYSTEMS IN TEXTILE MACHINES" authored by Ganapathy Nagarajan and Dr. G. Ramakrishnan is an authentic and in depth documentation of Mechanics of Textiles machines, which has not been attempted earlier by any textile expert. It is well recognised that any textile machine is the outcome of joint and balanced venture among mechanical, electrical and electronic engineering keeping the textile requirements in view. Even the basic fundamentals of textile machines are handled by mechanical engineers though the simple construction and working through synchronization of different parts is handled by textile people. From that point, this book is an example to take over the lead role in own hands.

In the starting chapters, conceptual information has been detailed such as transducers, potentiometers, strain gauges etc with practical equipment such as tachometer, stroboscope and so on. Though authors have put more weightage towards machines used in various stages in spinning and dedicated several chapters on that, adequate coverage has been made on control systems used in machines in testing, preparatory, weaving etc.

In textiles institutes, the fundamental aspects of textile machines is mostly taught by faculty from mechanical engineering, which invariably remains unconvinced to number of textile students. The information available in the book will enable textile faculty to take up the subject with better translational delivery to students.

Author's long experience in various capacities in textile mills, institutes and research organisations has been reflected in their write up. It is expected that the book will do reasonably well among textile fraternity and will enable them to learn basic concepts used in various textile machines ultimately leading to hassle free and effectively handling.

I wish a grand success of the book

Dr. J N Chakraborty
Professor
Department of Textile Technology
Dr. B R Ambedkar National Institute of Technology

1
Basic concepts – units and standards

1.1 Introduction

Measurement science is an important aspect in everyday life. In the present advanced technological world, everything requires to be measured before useful for practical life. Imagine any product which is made by human beings. It involves measurement at any stage of manufacture. In general, the operations of machineries have to be controlled either manually or automatically. Hence, measuring the concerned variable is the primary requirement for any control systems.

Numerous examples can be quoted in textile machineries. For example, before the invention of High Volume Instruments in textile testing laboratory, professionals and scientists used to measure the staple length of cotton fibre by hand stapling method. This requires skill to arrive at staple length of the cotton fibres which is the prime factor for determining spinnability. However, with the development in electronics and machine design, the fibre length can be measured accurately with the advanced equipments.

Examples like measuring of beater speed, pulley speed in Blow room lines, measuring of cylinder speed, Licker in speed, Doffer speed in carding, spindle speeds in Speed frames by stroboscope non-contact type instruments are very much required for running the textile mills efficiently and effectively.

1.2 Instrumentation

Instrumentation is one of the branches of engineering science which deals with the techniques involved in measurement, measuring devices used and the problems related to them. Most of the present day textile testing equipments works on the instrumentation. Examples like HVI, AFIS etc., utilise the concepts of instrumentation.

In general, measurement means to know about the physical quantities such as length, weight, temperature, pressure, force and so on. This physical quantity is called as measured variable which is the basic objective of the instrument.

Measurement is the outcome of a product in which the professionals normally set some standards which is acceptable by all concerned. Based on the output of the measured variable, engineers compare this with standard and takes decision to go for some corrections or not.

Examples like determining the hank or fineness of carding slivers or draw frame slivers, the mill managers decide whether the auto levelling equipment is working properly or not or whether it requires any calibration procedure.

1.3 Measurement system

Measurement is a process of comparing the input signal (unknown magnitude) with a pre-defined standard set by the engineers and comparing the measured variable with the standard and finally giving out the output signal or values. If the measured variable differs more than the standard values, necessary corrections can be made before giving the final desired output. This is shown in the Figure 1.1.

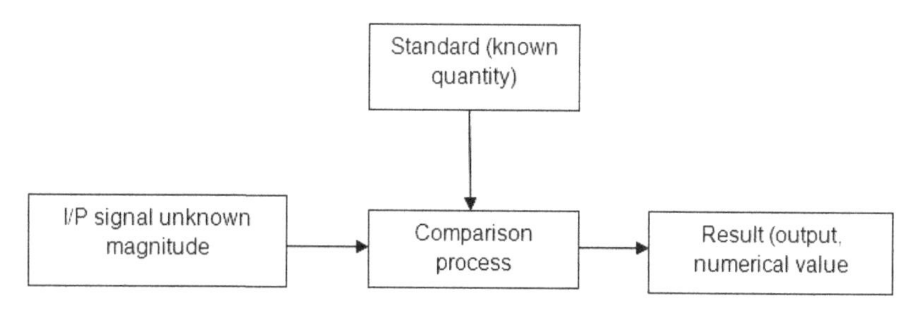

Figure 1.1 Measurement system

The result obtained should meet the following requirements:

a) The standard used for comparison should have common acceptability.
b) The procedure and the equipment used should be properly calibrated.

1.4 General concepts of measurement

The general concepts of measurements have been discussed in the following sections:

a) Measurement methods.
b) Generalised measurement system and its elements.
c) Three stages of generalised measurement system.
d) Applications of measuring systems/equipments.

1.4.1 Measurement methods

The methods of measurement have been classified as:

a) Direct comparison method.

b) Indirect comparison method.

a) Direct comparison method

Direct comparison methods can be used for measuring physical quantities like time, mass, length, etc. In this method, the measured quantity which is unknown is directly compared with the standard. The output result is a number and a unit.

Examples: Production of carding machines, draw frame, production of speed frames, ring frames, meters of warping material, production per loom in meters, etc.

b) Indirect comparison methods

Indirect comparison methods, human beings cannot make direct comparison with accuracy. Hence, in many applications indirect comparison methods are used.

Indirect comparison methods mean the use of a measurement system. These measurement systems use a transducer element. Transducer is one which converts the quantity to be measured from one form to another form (analogous signal) without changing the basic information content. The analogous signal is then processed and is sent to the end devices, which presents the result of measurement. In other words, in indirect comparison method, the input signal is converted to some other form and then it is compared with the standard reference values.

The methods of measurement are classified in to three types.

1. Primary measurement
2. Secondary measurement
3. Tertiary measurement

1. Primary measurement

In the primary measurement method, only subjective information is involved. These measurements are made by direct observation and they do not involve any translation of measurement.

Examples:
One vessel is bigger than the other.
One wire is longer than the other.

2. Secondary measurement

   In this secondary measurement, the output result is obtained by translation.

   This is shown in Figure 1.2.

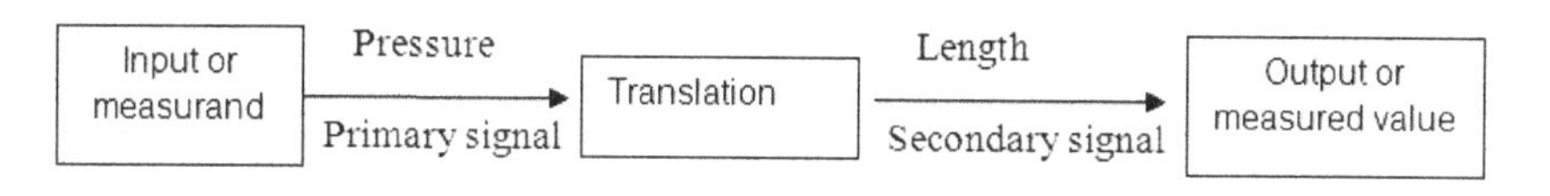

Figure 1.2 Secondary measurement

Conversion of measured variable or measurand in to length as shown in Figure 1.3 by means of Bellows, spring loaded with weight, Bourdon's pressure gauge.

3. Tertiary measurement

   In this tertiary measurement, the output result is obtained by two translations.

This is shown in Figure 1.3.

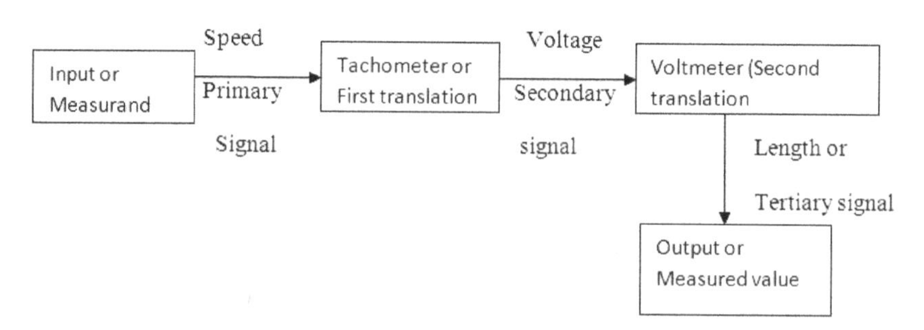

Figure 1.3 Tertiary measurement

Example: Electric Tachometer.
The input result (measurement) is converted in to voltage which is first translation. Then this voltage is converted in to length which is second translation.

1.4.2 Generalised measurement system and its elements

For any instrument the main functions are

a) To get information
b) Process the information
c) Presentation of data

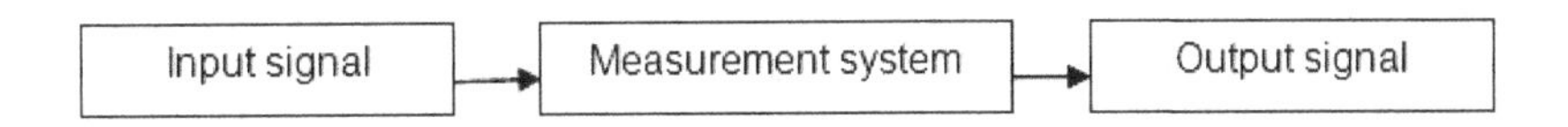

Any instrument is considered as a system. A system is defined as an assembly of components which are interconnected to do a specific function. Each component is called an element and each element performs a particular function during a measurement. The common elements of a generalised measurement system are listed in Table 1.1.

Table 1.1 Measurement system

Elements	Stages
a. Primary sensing element b. Variable convertor or transducer element	Detecting by transducer
c. Variable manipulation element d. Data transmission element e. Data processing element	Intermediate modifying stage
f. Data presentation element g. Data storage and play back element	Terminating stage

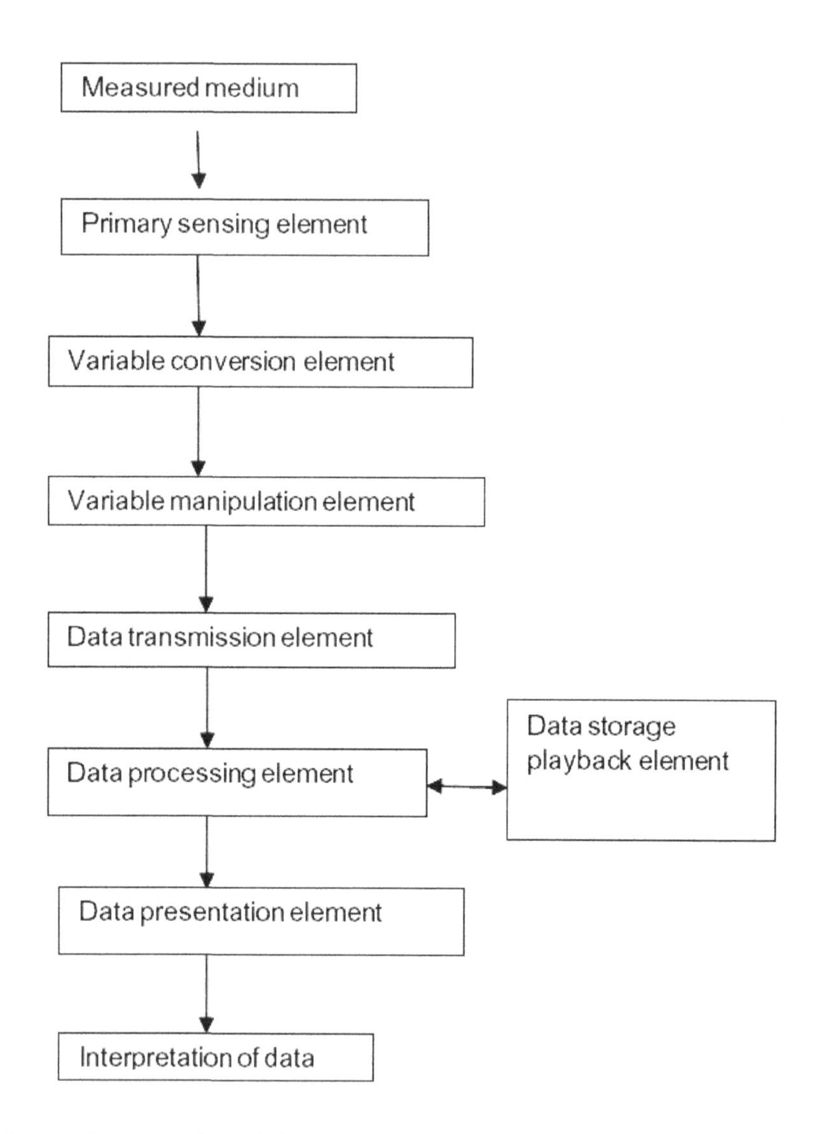

Now each process is explained below:

a. Primary sensing element

   The primary sensing element is the first element in the measurement system. This element utilises the energy from the measured medium and produces an output based upon the measured quantity or variable.

b. Variable conversion or Transducer element

   The primary conversion element gives the output which is some physical form like voltage or displacement. The transducer converts the signal from one physical form to another without changing the basic information content of the signal.

c. Variable manipulation element
The purpose of the variable manipulation element is to amplify the input signal in to desired output.

Example: Displacement amplifier

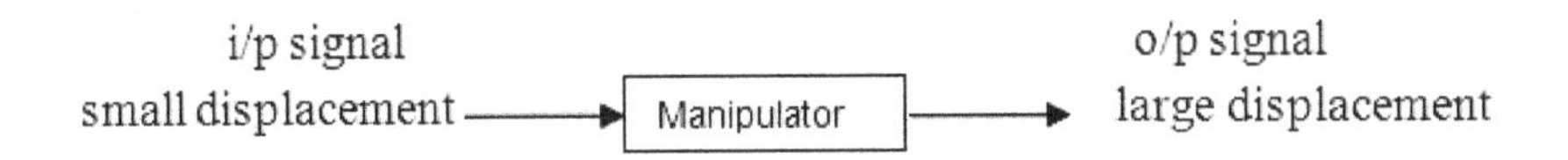

Hence, output signal = input signal × constant.

a) Data transmission element
b) Data transmission element transfers the signal from one place to another without causing any disturbance to the signal being transmitted. Example: From the shaft of the machine to gear.
c) Data processing element
d) The prime work of this element is to convert the data in to an acceptable or understandable form and alters the data before it is presented on the display.
e) Data presentation element
f) The information about the measured quantity is to be presented to the observer for monitoring, controlling or analysis purposes.
g) Data storage or play back element

   This element stores the data in computer or magnetic tape recorder and retrieves the information whenever required by the observer or engineers.

1.4.3 Three stages of a generalized measurement system

The three stages of a generalised measurement system is classified in to

a. Basic detector – Transducer stage
b. Intermediate modifying stage
c. Terminating stage or recorder /Integrator

This is depicted in the block diagram (Figure 1.4).

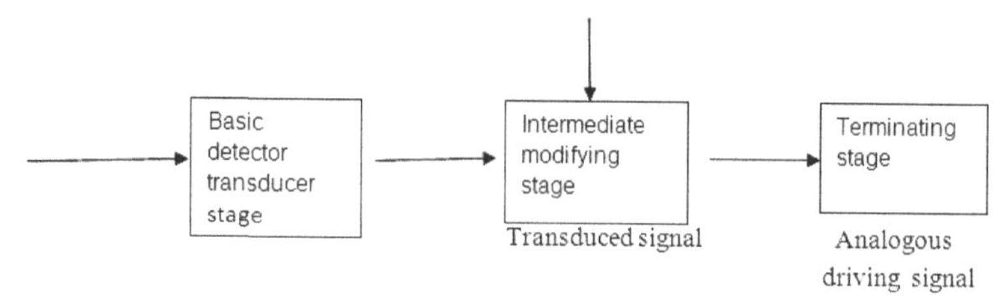

Figure 1.4 Generalised measurement system

1.4.4 Applications of measuring system instruments

Measuring system instruments are useful in the following ways:

a) Monitoring of processes and operations
b) Control of processes and operations
c) Experimental engineering analysis

a. Monitoring of processes and operations
Any measuring instrument has to monitor the operations and control the process so as to maintain the performance of the machinery.

Examples: Pressure transducer, Tachometer.

Pressure transducer employed in modern Blow room lines have to monitor the pressure inside the Multi mixer or Unimix. The function of this pressure transducer is that when there is no material inside the multimixer, the pressure inside the chamber is equal to atmospheric pressure. When the flow of material starts the multimixer will be gradually filled up with material. The pressure transducer monitors the pressure build up inside the chamber and cuts off the material build up when the required material is filled up in the chamber. This helps to prevent the overflow of material and also jamming of machinery parts.

Tachometer

Tachometer is a measuring instrument used to check the speeds of different parts of the machine. This is an easy method to check the speeds at any time and have a control over the process.

b. Control of processes and operations
Automatic control system of a measuring instrument is shown in the block diagram. (Figure 1.5)

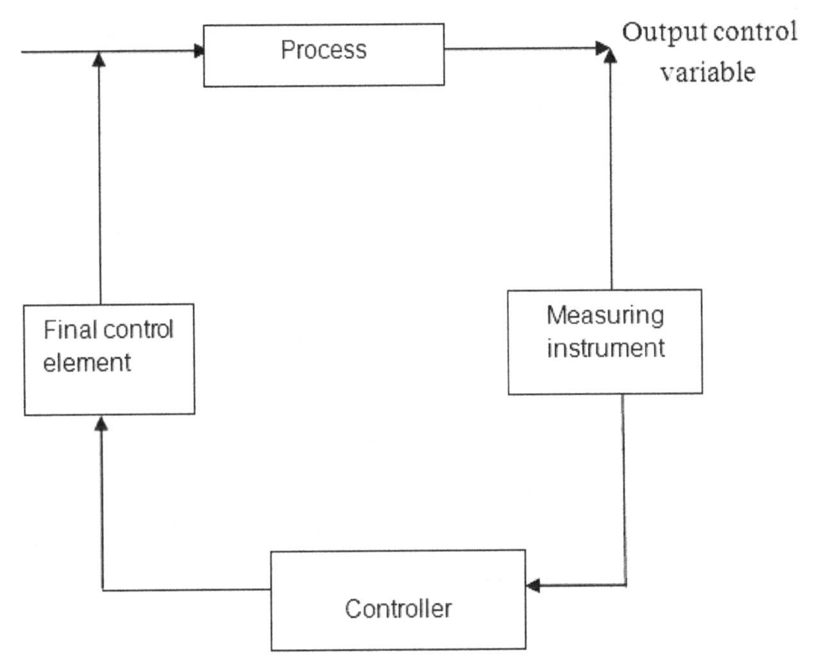

Figure 1.5 Automatic control system

c. Experimental engineering analysis
Basically there are two methods to solve engineering problems namely theoretical and experimental methods. In general, these two methods have to be adopted to solve engineering problems.

1.5 Programmable logic controllers (PLC)

A Programmable Logic Controller (PLC) is an industrial computer control system that continuously monitors the state of input devices and makes decisions based upon a custom program to control the state of output devices. Almost any production line, machine function, or process can be greatly enhanced using this type of control system. However, the biggest benefit in using a PLC is the ability to change and replicate the operation or process while collecting and communicating vital information.

Another advantage of a PLC system is that it is modular. That is, you can mix and match the types of Input and Output devices to best suit your application.

PLC structure

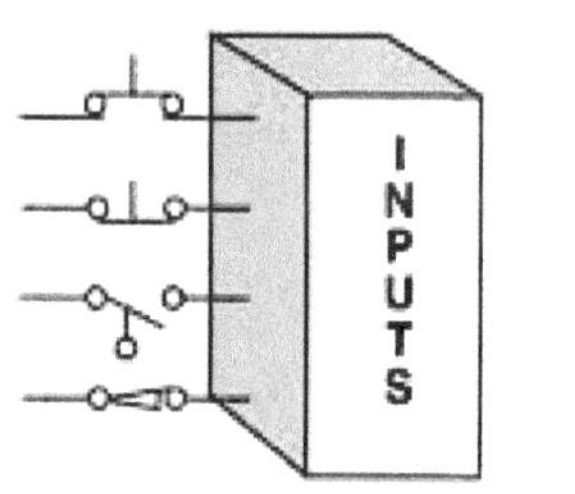

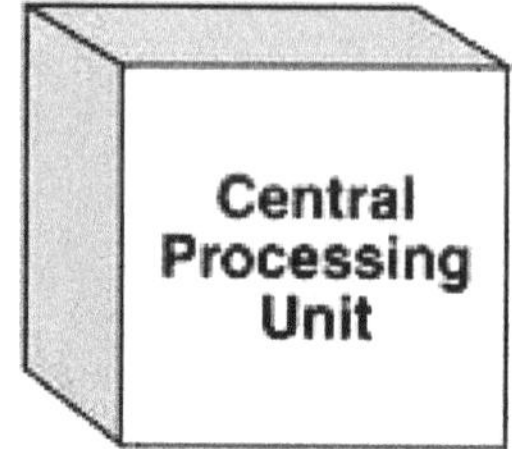

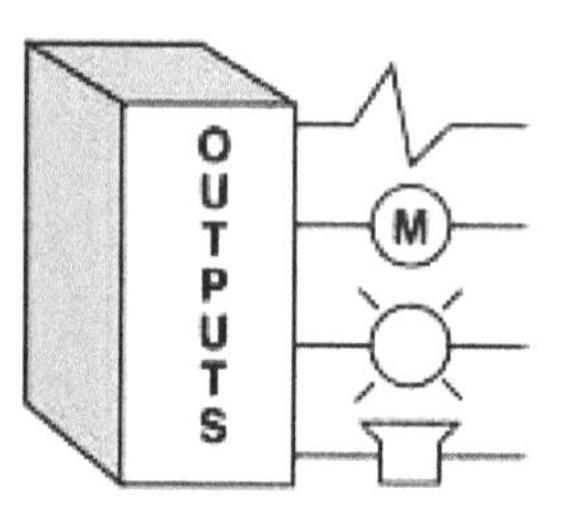

Figure 1.6 PLC structure

The Central Processing Unit (CPU), contains an internal program that tells the PLC how to perform the following functions (Figure 1.6):

- Execute the Control Instructions contained in the User's Programs. This program is stored in "non-volatile" memory, meaning that the program will not be lost if power is removed.
- Communicate with other devices, which can include I/O Devices, Programming Devices, Networks, and even other PLCs.
- Perform house-keeping activities such as communications, internal diagnostics, etc.

1.5.1 Four steps in the PLC operations

1) Input scan

- Detects the state of all input devices that are connected to the PLC.

2) Program scan

- Executes the user created program logic.

3) Output scan

- Energises or de-energise all output devices that are connected to the PLC (Figure 1.7).

4) House-keeping

- This step includes communications with programming terminals, internal diagnostics, etc.

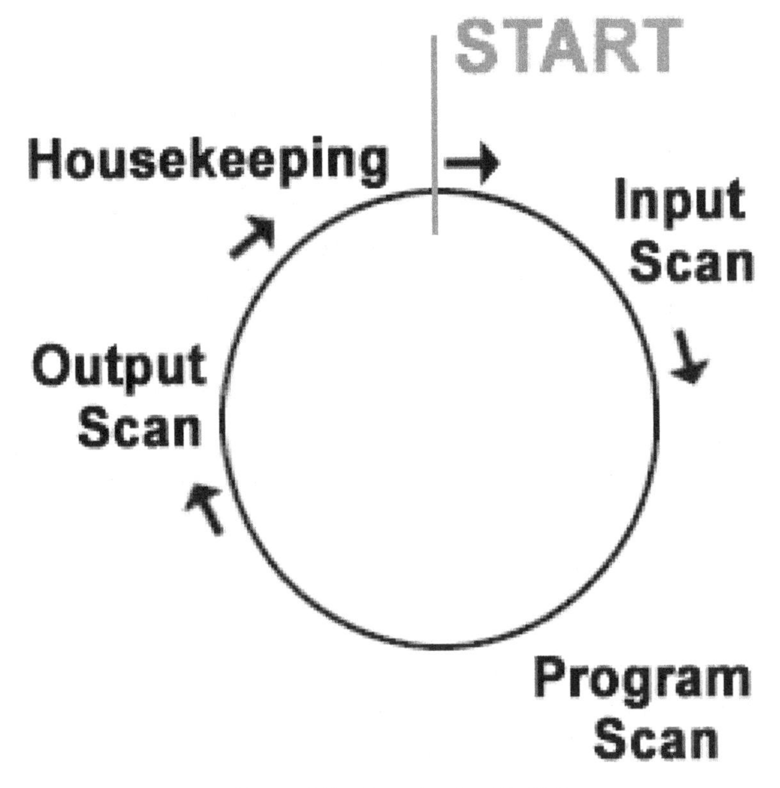

Figure 1.7 Steps involved in PLC.

1.5.2 Programming language in PLC

While Ladder Logic is the most commonly used PLC programming language, it is not the only one. The following table lists of some of languages that are used to program a PLC. (Figure 1.8).

Ladder Diagram (LD) – Traditional ladder logic is graphical programming language. Initially programmed with simple contacts that simulated the opening and closing of relays, Ladder Logic programming has been expanded to include such functions as counters, timers, shift registers, and math operations.

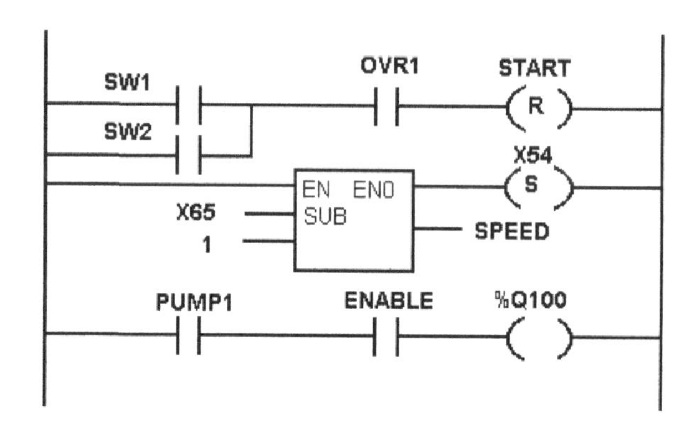

Figure 1.8 Ladder logic programming

Function Block Diagram (FBD) – A graphical language for depicting signal and data flows through re-usable function blocks. FBD is very useful for expressing the interconnection of control system algorithms and logic (Figure 1.9).

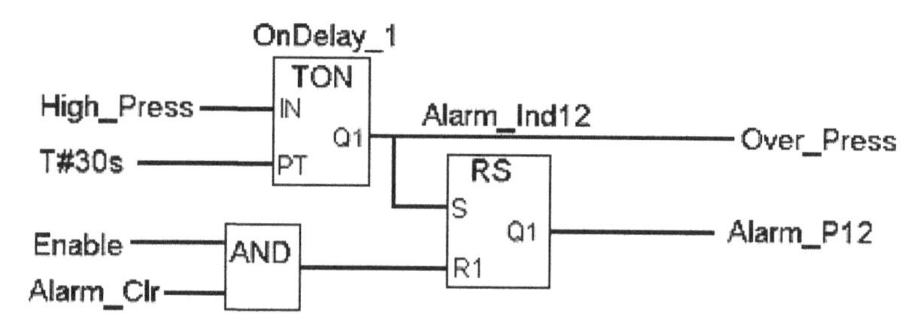

Figure 1.9 Function block diagram (FBD)

Structured Text (ST) – A high level text language that encourages structured programming. It has a language structure (syntax) that strongly resembles PASCAL and supports a wide range of standard functions and operators. For example:

```
If Speed1 > 100.0 then
   Flow_Rate: = 50.0 + Offset_A1;
Else
   Flow_Rate: = 100.0; Steam: = ON
End_If;
```

Instruction List (IL) – A low level "assembler like" language that is based on similar instructions list languages found in a wide range of today's PLCs.

```
      LD    R1
     MPC    RESET
      LD    PRESS_1
      ST    MAX_PRESS
  RESET:    LD   0
      ST    A_X43
```

Sequential Function Chart (SFC) – A method of programming complex control systems at a more highly structured level. A SFC program is an overview of the control system, in which the basic building blocks are entire program files. Each program file is created using one of the other types of programming languages. The SFC approach coordinates large, complicated programming tasks into smaller, more manageable tasks (Figure 1.10).

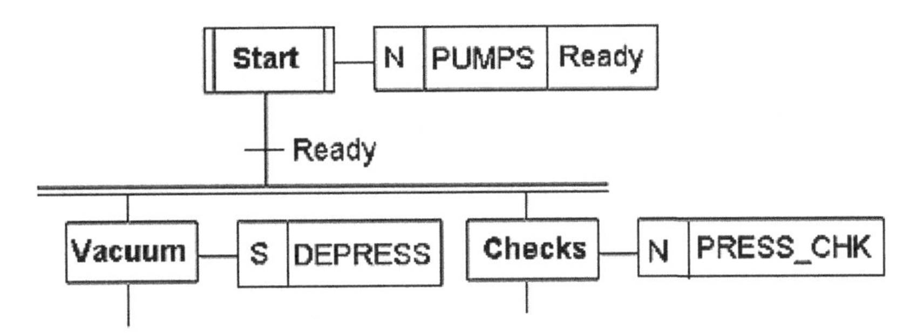

Figure 1.10 Sequential function chart (SFC)

1.6 Units and standards

There are four systems of units which are recognised universally.

a) C.G.S. units
b) F.P.S. units
c) M.K.S. units
d) S.I. units

a) C.G.S. units

The fundamental units of length, mass and time in this system of units are centimetre, gram and second, respectively. These units are also known as absolute units or Physicist's units.

b) F.P.S. units

The fundamental units of length, mass and time in this system of units are foot, pound and second, respectively.

c) M.K.S. units

The fundamental system of units of length, mass and time in this system of units are metre, kilogram and second, respectively. These units are also called gravitational units or Engineer's units.

d) S.I. units (International system of units)

In the S.I. system, there are seven fundamental units and two supplementary units. They are length, Mass, Time, Temperature, Electric current, Luminous intensity, Quantity of substance. Phase angle and Solid angle are two supplementary units.

2
Instrumentation

2.1 Introduction

Instrumentation plays an important role in all branches of engineering and science. Measurement forms an important basis for research and development. In the field of instrumentation, microprocessor significantly extends the capabilities and flexibility of the measuring system. With the development of electronics, mechanical control systems have slowly become obsolete and the microprocessor based operating of the machines have occupied due to its in-built flexibility and ease of operation. In textiles, right from the ginning to finishing of garments, textile testing lab equipments, the importance of instrumentation is felt. Microprocessor is a programmable electronic device which controls the interpretation and execution of instructions. The word "micro" refers to the small size and "processor" refers to the device that performs computational and control operations. Microprocessor is not an operational computer but it requires additional circuits for memory and input/output devices to be interfaced with the system to make a micro computer.

2.2 Instrumentation systems

Instrumentation systems are classified in to two main sections.

1. Analogue instrumentation systems
2. Digital instrumentation systems

Analogue systems deal with the measurement of information and signals in analogue form. It is a continuous function such as a plot of voltage against time or pressure.

Digital systems deal with the measurement of information in digital form. A digital quantity consists of number of discrete or discontinuous pulses whose magnitude or nature varies with time. Both these types will be explained in block diagram in the following sections.

2.3 Analogue instrumentation system

Any electronic instrumentation system consists of the following three important elements (Figure 2.1).

a) Input unit
b) Signal conditioning or processing unit
c) Output unit.

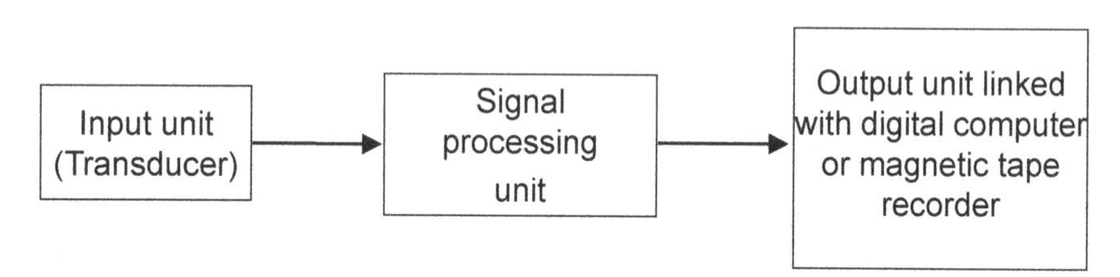

Figure 2.1 Block diagram of an analogue instrumentation system

a) Input unit (Transducer)
The input unit receives the quantity under measurement and converts in to proportional electrical signal such as voltage, current, etc. The electrical signal is fed to the next unit called signal processing unit.

b) Signal processing unit
The electrical signal thus received from the input unit (Transducer) is generally very weak. Hence, it is amplified and in addition to that, it is filtered, modified so as to make it acceptable by the output unit. The signal processing unit performs the following functions: Amplification, Filtering and Modification of form. It also forms an integral part of the input unit.

c) Output unit
The output unit measures the output of the signal processing unit and fed to the digital computer or indicating instrument or magnetic tape recorder for data manipulation or process control. Thus, the complete instrumentation system assumes various forms depending on what is to be measured and the way in which the output is presented.

2.4 Digital instrumentation system

The block diagram of the digital instrumentation system is shown in the Figure 2.2. In digital instrumentation systems, the basic functions are handling analogue signals, making measurements, converting and handling digital data, internal programming and control. The functions of each of the system are shown in Figure 2.2.

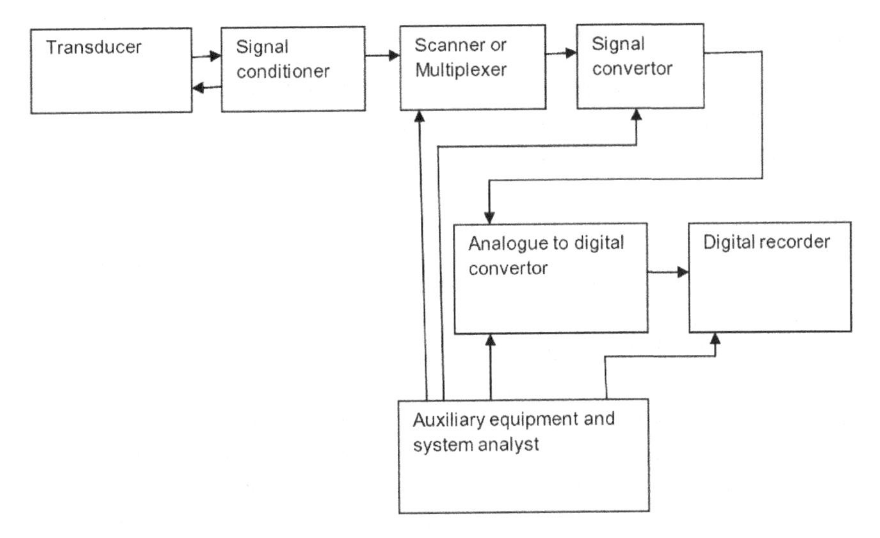

Figure 2.2 Block diagram of digital instrumentation system

a) Transducer: It converts the physical quantities such as temperature, pressure, displacement, velocity, acceleration, etc., in to electrical signal. Other electrical quantities such as voltage, current, etc., may be measured directly.

b) Signal conditioning It consists of supporting circuitry for the transducer. This circuitry provides excitation power, balancing circuits and calibration elements. Examples are strain gauge bridge balance and power supply unit.

c) Scanner or multiplexer It accepts multiple analogue inputs and sequentially connects them to one measuring instrument.

d) Signal convertor It translates the analogue signals in to suitable form acceptable by the (A/D) Analogue to digital convertor.

e) Analogue to digital convertor (A/D) It converts the analogue signals in to suitable digital form. The output of A/D convertor is suitably displayed visually as voltage output in discrete steps for further processing or recording on a digital recorder.

f) Auxiliary equipment This includes instruments for system programming functions and digital data processing.

g) Digital recorder It records digital signals on punched cards, magnetic tape, type written pages or combination of all these systems.

2.5 Transducers

Transducer forms an important element of a measuring system. Transducer converts the input physical variable in to convenient and usable forms of output. In general, it is essentially based on "cause and effect" relationship. For example, a spring when subjected to a force (input is cause) changes the length of the spring (output is effect). And this converts force in to displacement and forms the basis of spring balance – a force measuring device. In general, transducers provide an usable output in response to a specific input signal which may be a physical quantity, property or a condition. It may involve .a conversion of one form of energy to another according to specified relationship. Another term often associated with transducers is sensors. Sensor can be termed as the element, which first detects the measurand and it is the element in contact with the process. For example, non-contact type sensors like stroboscope used for measuring spindle speeds in ring frame in spinning machines, contact type sensor like tachometer used to find the speed of cylinder, doffer in carding machines. Mercury thermometer, thermocouples and photodiodes used in HVI testing instruments are sensors as well as transducers.

They are (a) Sensing element and (b) transduction element.

a) Sensing element: Sensing element is the part of a transducer which senses the physical quantity or the change in physical quantity.

b) Transduction element: It is the part of the transducer which transforms the response of the sensing element in to an electrical signal.

2.6 Classification of Transducers

Transducers are generally classified in to a number of ways as given below.

a) According to their application.

b) Depending upon the method of converting physical quantity in to an electrical signal.

c) According to the nature of the output signal.

d) According to the nature of the transducer whether it is active (self-generating or passive externally powered).

e) According to the electrical parameter or phenomenon being affected by the physical quantity under measurement.

2.6.1 Basic requirements of a transducer

The main function of a transducer is to respond only for the measurement under specified limits for which it is intended to be. It is therefore necessary to know the relationship between the input and output quantities and it should be fixed. Transducers should meet the following requirements.

- Linearity: It means it input–output characteristics should be linear and should produce those characteristics in a symmetrical way.
- Ruggedness: It should be capable of withstanding overload and some safety arrangement must be provided for overload protection.
- Repeatability: Under fixed environmental conditions, i.e. temperature, pressure and humidity, etc., it should produce same output signal when the same input signal is applied again.
- High output signal quality: The quality of the output signal must be good. It means the ration of the signal to the noise should be high and the amplitude of the output signal should be enough.
- High reliability and stability: It should give minimum error in measurement for temperature variations, vibrations and other various changes in the surroundings.
- Good dynamic response: Its output must be in accordance to the input when taken as a function of time. The effect is analyzed as frequency response.
- No hysteresis: It should not give any hysteresis during measurement while input signal is varied from low value to high value and vice-versa.

2.6.2 Selection of transducers

In any measurement system, selection of more appropriate transducer is the most important thing in getting accurate results. Selection of proper transducer depends upon the following factors.

a) Physical quantity to be measured
b) Appropriate transducer principle to be used for the given physical quantity and
c) Accuracy required.

2.6.3 Potentiometers

A potentiometer is a simple and cheap form of transducers and is widely used in converting linear or rotational displacement in to a voltage. The simplest and cheapest form is the linear potentiometer which consists of a single length of wire along which a slider or other form of moving device contacts the wire. The position of the slider determines the effective length of the conductor. Hence a change in electrical resistance or voltage drop is related to the position or displacement of the slider. This simple device is used mostly in laboratories and finds little application in industries. Figure 2.3 shows the basic arrangement of linear potentiometer.

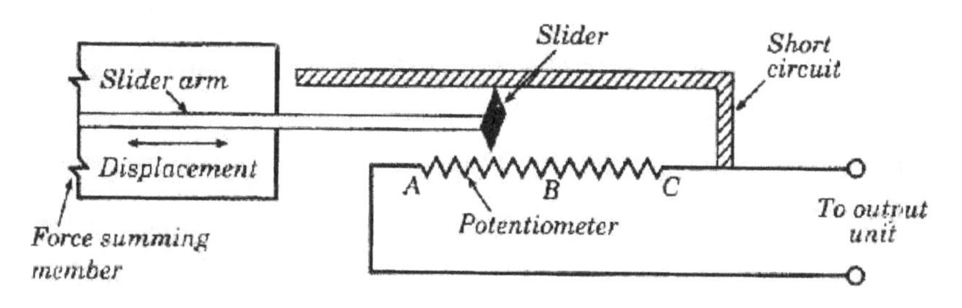

Figure 2.3 Linear potentiometer

2.6.4 Basic arrangement of a linear potentiometer

Potentiometer having translatory motion of the wiper is called translatory potentiometer. Such devices have strokes from 2 mm to 0.5 m.

Potentiometers having rotary motion are called rotary potentiometers, as shown in the Figure 2.3. These may have a full scale angular displacement as small as 10° and as much as full turns. When the motion of the wiper are both ways, i.e. translational and rotational, such potentiometers are called *helipot* as shown in the Figure 2.3.

2.6.5 Displacement transducers

Transducers basically convert an applied force in to a displacement. The mechanical elements used for converting applied force in to displacement are called force-operated or force summing devices. Some of the principal force-summing members are discussed below.

a) Flat or Corrugated diaphragm
b) Bellows
c) Bourden tube – circular or twisted
d) Straight tube
e) Mass cantilever – single or double operation
f) Pivot torque.

Figure 2.4. shows basic structure of these devices. Pressure transducers, in general utilize the first four force-summing devices mentioned above while accelerometers and vibration pick-ups generally use the last two devices.

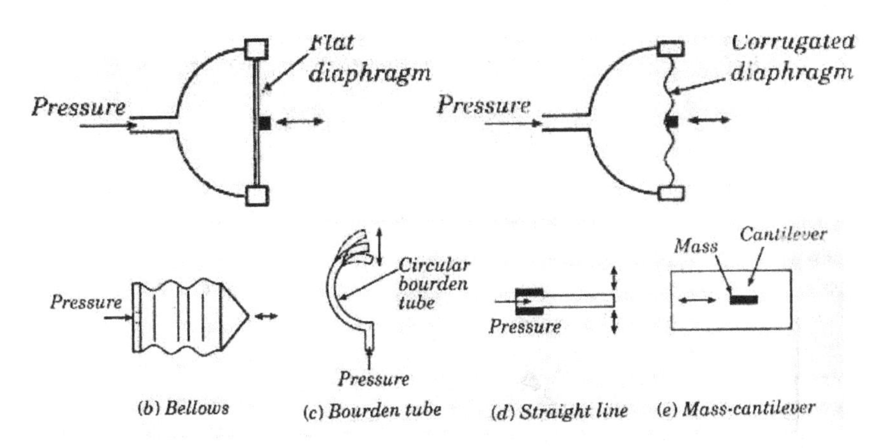

Figure 2.4 Basic structures of pressure transducers

The displacement produced by any force summing device is converted in to an electrical parameter involving one of the following transducers.

i) Linear potentiometer transducer
ii) Linear motion variable inductance transducer
iii) Proximity inductive transducer
iv) Capacitive transducer
v) Piezo-electric transducer
vi) Photo-electric transducer
vii) LVDT.

2.6.6 Linear variable differential transducer (LVDT)

It is the most popularly used inductive transducer used for translating linear motion in to an electrical signal. Figure 2.5 shows the basic structure. It consists of one primary winding (P) and two secondary windings S_1 and S_2 wound side by side on the same cylindrical former. The two secondary windings (S_1) and (S_2) have equal number of turns and are symmetrically placed on either side of the primary winding (P). A movable soft iron core is placed along the axis of the cylindrical former. The linear displacement under measurement is applied to an arm attached to the core.

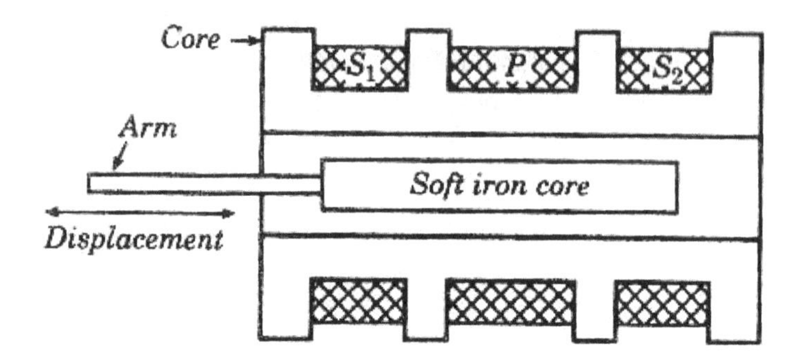

Figure 2.5 Structure of LVDT

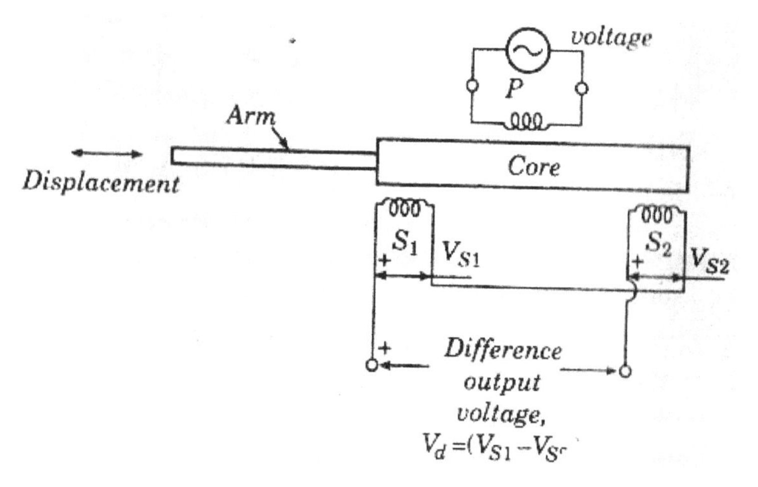

Figure 2.5 Linear variable differential transducer (LVDT)

An a.c. voltage of frequency 50 Hz or 400 Hz is applied to the primary (P) as shown in Figure 2.5 (A). This results in voltages V_{s1} and V_{s2} across the two secondary windings S_1 and S_2, respectively. The two voltages V_{s1} and V_{s2} are combined in series to give the difference output voltage ($V_{s1}-V_{s2}$) as shown in Figure 2.5 (B).

With the soft iron core in the central or null position, the magnetic flux linking with the two secondary windings (S_1) and (S_2) are equal thereby indicating equal voltages across them. Hence the resultant output voltage is zero.

As the core moves to the right of the null position, flux linkage with winding (S_2) exceeds the flux linkage with winding (S_1). The resultant output voltage ($V_{s1}-V_{s2}$) is in phase with V_{s2}. On the other hand, with core displaced to the null position, the flux linkage with (S_1) exceeds the flux linkage with winding (S_2) and the resultant ($V_{s1}-V_{s2}$) is in phase with V_{s1}.

Such LVDT principle is used in the high volume instrument (HVI) for determining the strength of cotton fibres.

Merits of LVDT

a) High resolution.
b) Small, light weight, easy to align and rugged construction.
c) Low hysteresis and hence easy repeatability.

Demerits of LVDT

a) Performance occasionally affected by vibrations.
b) Low power output.

Performance is temperature sensitive.

2.7 Measurement of force and pressure

Whenever a force is applied on an object, it undergoes certain displacement. This displacement may be measured making use of one of the various methods available for measuring displacement and hence the force is calculated. However, we may measure the force or pressure directly by making use of the proper transducer. Some of the transducers are discussed in the following sections.

a) Piezo-electric transducer
In any Piezo-electric transducer shown in the Figure 2.6 when it is subjected to pressure produces electric charges and hence potential difference across opposite faces. This potential difference may be measured or recorded. Piezo-electric transducers are used to measure gas pressure in gasoline engines from time to time during a cycle of operation. Figure 2.6 shows a Piezo-electric transducer.

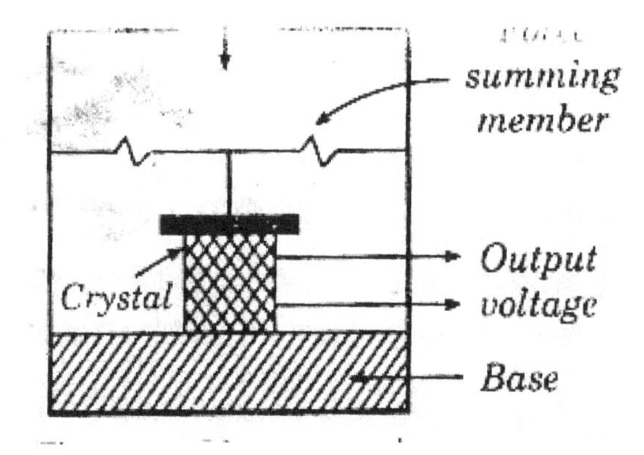

Figure 2.6 Piezo-electric transducer

In the Figure 2.6, a Piezo-electric crystal is placed between a solid base and the force summing member. Any external force applied through the pressure port exerts pressure to the top of the crystal. This produces a potential difference across the crystal which is proportional to the magnitude of the applied force. The main limitations of this type of transducers are:

a) It cannot measure static conditions directly.
b) Output voltage is affected by temperature variations of the crystal.

2.8 Measurement of temperature

Transducers have been designed to produce either changes in voltage or impedance whenever the temperature changes. Transducers using change in resistance are resistance thermometer and thermistors. Transducers using change in voltage are thermocouple and thermopile.

Resistance thermometers

Resistance thermometers are also called as thermo resistive transducers. The change of resistance with temperature in case of some materials form the basis of temperature measurement. Such type of materials falls in to two categories: one using metal conductors and the other semi-conductors. Commercial form of thermistors is shown in the Figure 2.7. The resistance of highly conducting materials (generally metals) increases with temperature. Examples are nickel, copper, platinum, tungsten and silver. A temperature measuring device using an element of this type is called resistance thermometers or resistance temperature detectors (RTD). In other materials like semi-conductor devices, the resistance of semiconductors generally decreases with increase in temperature and are called as *thermistors*.

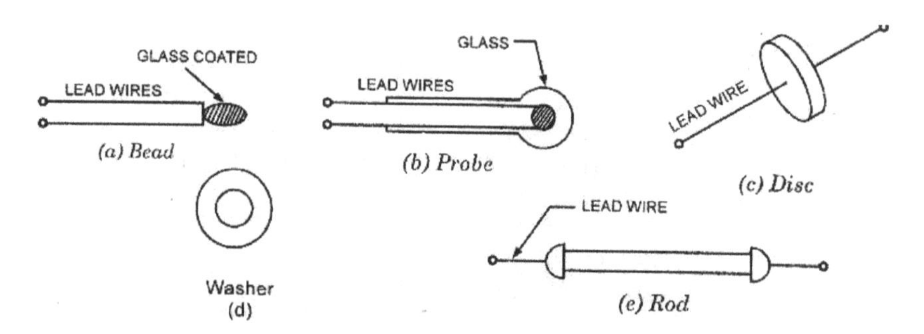

Figure 2.7 Commercial forms of ihermistors

2.9 Strain gauges

Strain gauges are a passive instrument which converts a mechanical elongation or displacement in to a change of resistance (R) or inductance (L) or capacitance (C). Thus strain gauges may be of the following three types.

i) Resistance type
ii) Variable inductance type and
iii) Variable capacitance type.

For measurement of elongation, or compression resistance type strain gauges are used. Such strain gauges are used in textile testing equipment. Here we will discuss only the resistance type strain gauge instrument which is applicable for testing of textile materials.

i) Resistance type strain gauge

As shown in the Figure 2.8. A resistance type strain gauge is in the form of a thin wire placed on a flexible paper tissue which can be attached or bonded to a variety of materials to measure strain in the material. The gauge is so placed that the wires of the gauges are aligned along the direction of the strain to be measured. This principle is also used in the fibre testing instrument using cantilever and strain gauge.

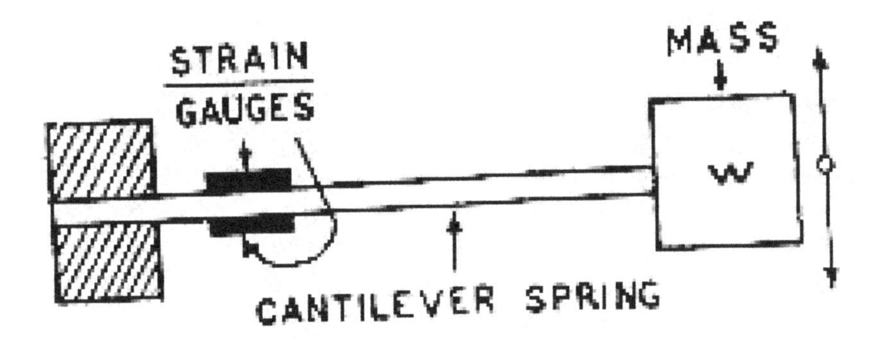

Figure 2.8 Resistance type strain gauge

Acceleration of shuttles

Metallic strain gauges are manufactured from small diameter resistance wires, such as Copper (60%) and Nickel (40%) alloy or etched from thin foil sheets. The resistance of the metal foil or wire changes with the change of length as the material to which the gauge is attached undergoes tension or compression. The change in resistance is proportional to the strain developed and is usually measured with the specially designed Wheat Stone Bridge.

2.10 Terms and definitions used in instrumentation

a. Drift

Drift is defined as the inability of the instrument to reproduce the same result at different times of measurement with the same input signal. If the instrument is able to reproduce the same result or perfect reproducibility, then it is said to have no drift.

b. Calibration

Calibration is defined as a known input signal is given to the instrument and the output signal is noted. If the system's output deviates with respect to the given input, then necessary corrections are made in the instrument which matches with the input.

In textile testing instruments like HVI (High Volume Instrument,), AFIS (Advanced Fibre Information System) and other yarn testing instruments calibration is performed in order to assess the expected result matches with our requirements.

In HVI, calibration cotton is used to check the output results matches with the prescribed values. If the output result deviates from the standard, the instrument will show "Fail" and necessary corrections have to be made before proceeding to regular testing of bulk cotton samples. Similarly, in AFIS, calibration materials are available to confirm the test results are satisfactory.

c. Hysterisis

Energy that was supplied in to the stressed component will not be recovered to the same level during unloading. Hence, there is slight deviation in the output result with the previous input signals and this is called as "Hysterisis Effect".

d. Sensitivity

$$\text{Sensitivity} = \frac{\text{Change in the output signal}}{\text{Change in the input signal}}$$

e. Threshold value

The minimum value of input signal that is required to make a change or start from zero is called "Threshold value".

f. Resolution

The minimum value of the input signal (non-zero value) required to case an appreciable change of an increment in the output is called resolution.

g. Accuracy

Accuracy in instrumentation refers to the closeness of the measured value with respect to its true value.

h. Precision

Precision is referred to the ability of the instrument to reproduce the same result again and again for the constant input signal.

3

Measuring devices in textile mills

3.1 Introduction

Measurement plays an important activity in all branches of engineering and science. Measurement can be defined "as a process of obtaining a quantitative comparison between a pre-determined standard and an unknown magnitude of a parameter". For example, the measurement of spindle speed in ring frame, speed frame, cylinder speed, etc. in textile machines. Speed of these components can be measured using contact or non-contact type instruments. The purpose is to ascertain that the components are running at the set speed according to the standard set on the machine display and to find out if there is any variation in the speeds. Measurement forms an integral part of any automatic control system for control action, discrepancy or error between the actual value and the desired value of a variable is to be determined or measured. Evaluation of the performance of the machines requires measurement of various parameters.

3.2 Angular motion measurement in textile machines

The measurement of the speed of various elements in the textile machines is important not only from the point of production but also to ascertain if there is any slippage in the spindle speeds and so on. Necessary actions have to be taken immediately or otherwise it will affect production and quality of the end product. In general, in textile mills, quality assurance department have the schedule of checking the speeds of the beaters, front roll delivery in speed frames and ring frames, spindle speeds in ring frames. This is very essential so as to have a control over the process and identifying the defective rotating elements and rectifying the same.

Measurement of the rotational speed of any working element is often required. It is generally expressed in revolutions per minute (RPM) or in metres/minute (m/min) or in radians/sec (rad/s). Essentially it requires a mechanism for counting the number of revolutions and a timing mechanism.

There are some types of transducers whose output is directly proportional to the angular speed. For example, the output voltage of an a.c. or d.c. generator is directly proportional to the angular speed of the armature. A device which measures the angular speed directly is generally called as tachometer. There are many types of tachometer available in practice and the common types still in use will be discussed in the following sections.

1. Mechanical type
   a) Revolution counter and timer
2. Electrical type
   a) Generator type (a.c. and d.c) Tacho generator
   b) Inductive non-contact type
3. Optical type
   a) Photo cell – Non-contact type
   b) Stroboscope

3.2.1 Revolution counter and timer

In this arrangement, a revolution counter is attached to and driven by the input shaft. A pointer or a direct counting digital display indicates the number of revolutions turned by the shaft in a certain period of time as indicated by the timer. Timer can be a simple stop watch. The average speed in RPM can then be calculated. The revolution counter and the timer can be mounted integrally as a single unit so that they work simultaneously.

A familiar example is the scutcher blow room line in spinning mills. In this Scutcher machine, the lap delivered from the calendar rollers is continuously counted by the electronic counters by counting the number of revolutions made by the calendar rollers and the length of the material delivered per minute. The electronic counter gets the information from the calendar roll shaft and as soon as the required length of material is wound, say, 45 m or 50 m depending upon the lap weight required, the machine stops. The full lap is removed from the machine and replaced with empty lap spindle which comes in to position automatically by pneumatic means and the winding of the lap starts again. In this way, a constant length of material can be wound on the lap spindle.

3.2.2 Electrical type

Generator (a.c and d.c.) type/Tacho generators

These are generally a permanent magnet type a.c. or d.c. electrical generators. The principle involved in this type is the output voltage from these generators is directly proportional to the angular velocity and can be measured by a voltmeter.

D.C. Tacho generators

D.C. type Tacho generators produce d.c. voltage which can be measured by a simple d.c. voltmeter. These devices require some form of commutation and present the problem of brush maintenance and are sensitive to the direction of rotation as shown in the Figure 3.1.

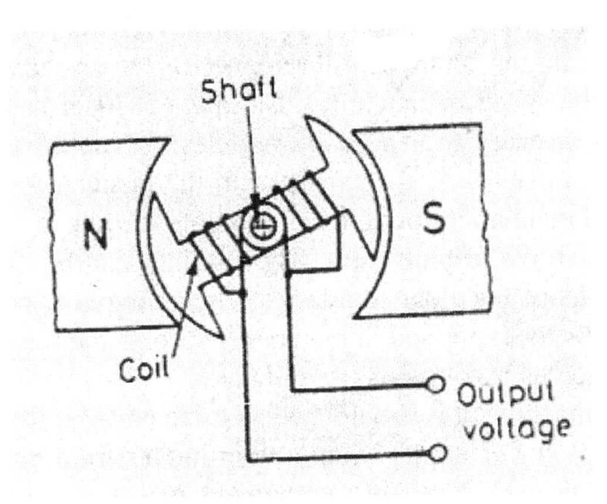

Figure 3.1 DC Tacho generator

A.C. Tacho generators

A.C. type Tacho generators shown in the Figure 3.2 produce a.c. voltage. A rectifier is required if simple d.c. voltmeter is used for converting the voltage for indication. These electrical devices produce a continuous indication of speed which can be recorded or displayed. Tacho generators are generally used in high speed draw frames in spinning mills.

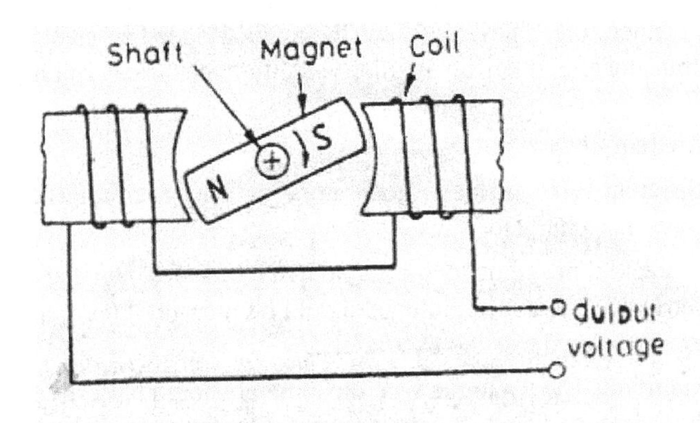

Figure 3.2 AC Tacho generator

3.3 Tachometers

A tachometer is a sensor device for measuring the rotation speed of an object such as the engine shaft in a car. This device indicates the revolutions per minute (RPM) performed by the object. The device comprises of a dial, a needle to indicate the current reading, and markings to indicate safe and dangerous levels. Historically, the first mechanical tachometers were designed based on measuring the centrifugal force. In 1817, it was adapted to be used for measuring the speed of machines. But after 1840, it has been predominantly used to measure the speed of vehicles. Advanced tachometers are being applied in novel uses, for example, in the medical field, a haema tachometer placed in an artery or vein can estimate the rate of blood flow from the speed at which the turbine spins. The readings can be used to diagnose circulatory problems like clogged arteries. There are two types of velocities. One is linear velocity expressed in meters per second and the other is angular velocity expressed as revolutions per minute or radians per second. For any type of rotating machine, it is usually necessary to measure the linear velocity or angular speed. For example, when a train moves from one station to the other, if the distance (x) is known and the velocity (v), then the time taken to reach the destination is calculated by

$$\text{Time } (t) = \text{Distance } (x) \,/\, \text{Velocity } (v)$$

In case of rotating machine

$$\text{Power } (P) = 2\pi NT/60$$

Thus, the power (P), speed (N) and torque (T) are related with each other. To know about the performance of the machine, sometimes torque speed characteristics are studied. The device used for measuring the speed of a rotating machine is called a *tachometer*. Electrical type of tachometers are called tacho generators. They represent the angular speed as electrical voltage. It is a transducer converting angular velocity to electrical voltage.

Characteristic requirements of Tacho generator

There are different types of tacho generators but they are required to meet certain performance characteristics.

- Accuracy: The input to the tachometer is speed and the output is electrical voltage. The voltage should be measured correctly so that it can indicate the speed accurately.
- Resolution: The analogue type tachometer should have divisions and sub-divisions of scale small enough calibrated to indicate the smallest variation in speed. In case of digital indicator, the resolution can be improved by increasing the digits displayed.

- System voltage: If the system voltage used for excitation is fluctuating, the output relation may be affected.
- System frequency: In case of ac tacho generator, if input frequency is fluctuating, the output relation is affected.
- Temperature effect: Armature and field winding resistances are affected by temperature. Magnetic property of the core is also temperature dependent. If the temperature variations are excessive, physical expansion co-efficient may affect the dimensions. In case of small air gap between fixed field system and moving armature system, the change of dimension may create magnetic distortion and mechanical problems.

3.3.1 Types of tachometers

The types of tachometers commonly found are mentioned below:

- Analogue tachometers – Comprise a needle and dial-type of interface. They do not have provision for storage of readings and cannot compute details such as average and deviation. Here, speed is converted to voltage via use of an external frequency to voltage converter. This voltage is then displayed by an analogue voltmeter.
- Digital tachometers – Comprise LCD or LED readout and a memory for storage. These can perform statistical operations, and are very suitable for precision measurement and monitoring of any kind of time based quantities. Digital tachometers are more common these days and they provide numerical readings instead of using dials and needles.
- Contact and non-contact tachometers – The contact type is in contact with the rotating shaft. The non-contact type is ideal for applications that are mobile, and uses a laser or optical disk. In the contact type, an optical encoder or magnetic sensor is used. Both these types are data acquisition methods.
- Time and frequency measuring tachometers – Both these are based on measurement methods. The time measurement device calculates speed by measuring the time interval between the incoming pulses; whereas, the frequency measurement device calculates speed by measuring the frequency of the incoming pulses. Time measuring tachometers are ideal for low-speed measurements and frequency measuring tachometers are ideal for high-speed measurements.

3.3.2 Inductive tachometer (Non-contact type)

The inductive tachometer of non-contact type is shown in Figure 3.3. In this type of device, a small toothed wheel is to be attached to the shaft whose speed is to be determined. A permanent magnet with a coil wound round it is placed near the rotating toothed wheel. As the wheel rotates, the magnetic flux produced with the magnet and the coil changes. As a result of this, a voltage is induced in the coil. The frequency or the number of pulses depends upon the number of teeth on the wheel and

$$\text{Speed of the shaft} = \frac{\text{Pulses per second}}{\text{Number of teeth on the wheel}}$$

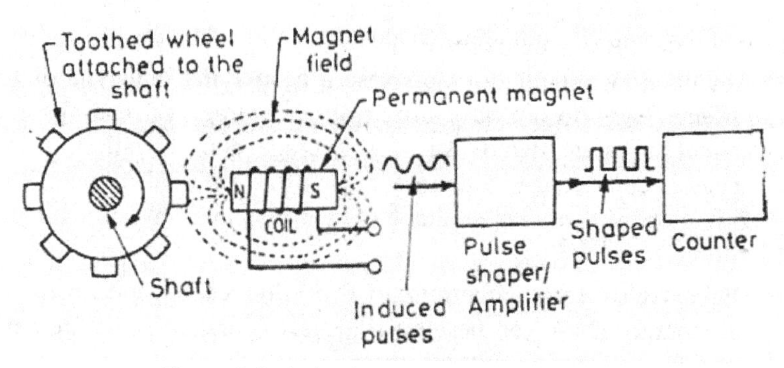

Figure 3.3 Inductive non-contact type tachometer

The pulses produced are not uniform and well shaped and need to be shaped and amplified. The output that is the number of pulses can be seen on a frequency measuring unit or can be converted in to proportional voltage.

3.3.3 Photoelectric type tachometer (non-contact)

The working principle of the photoelectric type tachometer is shown in the Figure 34. This is also a non-contacting type of speed measuring device. It consists of an opaque disk with evenly spaced holes on its periphery. It is attached to the shaft whose speed is to be measured. A light source is placed on one side and a light sensitive transducer is placed on the other side of the opaque disk. It should be noted that both the light source and light sensitive transducers should be in proper alignment with the holes in the disk. As the opaque disk rotates, the intermittent light falling on the photocell produces voltage pulses whose frequency is the measure of the speed of the shaft. From this, the output (number of pulses) can be seen on a frequency measuring unit or can be converted in to proportional voltage.

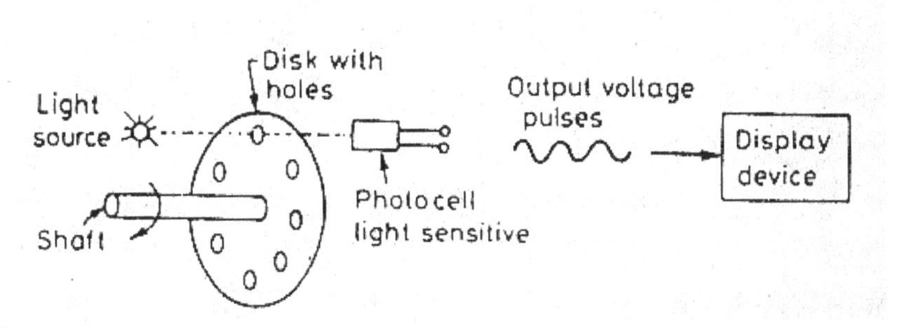

Figure 3.4 Photoelectric type tachometer

3.3.4 Contact type tachometer

Contact type tachometer is generally used in textile mills to measure the speed of the rotating shaft in RPM. Some of its applications are speed measurement of cylinder, Doffer, lickerin in carding machine, Front roll delivery speed in speed frames and in ring frames, etc. The working principle of the contact type tachometer is:

The contact type is in contact with the rotating shaft. The non-contact type is ideal for applications that are mobile, and uses a laser or optical disk. In the contact type, an optical encoder or magnetic sensor is used.

3.4 Stroboscope

Stroboscope is a measuring device for measuring the speeds of the objects having repeating or cyclic motion. For example, to measure speed of the rotating shaft and reciprocating mechanisms. In textile spinning machines, stroboscope is generally used to measure the speed of the spindles. Since the spindle speed in a ring frame is of high order, say from 15,000 rpm to 22,000 rpm, contact type tachometer is not possible. However, non-contact tachometer can also be used to measure the spindle speeds depends upon its measuring speed range.

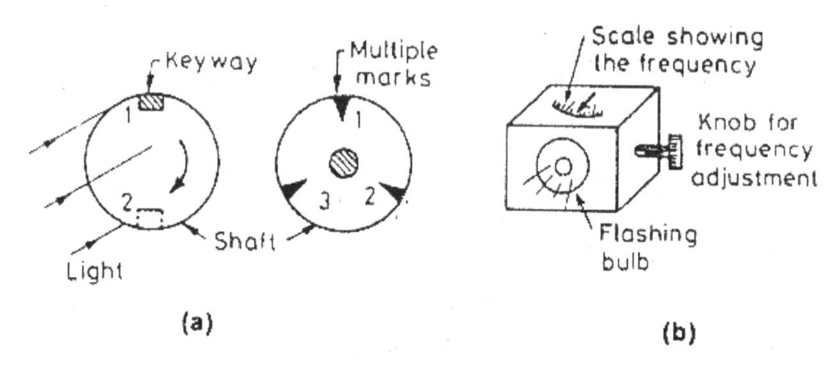

Figure 3.5 (A) & (B) Stroboscope

Working principle

The principle of working of stroboscope instrument is based on the fact that making the object to appear motionless (freezing the motion) whose speed is to be measured. It is done by adjusting the timing of the flashing light whose frequency matches with the frequency of the rotating body.

This can be explained by an example (A) and (B) shown in the Figure 3.5 (A) and (B).

Consider a shaft with a keyway rotating at 1 revolution per second. It means it is the frequency of the object. If a light source flashing once every second (flashing frequency) is made to illuminate the shaft at a particular frequency, the keyway appears to be in stationary position to the eye. When the keyway appears to be stationary for the eye, the speed can be read from the display which is the rotating speed of the shaft. The time taken by the keyway to occupy the position (1) after one revolution is the same as the time taken by the flashing light to flash again and illuminate the key way that is 1 second. If the flashing frequency is doubled to 2 flashes per second then, two stationary images will be seen one in position (1). And another in position (2) because, the keyway after ½ second, will occupy position(2), just at this time when flash of light occurs and illuminates it.

Stroboscope is a flashing light source which provides repeated short duration light flashes. The frequency of these flashes can be controlled by a variable frequency electronic oscillator which operates the flashing bulb. The frequency of the light can be adjusted by a knob and the measured value (speed) can be read off on the display. The flashing frequency is normally variable from 1 to 2500 Hz. The working of the instrument requires ambient light to get sub-dued and should not bright. Stroboscope does not require any physical contact with the object for measurement. A simple method is to work with a single mark or sticker for the frequency to be adjusted so that two images are seen. Frequency setting is then halved and adjusted till one stationary image is seen, giving the correct value of the frequency and the speed.

3.5 Flow measurement

Accurate measurement of fluid flow is very important in many situations from domestic water supply, canals and in engineering the velocity of air flow in ducts. This is very much required in the present day advanced textile machineries. Most of the textile spinning units has air engineering for their humidification systems in the mills. Furthermore, in spinning preparatory machines, the material is conveyed from one machine to another by means of air through ducts. The design of the ducts is also important for proper air flow without any abrupt bends. Hence, the instrumentation has developed an instrument to measure the flow of air velocity through the ducts. One such instrument is Hot Wire Anemometer.

Hot wire Anemometer

The hot wire Anemometer device is mainly used in industries and in research applications to study about the rapidly varying flows i.e., to study about the mean and fluctuating components of velocities. The principle of hot wire anemometer is rather simple. It consists of a thin wire of 5 μm in diameter of resistance (R) usually made of platinum, nickel or tungsten is heated by passing current and placed in the flowing fluid. Due to the heat applied to the wire, the convective heat transfer characteristics of the heated wire become a measure of fluid velocity.

The hot wire anemometer probe is shown in the Figure 3.6 (A), (B) and (C)

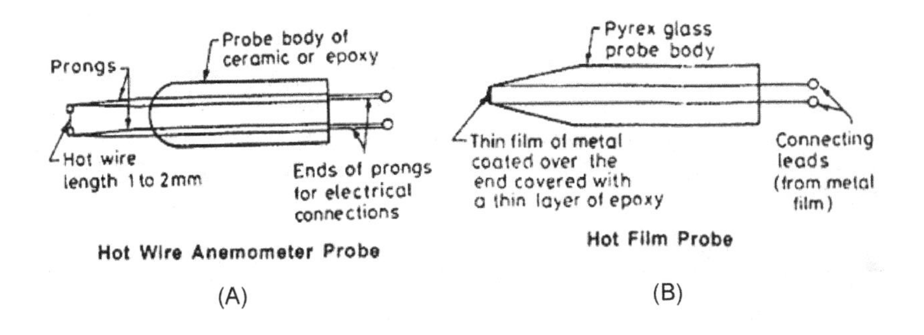

(A) (B)

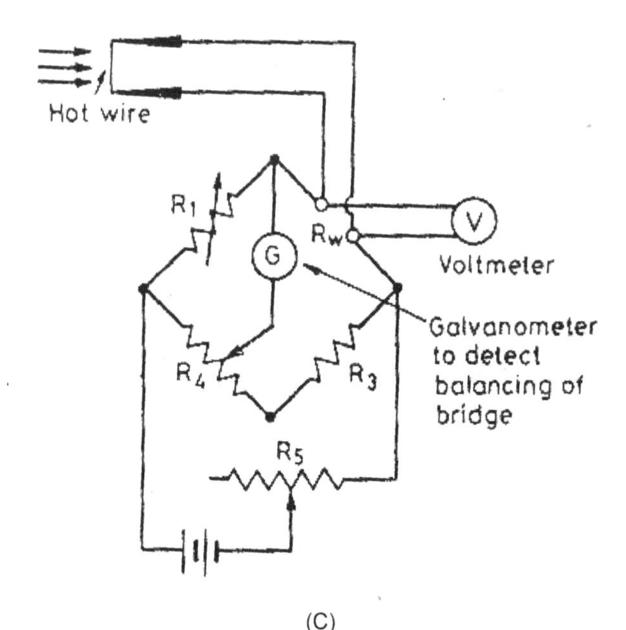

(C)

Figure 3.6 (A) Hot wire Anemometer probe, (B) Hot film probe, (C) Schematic circuit diagram of hot wire Anemometer

The arrangement shown works on Constant Temperature Mode. In this arrangement, hot wire resistance and hence its temperature is kept constant by continuously adjusting the current through the hot wire using a suitable servo system. The current or the voltage across the hot wire is the measure of the flow velocity. The constant temperature design has the advantage that the wire is protected from burn out. The hot wire probe can be calibrated against a pitot static tube in a wind tunnel. Accumulation of dirt on the wire should be avoided or otherwise, it gives serious heat transfer errors.

3.5.1 Laser Doppler Anemometer

Doppler effect at optical frequencies has been used for a quite a long time by our astronomers for measuring the velocity of stars. The same principle is used in fluid flow measurement by using laser Doppler Anemometer.

Principle

When a beam of light is from a laser is focused in to a moving fluid, a frequency shift in the light scattered by the minute particles present in the fluid or Doppler effect is observed. This frequency shift is proportional to the velocity of the scattering particles. As the slip velocity between particles and fluid is negligibly small, this frequency shift is an indication of fluid velocity. In most cases, the frequency shift is so small compared to the optical frequency of the laser light used. If a He–Ne laser having a frequency of approximately 5×10^{14} Hz is used, a change in frequency of the order 5×10^{6}Hz is difficult to measure. This problem is solved by heterodyning.

Heterodyning is a familiar technique used for frequency measurement. When two waves of equal amplitude and nearly equal frequency are added the resultant waves undergoes "slow beats". It means the amplitude of the resultant wave rises and falls with a cyclic frequency which is equal to half the frequency difference between the waves.

In the Figure 3.7 the laser light is split in to two beams of equal intensity and focused in to the flow. Both the beams undergo Doppler shift. But for each beam it is different because of their different angles of incidence. The resulting beat frequency is the difference in Doppler shift of two beams. This is called differential mode and is more commonly used as it is easy to align and use.

The schematic diagram consists of a laser light source, transmitting optical unit, receiving optical unit, photo multiplier tube and signal processing unit. The instrument is a complex combination of laser optics and electronics. The laser light is focused in to the transparent test section by transmitting optical unit. The light from the test section is received by the receiving optical unit which then focuses it on to a photomultiplier unit. Photomultiplier unit converts light signals to electrical signals to be processed by signal processing unit.

In certain occasions, if the signal is weak, it is necessary to introduce particles in to the flow or seed the flow with light scattering particles.

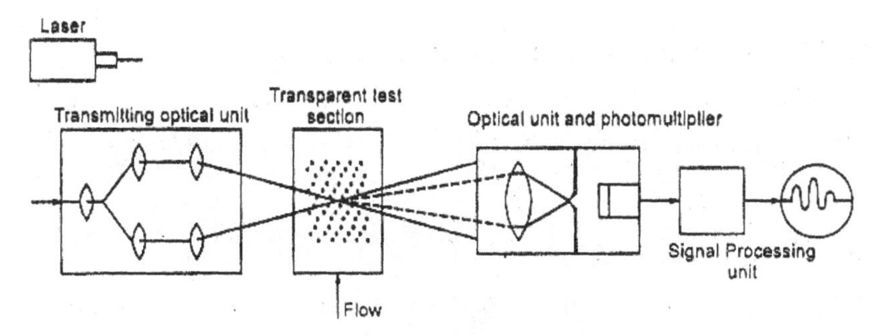

Figure 3.7 Laser Doppler Anemometer

3.6 Torque measurement

Torque measurement in rotating devices and machines are needed to determine the power required to operate a machine or power developed by a machine. Torque and power are related as

$$P = \omega T = 2\pi NT$$

where P = power in watt
T = Torque in N-m
ω = angular speed in radian/second
N = angular speed in revolutions per second

Thus, if the torque and angular speed are known, power can be determined. Torque "T" is the turning effect of a force "F" and is given by $T = F \times r$, where r (arm of the force) is the perpendicular distance of the moment centre from the line of action of the force. Torque measuring devices are commonly known as dynamometers and are classified as

- Absorption dynamometers
- Driving dynamometers
- Transmission dynamometers.

3.6.1 Absorption dynamometers

The principle used in these types of dynamometers is the mechanical energy is absorbed and dissipated (generally as heat) as torque of the machine is measured. However, these types of dynamometers cannot drive a test machine. These are particularly useful for measuring power or torque developed by engines and motors. Absorption dynamometers can be of the following types:

- Mechanical type
- Hydraulic type
- Eddy current type
- Electrical generator.

Mechanical / Prony Brake dynamometer

In this type, frictional resistance is created between a rotating flywheel attached to the test machine and a rope or band or brake shoes. The heat thus generated is dissipated by circulating cooling water. The most familiar type is prony brake dynamometer shown in the Figure 3.8.

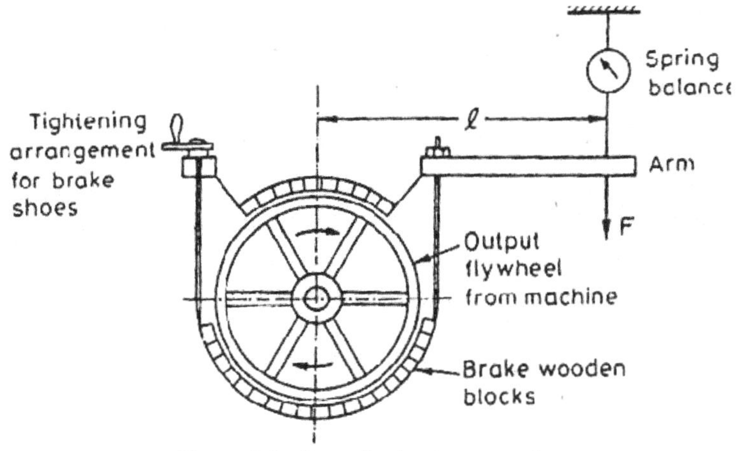

Figure 3.8 Prony brake dynamometer

Figure 3.8 shows the dynamometer and the frictional resistance can be varied by tightening of rope or wooden brakes. The force "*F*" can be measured by any force measuring device. In dynamometers involving small forces, an arrangement involving dead weights and spring balance can be used.

Torque (T) = $F \times 1$
Power (P) = = $2\,\pi\,NT/\,60$ where N is the speed in RPM.
$P = 2\,\pi\,NF.1\,/\,60$

3.6.2 Hydraulic dynamometer

Cradle type dynamometer of hydraulic type is shown in the Figure 3.9 in which the stator and rotor are coupled hydraulically. The stator is in two halves and placed on either side of the rotor. Semi-elliptical grooves in the rotor match with that in the casing through which a flow of water is maintained.

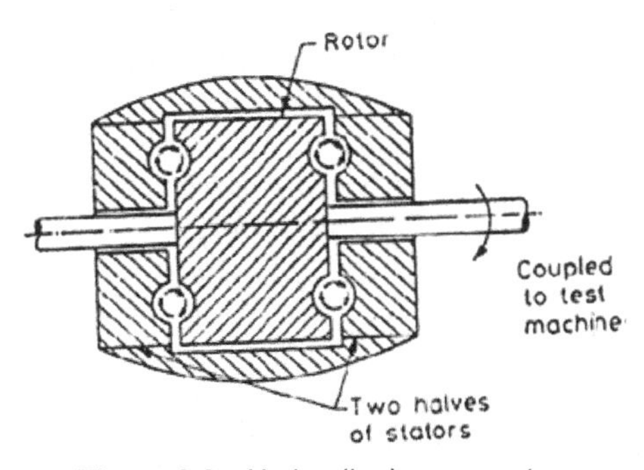

Figure 3.9 Hydraulic dynamometer

Water flows through helical paths creating eddy currents resulting in dissipation of energy. At the same time, the stator tends to rotate, which is freely mounted on bearings is opposed by weights applied to the moment arm on the stator. The braking effect can be varied by controlling the amount of water flowing by some mechanism between rotor and stator. Since water itself acts a coolant, it is not necessary to provide any cooling arrangement in this type of hydraulic dynamometer.

3.7 Pressure measurement

Pressure can be defined as the average force exerted by a fluid on a unit area.

$$P = \rho g H$$

where P = pressure in N/m^2
ρ = density of fluid in kg/m^3
H = height of the fluid column in m.

Pressure can be measured either as a force acting on a unit area (N/m^2) or by the height of a liquid column (generally water and mercury) it can support in metre or millimetre. Standard atmospheric pressure is 1.013×10^5 N/m^2) can support a mercury of 760 mmHg or 10.3 m of water column.

Pressure Terms

Absolute Pressure

It represents the pressure with absolute zero pressure or perfect vacuum taken as reference.

Gauge Pressure

It represents the pressure with atmospheric pressure as reference pressure.

Vacuum Pressure

It refers to pressures those are below atmospheric pressure and can be expressed in terms of both absolute or gauge pressure.

3.7.1 Differential pressure

It represents the difference between two pressures. Measuring devices generally measure pressure as a differential quantity shown in the Figure 3.10 (A) and (B).

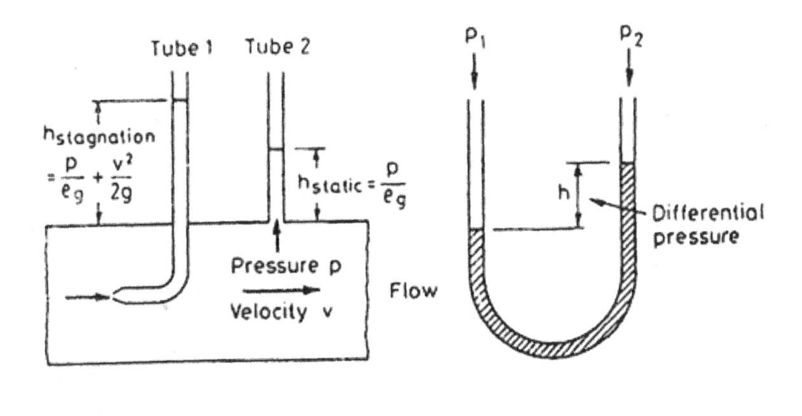

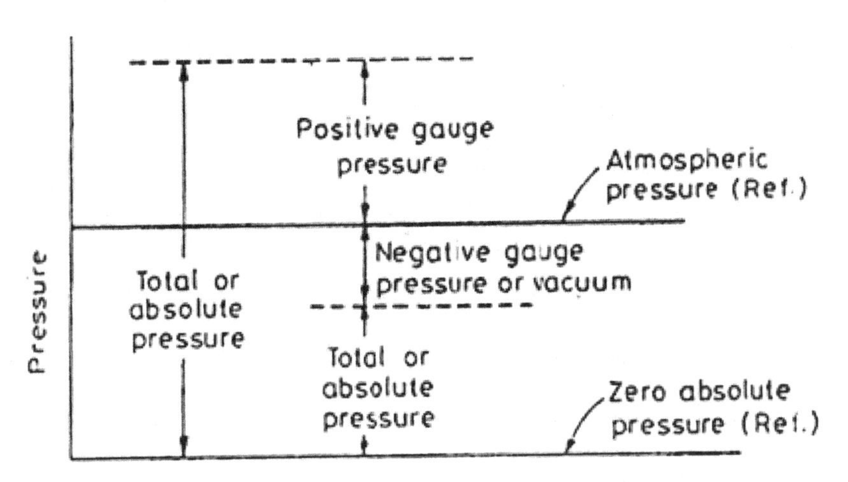

Figure 3.10 (A) and (B) Relationship between pressure terms

If one of the pressure is atmospheric pressure then that differential pressure can be termed as gauge pressure. In the Figure 3.10 (a), if pressure (P_2) is atmospheric pressure then the height of liquid column "h" represents gauge pressure (P_1).

3.7.2 Piezometer

Piezometer is a pressure measuring device in which the pressure in a pipe or a vessel can be measured using piezometer tube. It consists of vertical transparent tube connected to the pipe, the other end of which is open to the atmosphere as shown in the Figure 3.11.

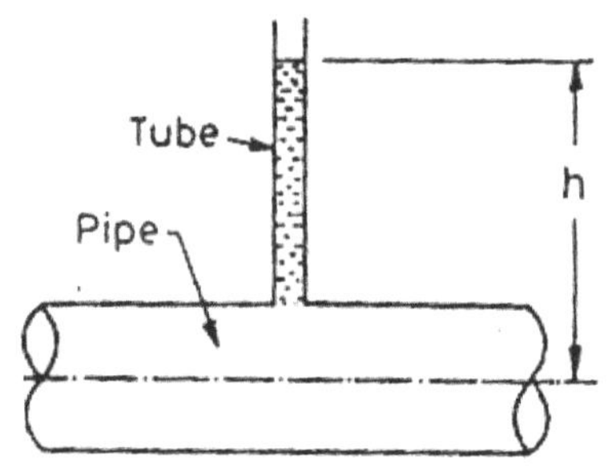

Figure 3.11 Piezometer

The rise '*h*' of the fluid in the piezometer tube is a measure of pressure in the pipe.

The absolute pressure in the pipe, is given by

$$P = P_a + \rho gh$$

where "ρ" is the density of the fluid and "pa" is the atmospheric pressure.

This instrument uses the working fluid for the measurement of pressure and hence it is not suitable for gases. Furthermore, it is not suitable for pressures which are below the atmospheric pressure as the atmospheric air will flow in to the pipe or container. Also, it is not suitable for pressures which are larger than atmospheric pressure.

3.7.3 Manometers

The difficulties associated with piezometer device are overcome by a simple U-tube manometer (double column) and its various modified versions are used for measuring pressure in the range of 10^{-1} mmHg to 10^4 mm of Hg.

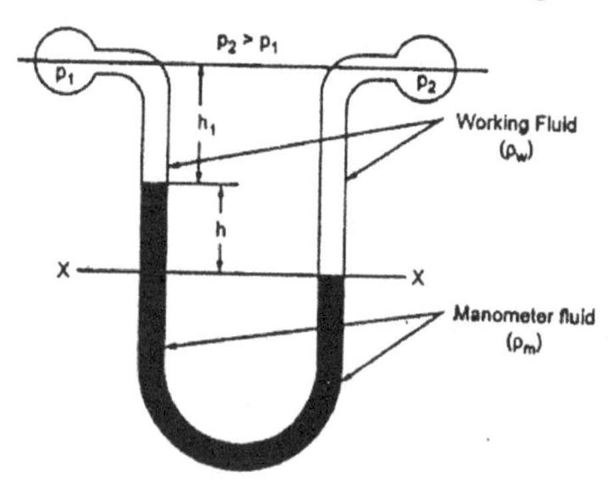

Figure 3.12 U- tube manometer

A simple U-tube manometer is shown in the Figure 312 uses water, mercury or any other suitable fluid as manometer fluid. The difference in levels

"h" between the limbs is an indication of pressure difference (P_2–P_1) of the pressures applied to the two limbs.

3.7.4 Elastic transducers used for pressure measurement

Any elastic elements when subjected to a pressure it gets deformed. There are wide variety of metallic elements that could be used as pressure transducers. The resulting deflection/deformation operates a pointer on a scale with some magnification obtained by gears or any linkages. The deformation can also be transduced to an electrical signal using displacement transducers such as strain gauges, LVDT, piezo electric crystal, capacitance and inductance transducers. Pressure measurement using elastic transducers is shown in the Figure 3.13.

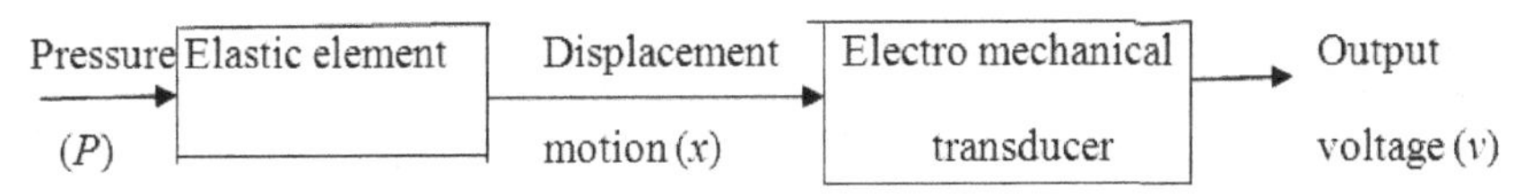

Figure 3.13 Pressure measurement with elastic transducers

Various elastic elements used are

- Bourdon tube
- Bellow
- Diaphragm and
- Capsule.

3.8 Sound measurement

Sound waves are vibratory phenomenon and produce pressure fluctuations in the liquid or gaseous medium. In addition, it produces sensation of hearing. Acoustic studies and sound measurements find wide application in the present scenario. Few examples are:

- In the development of machinery and equipment which should be less noisy.
- Diagnosis of vibration problems in machineries.
- In the testing and designing of sound proofing, sound recording and sound reproducing equipment.
- Detection devices for underwater objects.

- For checking and controlling noise pollution in machineries.

Sound pressure level

Sound waves produce pressure fluctuations. Sound pressure level is measured in terms of "decibels" (dBA) and described in terms of power ratio. The basic definition of sound is in terms of root mean square value of the fluctuating component of pressure called "sound pressure".

Equipments to measure sound

Sound level meter

Sound measuring systems have a transducer. It converts acoustic pressure in to proportional voltage. The output device can be a CRO (Cathode Ray Oscilloscope) or a simple meter.

Microphones

Microphones are transducers that convert sound pressure variations in to analogous electrical signal. In microphones, a thin diaphragm converts pressure variations in to mechanical movement. This mechanical movement is converted in to an electrical signal using and appropriate secondary transducers. Microphones are classified based on the secondary transducer are as follows:

- Capacitor or condenser microphone
- Piezo electric crystal microphone
- Electro dynamic microphone

a) Capacitor or condenser microphone

A capacitor or condenser microphone is shown in the Figure 3.14. In this device, an air dielectric capacitor is used in this arrangement. One of the plates of the capacitor acts as diaphragm. A constant charge is maintained on the plates. If the gap between the plates changes, it causes change in capacitance and hence the voltage of the capacitor changes. When sound impinges on the diaphragm, it produces an output voltage due to the change in air gap between the plates of the capacitor. The electrical output produced is the measure of sound.

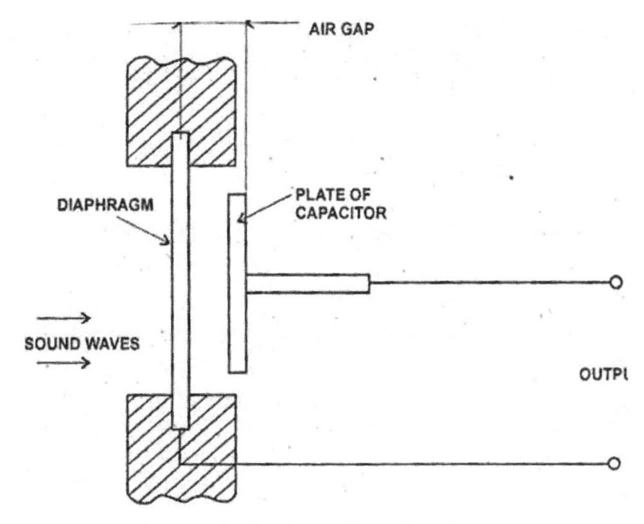

Figure 3.14 Capacitor microphone

(b) Piezo electric crystal microphone
In this type of microphone, a piezo electric element is activated by bending. The diaphragm is placed on a piezo electric crystal. When sound impinges on the diaphragm, the piezo electric crystal gives an electrical signal output which is a measure of sound

(c) Electro dynamic microphone
In this type, the principle of moving conductor in a magnetic field is used in these microphones. The device consists of a diaphragm that carries a coil (Figure 3.15).

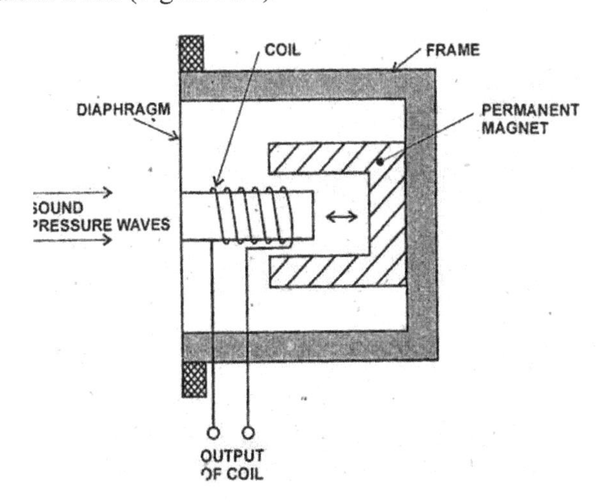

Figure 3.15 Dynamic microphone

There is a permanent magnet in which the coil moves to and fro in a permanent magnet whenever the sound pressure waves impinges on the diaphragm.

3.9 Measurement of vibration

Mechanical motion measurement generally concerns determination of displacement, velocity and acceleration. It includes measurement of displacements of elastic transducers when subjected to force, pressure, temperature, etc. and also vibratory motion of structures and machines. A time varying displacement which is continuous and has *some degree of repetitive nature is generally termed as vibration.* Motion measurement includes both shock and vibration measurement.

Vibration measurement

In vibrating systems, the quantities required to be measured are amplitude of displacement, velocity and acceleration. Furthermore, it may be noted that displacement, velocity and acceleration are related to each other, since velocity is the rate of change of displacement with time and acceleration is the rate of change of velocity with time. Each quantity may be obtained by using differentiating or integrating circuits if one of the quantities has been measured.

3.9.1 Simple vibration instruments

When the amplitudes of motion are greater than 1 mm a simple device is vibrating wedge of paper, or thin material, attached to the surface of vibrating member.

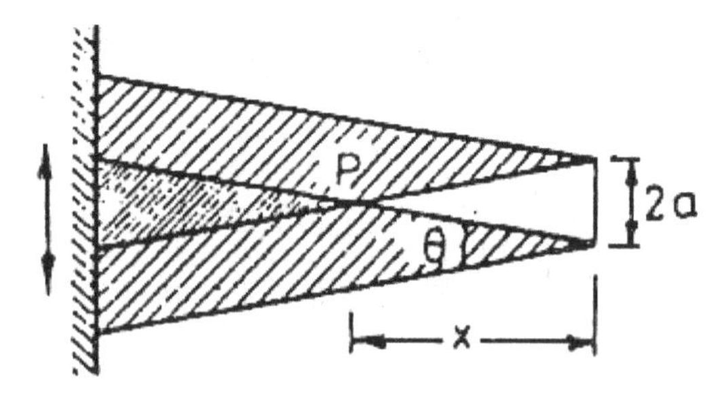

Figure 3.16 Vibrating wedge for amplitude measurement

In the Figure 3.16 the axis of symmetry of the wedge is placed at right angles to the motion. As the member vibrates, the wedge successively assumes two extreme positions. By carefully observing the point "P", where images overlap and measuring the distance "x" the amplitude of motion can be determined. At this distance, the width of the wedge is equal to the double amplitude "a" of the motion.

$$a = x \tan \theta/2$$

where, θ is the included angle of the wedge and "a" is the amplitude of motion.

Another simple device frequency consists of a small cantilever beam mounted on a block as shown in the Figure 3.17 (A) and (B) is placed against the vibrating surface.

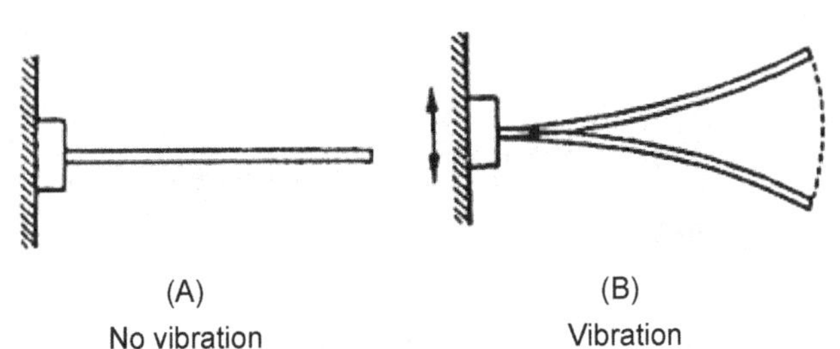

(A) No vibration (B) Vibration

Figure 3.17 (A) and (B) Cantilever beam for frequency measurement

The beam length can be varied by some appropriate arrangements provided. The beam length is slowly adjusted observing the length of the beam in which the resonance occurs and its natural frequency equals the frequency of the vibrating surface. The length of the beam can be calibrated in terms of frequency.

3.10 Seismic instrument

A schematic diagram of Seismic instrument is shown in the Figure 3.18. A mass (m) is connected through parallel spring ad dashpot device to the housing frame. The housing frame is connected to the vibration source

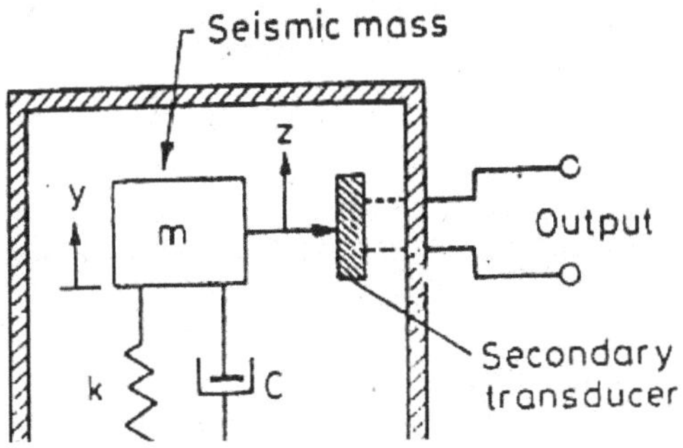

Figure 3.18 Seismic instrument

where the characteristics of vibration is to be measured. The relative motion between the mass (m) and the housing can be used for either displacement or acceleration measurement, depending upon the selection of mass, spring and dashpot combinations. In general, a large mass and soft spring designed for low un damped natural frequency is displacement measurement, while a relatively small mass and stiff spring designed for high un damped natural frequency is used for acceleration measurements. The relative displacement between the mass and the housing can be measured by various secondary transducers as given in the succeeding sections:

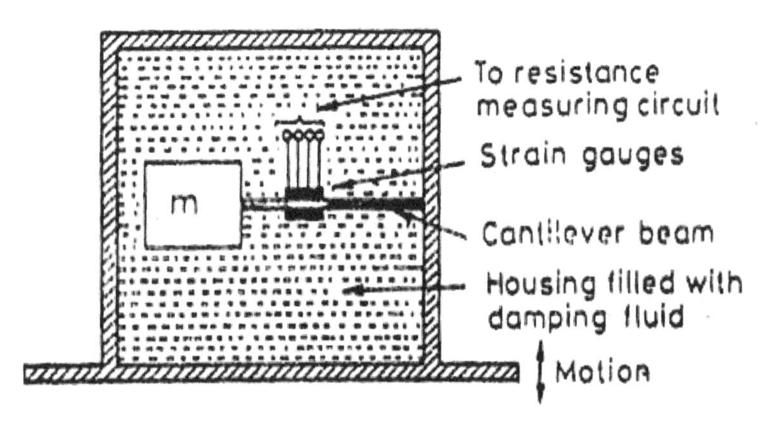

Figure 3.19 Bonded and unbonded strain gauges

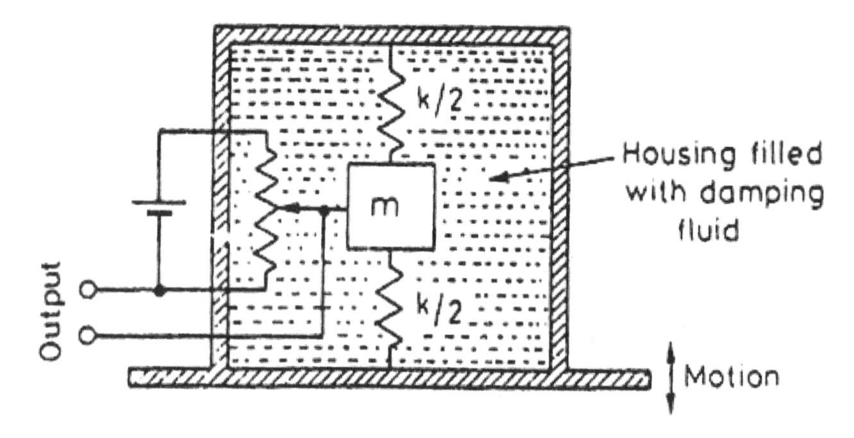

Figure 3.19 Resistance potentiometers

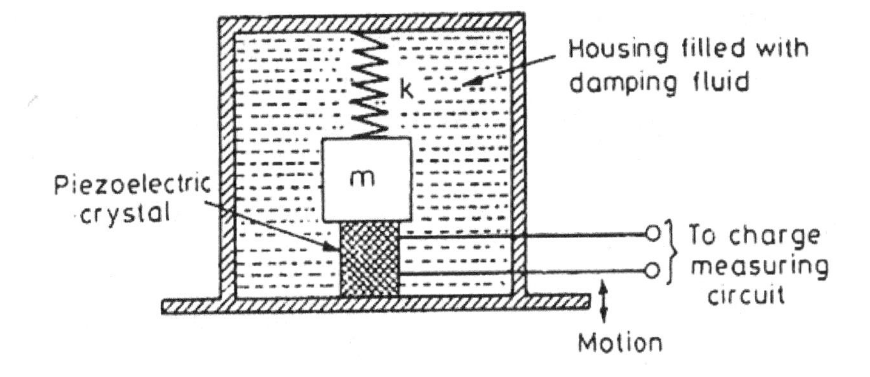

Figure 3.19 Piezo electric transducer

All the above listed transducers are displacement transducers.

If the secondary transducer used is a "velocity pickup" type, the output becomes a function of velocity. These transducers, in general, incorporate a permanent magnet and a coil. The output voltage induced due to the relative motion between the two is proportional to the rate of change of magnetic flux and hence to the rate of change of displacement or velocity as shown in the Figure 3.19 (D).

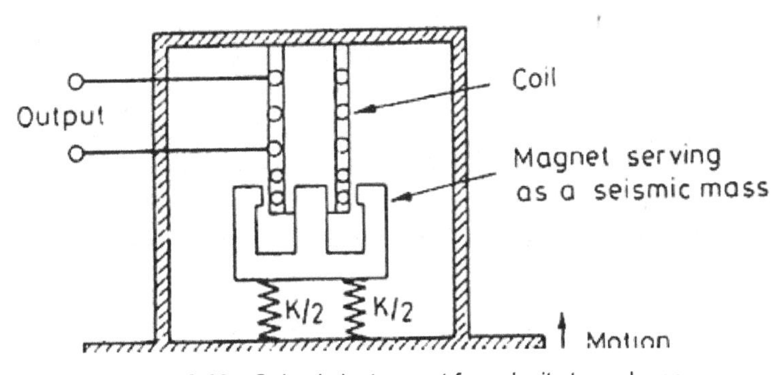

Figure 3.19 Seismic instrument for velocity transducer

3.11 Instrument transducers: principle and application in textile machines

Electronic instruments are powered by electrical energy and the information (output signal) they process should be in the form of electrical energy. This information after processing is generally obtained as a deflection of a pointer in a meter. This implies that in an electronic measuring system, a sensing or detecting element is necessary to convert the quantity being measured in to a corresponding electrical quantity to process it by the system. Such sensors or detectors are called *Instrument Transducers*. It is thus the first element in a general electronic measuring system as shown in the Figure 3.20. The block diagram shows three functional blocks.

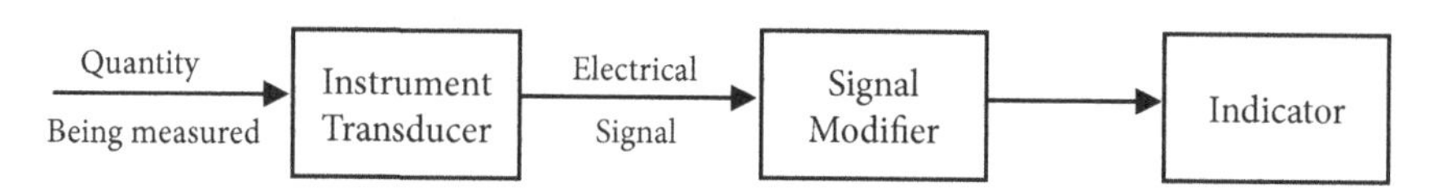

Figure 3.20 Block diagram of Instrument transducers

1. Detection or sensing by an instrument Transducer to produce an electrical signal.
2. Modifications of the electrical signal from the transducer by a signal modifier in to suitable waveform.
3. Indication to provide a read out of the quantity measured on a meter, cathode ray oscilloscope screen, recorder chart, counters, etc.

Although the instrument transducers convert the quantity being measured in to an electrical quantity, some of the transducers such as variable resistance transducers require an auxiliary source of electrical energy and are then strictly energy controllers. The term instrument transducers are generalised to include such energy controllers and are normally called as *Passive transducers*. Transducers which do not require any auxiliary electrical energy but generate the energy proportional to the quantity being measured are called *Active transducers*.

3.11.1 Active instrument transducers

In the active instrument transducers, some of the energy conversion principles are used. They are

- Electro-magnetic
- Piezo electric
- Thermo-electric and
- Photo-voltaic type.

Electro-magnetic type

Principle

Electro-magnetic type of transducers work on the principle of Faraday's law of induction which states that an e.m.f. developed in a loop of wire placed in a magnetic field is proportional to the rate of change of magnetic field. In other words, a relative motion between the magnetic field lines and the loop of wire is necessary for the e.m.f. to be developed in the loop of wire.

Figure 3.21 (A) shows the loop of wire placed in a magnetic field.

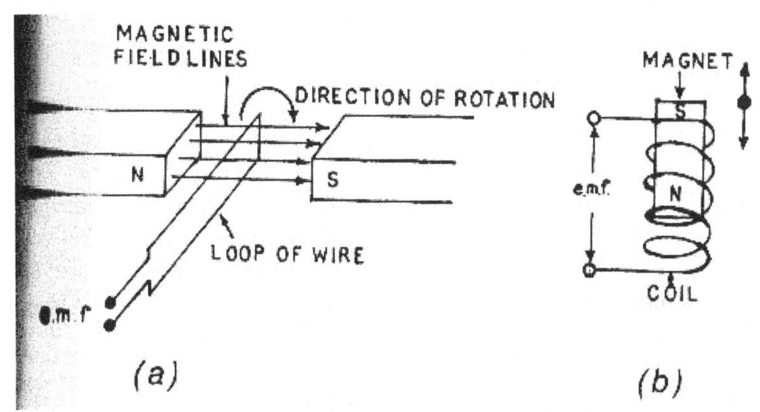

Figure 3.21 (A) Loop of wire moving in a magnetic field. (B). Magnet being moved in the stationary coil.

If the loop is rotated, a voltage will be generated in it. Alternatively, if a magnet is involved in the stationary coil, a voltage will be developed as shown in Figure 3.21 (B)

Applications

This type of transducers is suitable only for dynamic measurements like RPM, vibrations, velocity and acceleration.

Tacho generator

Tacho generator is an electrical tachometer used for measuring speeds in terms of RPM of the rotating shafts by means of generated voltage which is directly proportional to the speed.

Vibration transducer

The vibration transducer shown in the Figure 3.22 is similar in construction to a microphone.

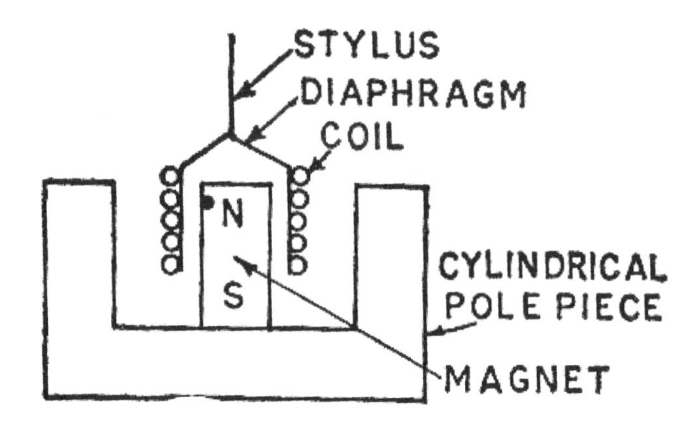

Figure 3.22 Vibration transducer

A coil wound on core of insulating material is placed in radial field arrangement. A diaphragm and stylus are attached to the coil. The stylus when kept in contact with a machine member transfers the vibrations to the coil and makes it to move to and fro in the magnetic field. Hence a voltage is developed in the coil at a frequency equal to that of the vibrations.

Measurement of rate of flow of conducting liquids

Faraday's law of induction is used to measure the rate of flow of electrically conducting liquids. The schematic arrangement is shown in Figure 3.23.

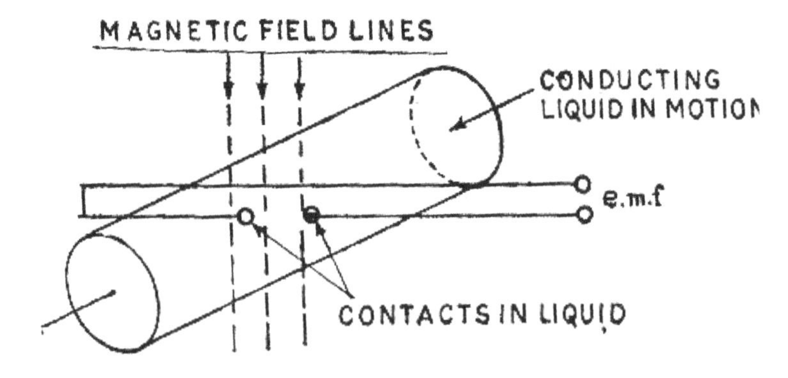

Figure 3.23 Measurement of rate of flow of conducting liquids

In this Figure 3.23, two contacts are placed in the conducting liquid in a direction mutually perpendicular to the direction of flow of the liquid and also to the lines of magnetic field. The liquid as it flows, cuts the lines of magnetic field and hence voltage generated in the liquid is picked up by the two contacts in the liquid. The amplitude of the generated voltage is proportional to the rate of flow of liquid.

Piezo electric type

Principle

Mechanical strains in crystals of certain classes produce piezo electricity an electrical polarisation, the polarisation being proportional to the strain and changes sign with it. Certain crystals like those of quartz, Rochelle salt, barium fitanate compounds exhibit piezo electric effect (Figure 3.24).

Applications:

Piezo electric transducers are force sensitive devices and there are useful in the measurement of physical quantities which can be reduced to forces such as pressure, stress or acceleration. Transducers of this type utilize two piezo electric elements cemented together called *bimorph*, to increase the sensitivity.

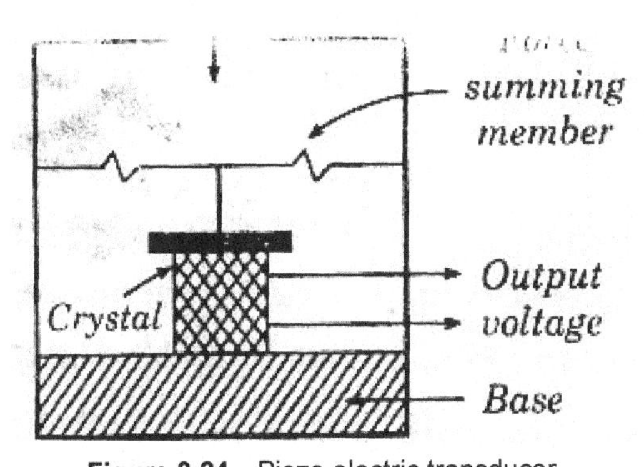

Figure 3.24 Piezo electric transducer

Bimorph is cemented to a machine member and the distortions produced in the machine member are transferred to the *bimorph*. The stress thus developed generates a voltage in the bimorph which is proportional to the amplitude of vibrations.

Acceleration Transducer

Figure 3.25 shows the arrangement of acceleration transducer.

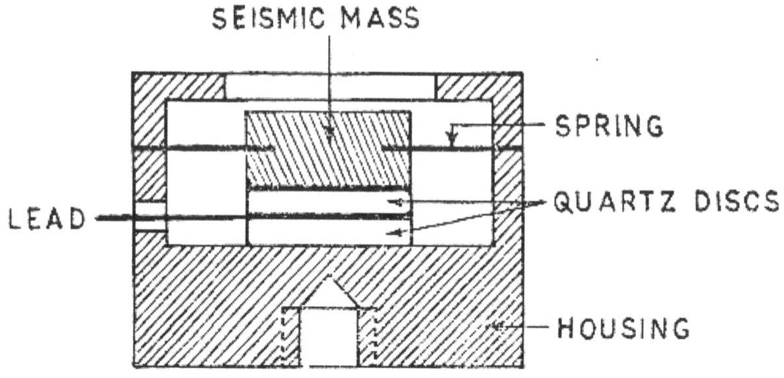

Figure 3.25 Acceleration transducer

In this type of transducers, the arrangement consists of two quartz discs for measuring acceleration. Seismic mass supported by springs rests on two quartz discs. The seismic mass under acceleration exerts pressure on quartz discs which causes a voltage to be developed across the two quartz discs. Thus the voltage developed is proportional to the acceleration. As the size of the transducers is small, it is possible to measure the acceleration of shuttles in shuttle looms with transducer fixed in it.

Thermo-electric type

Principle

When junctions of two dissimilar metals, iron and copper are kept at different temperatures, electric current flows from one junction to another which is proportional to the temperature difference between the two junctions.

Applications

Thermocouples are suitable for measuring high temperatures. Different sensitivities and temperature ranges can be obtained by using combination of various metals and alloys. A basic thermocouple is shown in the Figure 3.26.

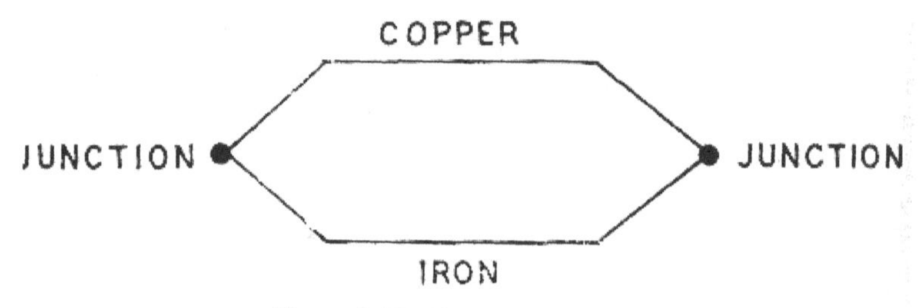

Figure 3.26 Thermo-electric type

Photo-voltaic type

Principle

In this type, light is directly converted in to electrical energy. It was observed that when light falls on oxides deposited on certain metals, a voltage is developed across the two surfaces. Construction of such type of voltaic cell is shown in the Figure 3.27.

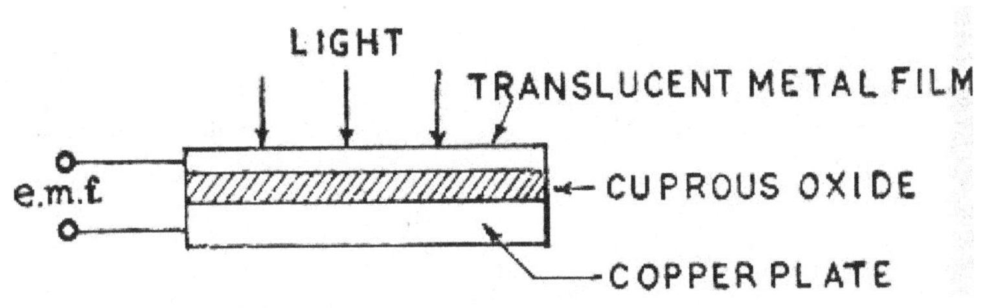

Figure 3.27 Photo-voltaic type

Cuprous oxide deposited on a copper plate is covered with a thin translucent metal film. This thin film serves as electrical contact and at the same time protects the oxide. Light falling on this oxide generates a voltage between the copper plate and oxide.

Applications

Photo-voltaic cells are useful in measuring light intensity, temperature and also are commonly used in spectroscopy and lustre measurements.

3.12 Passive instrument transducers

Some of the physical effects used in passive transducers are:

- Resistance type (Requires AC or DC auxiliary supply)
- Capacitance type (Requires AC auxiliary supply)
- Inductance Type (Requires AC auxiliary supply)

Variable resistance type

Principle

There are many versions of this type though they differ considerably in their design characteristics they work on the same basic principle change in resistance.

$$\delta R = \delta l / \delta A$$

Where

δR = Change in resistance

δl = Change in length of resistance wire

δA = Change in cross-sectional area of wire.

The resistance can be changed by either length, area or by both. In general, when the length is changed, the cross-sectional area also varies. The changes are such that it changes the resistance value. This principle is used in Strain Gauges in the strength testing of textile materials in textile testing equipments.

Strain gauge

Strain gauge is a commonly used transducer to sense the strain or effects of forces like weight, pressure, displacement and similar quantities. There are two types of strain gauges.

- Bonded
- Un bonded

The construction of bonded type strain gauge is shown in the Figure 3.28.

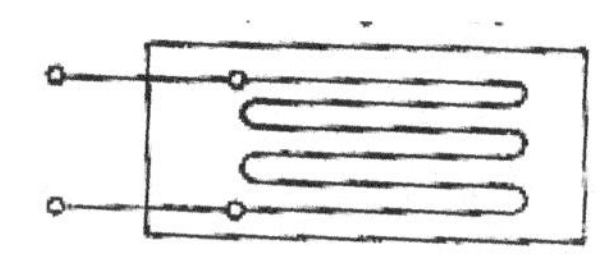

Figure 3.28 Bonded strain gauge

Applications

Figure 3.29 shows the applications of bonded strain gauges.

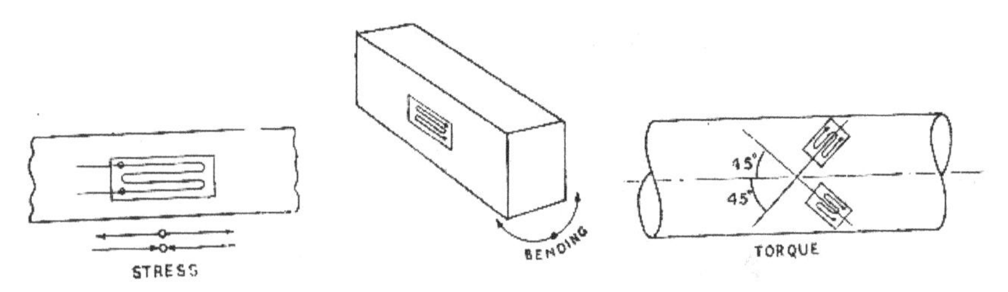

Figure 3.29(A) Stress, strain, bending and torque

Figure 3.29 (A) shows the application of bonded strain gauges for measurement of linear or bending strain of machine members and torque of rotating shafts.

Figure 3.29 (B) shows the pressure of fluids imparts strain on the cantilever spring which is sensed by the strain gauge.

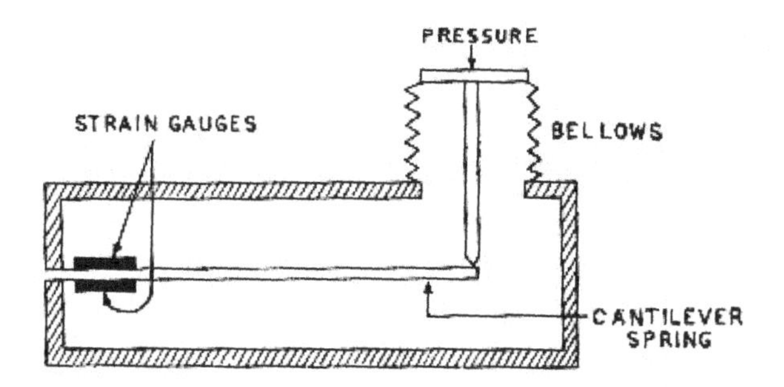

Figure 3.29(B) Pressure

A fine wire element is looped back and forth and bonded on a mounting plate like stiff paper or woven in to a fibre-glass fabric. The mounting plate is connected to a member which undergoes strain. Due to the loading, the extra length obtained by a hair pin looping increases the effects of stress applied in the direction of length. Hence a tensile stress would stretch the wire, increasing its length and reducing the cross-sectional area and thus increasing its resistance.

Figure 3.29 (C) shows a load cell comprising a ring and four strain gauges bonded on the inner surface of the ring. Stress strain curves of fibres, yarns, etc., can be obtained by using such type of load cell.

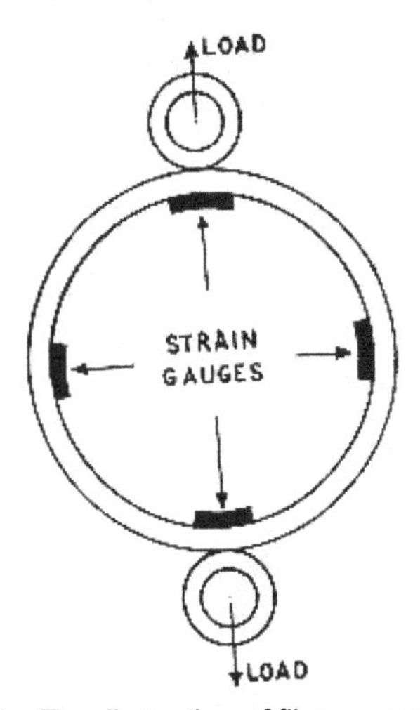

Figure 3.29(C) Tensile testing of fibres, yarns and fabrics

Figure 3.29 (D) shows an arrangement of acceleration of shuttles. The cantilever spring is strained by the known weight (W) which is forced to move in one direction due to acceleration.

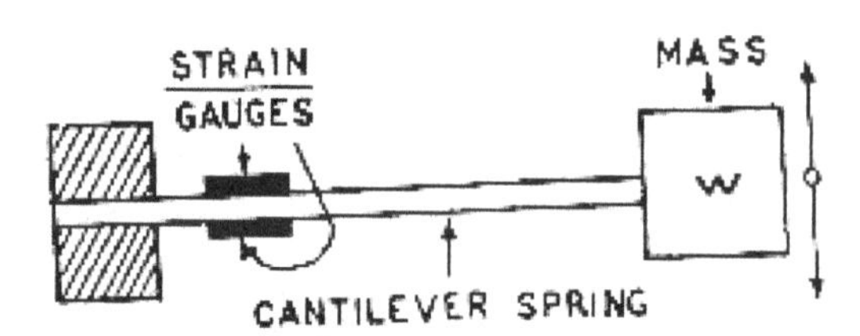

Figure 3.29(D) Acceleration of shuttles

3.13 Humidity measurement

The amount of water vapour in the atmosphere or humidity/moisture is very important in a number of processes. Humidity or amount of moisture in the atmosphere affects the behaviour of many materials like textile yarns, fabrics, paper, leather and wood products. Hence, measurement of humidity and control is important for satisfactory running of machines to achieve required productivity, quality besides human comfort. In many textile mills, air conditioning is installed to control the temperature, relative humidity and air cleanliness. Atmospheric air consists of mixture of dry air and water vapour. The amount of water vapour in the atmosphere varies depending upon the temperature and pressure. Many instruments are there to measure the relative humidity of the atmosphere as well in the departments. In textile mills, hygrometers and sling hygrometers are most commonly used for measurement of humidity. One such humidity sensor works on electrical method is discussed below.

Electrical method (Electrical Humidity Transducer)

Humistor is a humidity sensitive resistor. The principle is the electrical resistance of these resistors vary with the changes in the humidity of the surrounding air. The transducer consists of a resistance element constructed by winding dual noble metal elements on a plastic sheet with a controlled spacing between them. The windings are then coated with moisture sensitive salts like lithium chloride which forms a conducting path between the wires. When the humidity increases in the atmosphere, the coating becomes conductive and the electrical resistance get decreased. The variation of resistance is calibrated in terms of relative humidity (RH) units. It is very suitable for industry because of its, speed, accuracy and high sensitivity.

Humidity sensors

Humidity sensors have gained increasing applications in industrial processing and environmental control. For manufacturing highly sophisticated integrated circuits in semiconductor industry, humidity or moisture levels are constantly monitored in wafer processing. textile production, etc.

Classification of humidity sensors

Humidity Sensing – Classification and Principles (Figure 3.30).

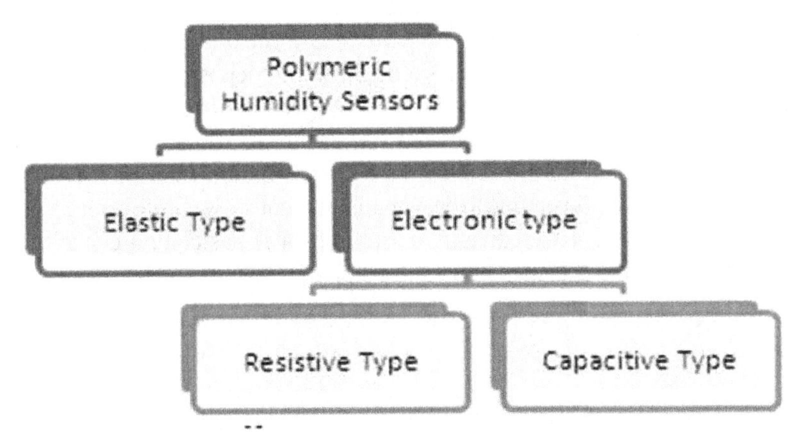

Figure 3.30 Classification of humidity sensors

Sensing Principle

Humidity measurement can be done using dry and wet bulb hygrometers, dew point hygrometers, and electronic hygrometers. There has been a surge in the demand of electronic hygrometers, often called humidity sensors. Electronic type hygrometers or humidity sensors can be broadly divided into two categories: one employs capacitive sensing principle, while other uses resistive effects.

3.13.1 Sensors based on capacitive effect

Humidity sensors relying on this principle consists of a hygroscopic dielectric material sandwiched between a pair of electrodes forming a small capacitor. Most capacitive sensors use a plastic or polymer as the dielectric material, with a typical dielectric constant ranging from 2 to 15. In absence of moisture, the dielectric constant of the hygroscopic dielectric material and the sensor geometry determine the value of capacitance. At normal room temperature, the dielectric constant of water vapour has a value of about 80, a value much larger than the constant of the sensor dielectric material. Therefore, absorption of water vapour by the sensor results in an increase in sensor capacitance.

At equilibrium conditions, the amount of moisture present in a hygroscopic material depends on both the ambient temperature and the ambient water vapour pressure. This is true also for the hygroscopic dielectric material used on the sensor. By definition, relative humidity is a function of both the ambient temperature and water vapour pressure. Therefore there is a relationship between relative humidity, the amount of moisture present in the sensor, and sensor capacitance. This relationship governs the operation of a capacitive humidity instrument.

Basic structure of capacitive type humidity sensor is shown in the Figure 3.31.

On alumina substrate, lower electrode is formed using gold, platinum or other material. A polymer layer such as PVA is deposited on the electrode. This layers senses humidity. On top of this polymer film, gold layer is deposited which acts as top electrode. The top electrode also allows water vapour to pass through it, into the sensing layer. The vapours enter or leave the hygroscopic sensing layer until the vapour content is in equilibrium with the ambient air or gas. Thus capacitive type sensor is basically a capacitor with humidity sensitive polymer film as the dielectric.

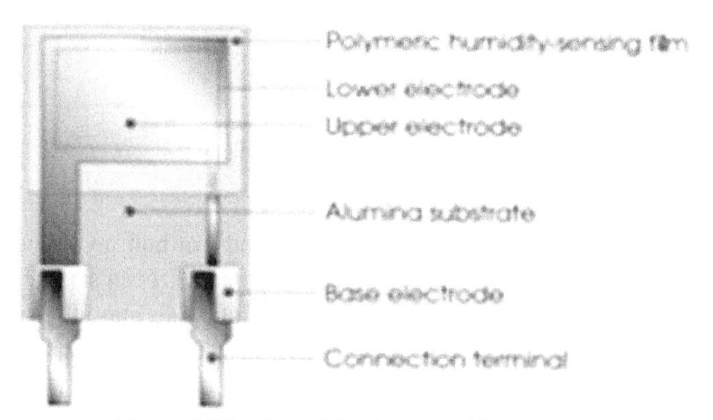

Figure 3.31 Capacitive type humidity sensor

3.13.2 Sensors based on resistive effect

Resistive type humidity sensors (Figure 3.32) pick up changes in the resistance value of the sensor element in response to the change in the humidity. Basic structure of resistive type humidity sensor is discussed below. Thick film conductor of precious metals like gold, ruthenium oxide is printed and calcinated in the shape of the comb to form an electrode. Then a polymeric film is applied on the electrode; the film acts as a humidity sensing film due to the existence of movable ions. Change in impedance occurs due to the change in the number of movable ions.

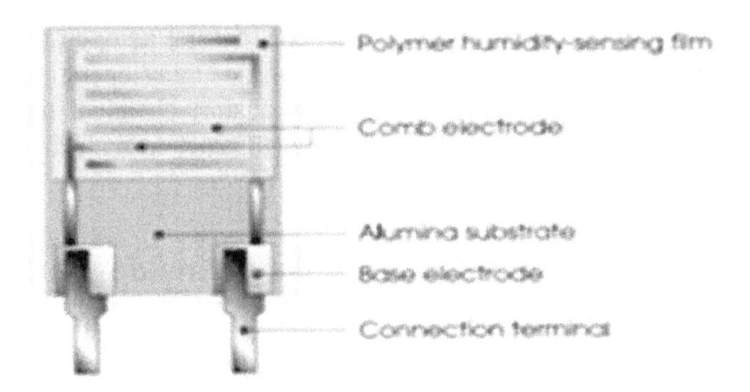

Figure 3.32 Humidity sensitive resistors

Semi conducting materials like silicon change their resistance with change in temperature. Such temperature sensitive resistors are called thermistors which are suitable for measurement of temperature, rate of flow, etc. In certain salts like lithium chloride, the changes in humidity cause changes in resistance. A typical transducer using lithium chloride of 3% concentration dope filled on a ploy acetate film changes resistance from 20 million ohms at 5% R.H. to 1000 ohms at 100% R.H. Semi-conductors like titanium dioxide and some conducting plastics are also suitable for measurement of humidity.

3.13.3 Pros and cons of these two types of sensors

- Capacitive type sensors are very linear and hence can measure RH from 0% to 100%, but require complex circuit and also need regular calibration.
- Resistive type sensors find difficulty in measuring low values (below 5% RH), the change is impedance is too high and hence it is difficult to control the dynamics, temperature affects the properties significantly.
- However, advances in electronics can mitigate the problems of temperature effects and high impedance change.
- Capacitive RH sensors dominate both atmospheric and process measurements and are the only types of full-range RH measuring devices capable of operating accurately down to 0% RH.
- Because of their low temperature effect, they are often used over wide temperature ranges without active temperature compensation.

- Thermo set polymer-based capacitive sensors, as opposed to thermoplastic-based capacitive sensors, allow higher operating temperatures and provide better resistivity against chemical liquids and vapours such as isopropyl, benzene, toluene, formaldehydes, oils, common cleaning agents, etc.

3.14 Measurement of air pressure inside the duct in spinning mills

Measurement of air pressures is the method of knowing the air velocity and quantity of air flowing through the ducts. It is important to know about the pressures inside the duct so as to identify if there is any pressure loss in the duct. In modern spinning technology, the technological air combined with air engineering concepts play a major role in transporting the opened fibres from one machine to another machine. A knowledge of this is required since the air engineering concepts also involve the removal of waste from the machines. In any duct, the total pressure is the sum of static pressure + dynamic pressure.

Static pressure (Ps)

The air delivered by the fan in to the duct system exerts a pressure on the sides of the wall of the duct. Static pressure inside the duct is the pressure inside the duct which resists the air flow. This is the pressure that is exerted on the surface of the duct in the direction parallel to the air flow. The static pressure in the duct is measured by making a hole perpendicular to the duct and connecting a pressure measuring instrument to it. Static pressure is generated by forcing air in through the one end of the duct with the other end sealed. In this case, there is no air flow. This is known as static pressure and this is the pressure exerted by the air even when it is static (air is not flowing).

Dynamic pressure (Pdyn)

The dynamic pressure developed in the duct when the outlet is open and the air in the duct is set in motion. This can be explained by an example. If a plate is suspended in the duct perpendicular to direction of flow of air, it will tilt in the direction of air flow due to the impact of air in motion. The plate would remain inclined at the same angle as long as the velocity of the air does not change. This is because the pressure exerted on the surface of the plate causing it to tilt would increase/ decrease with the velocity of air moving air. This is known as Dynamic pressure.

P dynamic = Pdyn = $pv^2/2$

where p = density of air 1.25 kg/m^3

and v= velocity of air in m/sec.

The dynamic pressure cannot be measured directly. It is calculated from the difference in total pressure and static pressure.

Total pressure (Pt)

Total pressure may be explained with an example. Consider the example of the same plate for dynamic pressure. In this case, the plate would be hung with its surface parallel to the air flow, hence it would not affect the position of the plate and the plate would remain hanging vertically. The static pressure exerted on the plate will be equal on both sides of the plate and will not affect the position of the plate. Thus, the air in motion inside the duct exerts both static and dynamic pressure. If the probe of the measuring instrument, say, manometer, is inserted in the duct with its mouth perpendicular to the air movement, the manometer will indicate the static pressure. If probe of the manometer is inserted opposing the movement of the air, the manometer will show the total pressure since the probe is subjected to static and dynamic pressure.

Measurement of pressure

The pressure is generally measured using an instrument, manometer. The principle of this instrument is to measure the air pressure in a duct by forcing a column of water upwards in a tube. This is called water manometer. The instrument consists of a U-tube which is transparent filled with water about the half of the height of the two legs. The water level in both the legs will be same when both ends of legs open at the top exposed to the atmosphere or to the same pressure in both the legs. A flexible measuring tube is connected to one leg. The end of the flexible tube is inserted in the duct perpendicular to the air flow as shown in Figure 3.33.

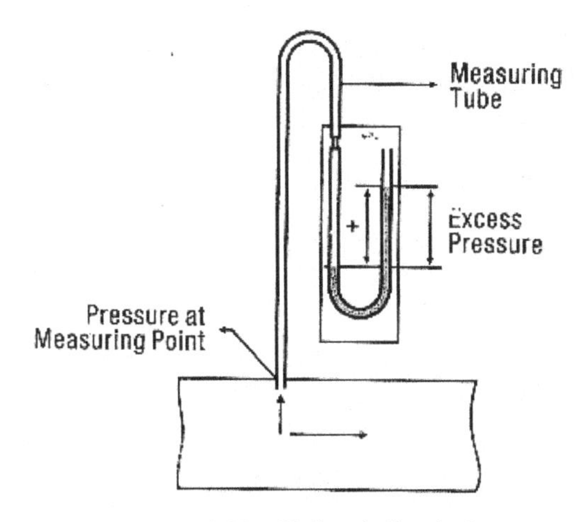

Figure 3.33 Air flow in the duct

Depending upon the air pressure in the duct, the water level is pushed down and up in the left leg of the manometer. The difference in water level in the two legs measured in millimetres is the static pressure of air.

3.14.1 Pressure measurement using Pilot tube

Another method of measuring the static and dynamic pressure is by using the instrument Pilot tube. The construction of this Pilot tube is, it consists of two tubes one within the other.

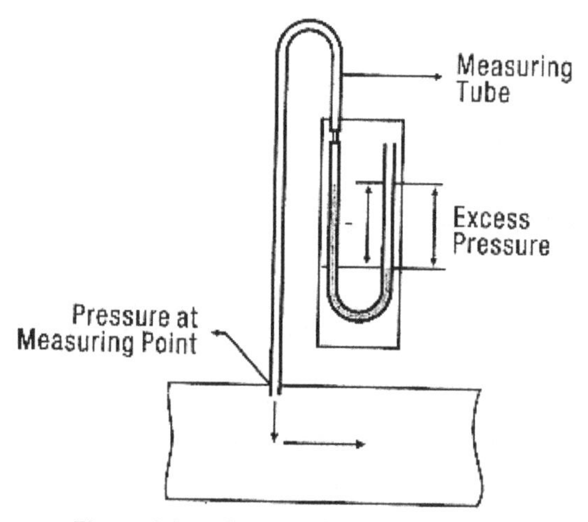

Figure 3.34 Suction of air in the duct

The inner tube is used to measure the total pressure. The outer tube has a number of small holes usually eight in number and each hole is 1 mm in diameter drilled on the sides so as to measure the static pressure. Connection points are provided on the other end of the pilot tube where a manometer is to be connected for pressure measurement (Figure 3.34).

3.14.2 Measurement of dynamic pressure using Pilot tube

- Connect the manometer to the two open ends of the Pilot tube as shown in Figure 3.35 (A) and (B).
- Drill a 6 mm hole in the duct or pipe where the dynamic pressure is to be measured.
- Insert the pilot tube in the duct or pipe opposing the air flow.

- The difference in the water levels in the two legs measured in millimetres is the dynamic pressure of the air flowing in the duct.

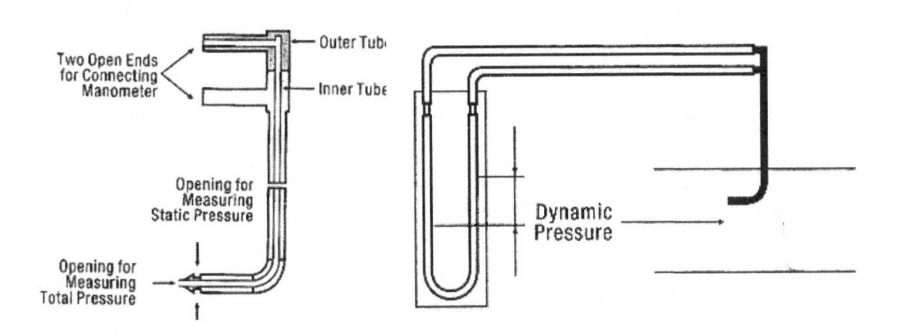

(A) Pilot tube (B) Dynamic pressure measurement

Figure 3.35 (A) Pilot tube (B) Measurement of dynamic pressure

3.14.3 Measurement of static pressure

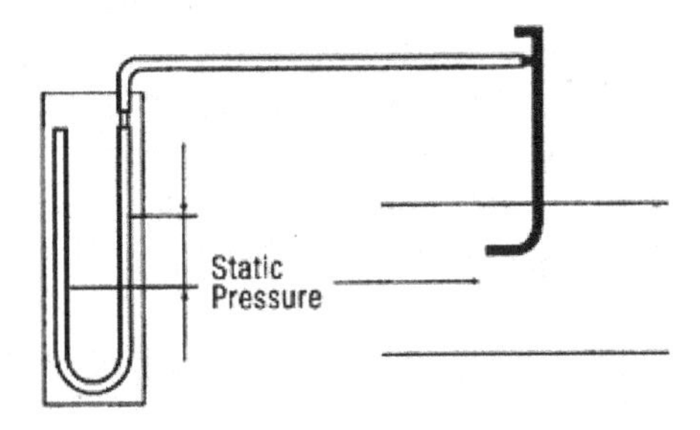

Figure 3.36 Measurement of static pressure

- Connect the manometer to the outer open ends of the Pilot tube as shown in the Figure 3.36.
- Drill a 6 mm hole in the duct or pipe where the static pressure is to be measured.
- Insert the pilot tube in the duct or pipe opposing the air flow.
- The difference in the water level in the two legs measured in millimetres is the static pressure of the air in the duct.

3.15 Measurement of viscosity of size paste in sizing machines

Viscosity is a measure of size pickup, since other factors in pickup are constant. Laboratory checks on viscosity are inadequate, mills are finding, and a viscometer on the slasher is being used more and more.

The size content or pickup of a warp is a function of

a) properties of the sizing solution, which can be measured by viscosity,
b) character of the material sized,
c) time of applications, and
d) the effect of the squeezing process.

Because items b, c, and d are normally constant, viscosity is also a measure of the size pickup in the warp, and can be used in finding the best size formula for particular warps.

Properties of the size solution

The properties of the sizing solution depend upon the type and quantity of size used, how it is prepared, and the operating temperature. Viscosity is a measure of each of these factors because viscosity changes with different types of size. If the quantity of size used is increased, viscosity normally increases. If the operating temperature is increased, viscosity normally decreases. If the size is prepared by cooking, viscosity is low in the early part of the cooking process. The viscosity increases to very high values during the paste state, and then gradually decreases to the correct value when the process is terminated. If the size is under-cooked, the viscosity will be high. If it is overcooked, it will be too low.

Viscosity is also an important factor when starches are mechanically converted by homogenization, and the pressures applied during the homogenization can be adjusted to obtain the desired viscosity. Size held in the storage kettle at high temperature decreases in viscosity. Furthermore, condensation from the steam used to maintain the proper temperature in the size box dilutes the solution and lowers the viscosity. Mechanical agitation of the size in the box further breaks down the starch and also lowers the viscosity. Viscosity of the sizing solution as it is used is therefore an overall measurement of all of factors affecting the properties of the solution, and the viscosity will remain constant if all of these factors are not permitted to vary. The properties of the textile material being sized and its previous treatment will of course influence size pickup to some extent, but normal variations in the character of the material do not appreciably affect pickup.

3.15.1 Viscometer

Viscometers have been perfected for use on slashers for process control. Such instruments automatically measure the viscosity of the solutions in the actual process where they are used, and eliminate the necessity of removing samples for laboratory tests. As a result, accurate control of viscosity can be maintained under actual operating conditions.

The measuring element of one type of viscometer that is in use in the industry is shown in the accompanying photograph as applied to a cotton slasher. In this particular type, a plunger or piston is raised by a motor cam mechanism and then allowed to drop by gravity. A sample of the solution whose viscosity is to be measured is drawn in through orifices as the piston is raised, and expelled when the piston is dropped. The time required for the piston to drop is a measure of the viscosity. This measurement is transmitted to a recorder through a suitable multi-conductor cable.

Measurements are then converted into chart readings by means of motors, relays, and other associated parts. The recorder can be set to sound an alarm if high and low limits of viscosity are reached. The viscometer records viscosity in convenient arbitrary units from 0 to 100. For such applications, it is not necessary to have the instrument calibrated into absolute viscosity units, because it is not feasible to compute the optimum value of viscosity to be used for any particular warp.

3.16 Stepper motors

As the name *step* implies that the stepper motor is different from electric motors of a.c. and d.c. type. The electric motors like shunt, series and compound motors give continuous motion. Consider the movement of snakes, they have no feet. They just slide or move smoothly. Human beings and animals move with two and four feet. It means they traverse the distance in discrete steps.

Technically speaking, it can be stated that it is a motor whose rotor rotates in discrete angular steps in response to planned sequence or programmed sequence of excitation and energisation. The movement I basically due to magnetic interaction between salient rotor poles and electromagnetic poles created by excitation of stator windings in a specified sequence. In the present context, computerized control, digital control, microprocessor and micro controller based control systems are becoming more popular. In the conventional days, spring wound watches correspond to a continuous motion and digital watch correspond to continuous motion and digital watch with analogue display is the best example for discrete stepped motion.

In general, there are two types of stepper motors are in use. One type uses permanent magnets and the other type is based on variable reluctance.

Permanent type motors are very simple in construction and cheaper in cost. In computer peripherals the line printers, floppy drives, reading heads and scanner heads require discrete stepped motions. For this type of applications permanent magnets are used. The rotor has permanent magnets and excitation system is changed on stator side. In case of variable reluctance type stepped motor, permanent magnets are not used on either stator or rotor side. Due to particular pattern of stator excitation the rotor teeth will try to align for minimum reluctance position. The hybrid stepper motor takes advantage of both permanent and variable reluctance.

Permanent magnet stepper motor

In this permanent type stepper motor, there are two cylindrical permanent magnets on rotor side with north (N) and south (S) poles which are excited by winding and excitation. One general core construction is considered on stator and rotor side. The construction and working principle of a simple two phase permanent magnet stepper motor is shown in the Figure 3.37. The rotor is considered to have two permanent magnetic poles north (N) and south (S), respectively. The stator windings are excited with supply in such a way that windings (1) and (2) establish South pole. Windings (3) and (4) are excited in such a way that they establish combined North Pole. By definition from North pole flux comes out towards the rotor and for South Pole the flux enters from the rotor side. The windings can be connected in series or in parallel but proper direction of current is to be established as shown in Figure 3.36. Due to the

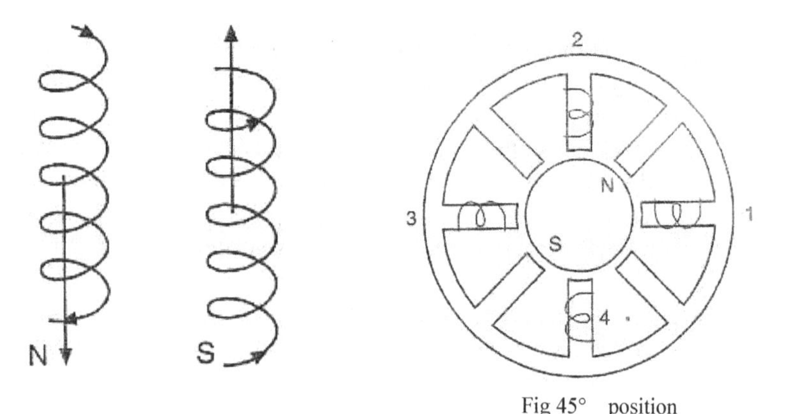

Figure 3.37 Permanent magnet stepper motor

attraction between opposite poles and repulsion between similar poles if windings (1) and (2) establish (S) pole windings and windings 3 and 4 establish the (N) pole the rotor (N) and (S) poles will align in position at 45° with respect to vertical direction as shown in the Figure 3.37.

After some time, if windings 1 and 2 are excited to make (N) pole and windings 3 and 4 are made to establish (S) pole the flux axis will shift by 180°. The rotor poles will rotate. Rotor (N) pole will be at 225° and (S) pole will be at 45° as shown in the Figure. In other words, with two stator poles and two rotor poles the step size is 180°. If same excitation sequence is continues, it continues to move with a step size of 180°.

Alternatively, if polarity of current for alternate coils is reversed instead of keeping same coils 1.2.3.4 will establish S,N,S,N poles, respectively. The rotor is having the same two poles. So, if excitation sequence and polarities are properly adjusted the stator poles will be N,S,N,S then S,N,S,N and again N,S,N,S. Each time the magnetic axis will shift by 90° and rotor will follow it resulting in four equal steps of 90°.

3.16.1 Variable reluctance type stepper motors

In this type, three phase winding is arranged in 12 stator slots and it is excited by a particular sequence. The rotor has eight teeth and slots but they are not permanent magnets as shown in the Figure. Each of three phase winding occupies 120° of the periphery and consists of four coils forming four poles. Hence, the adjacent poles are at 30°distance. The rotor has eight teeth at 45° (360/80) each. It means that at any time, the distance between the nearest stator pole and rotor slots is 15°.

As shown in the Figure 3.38 centre line of one rotor slot (1) is just against the centre line of stator pole (1) excited with supply phase (R). After that, if phase (Y) is excited stator pole (2) is energised, the centre line of rotor slot (II) is at a distance of 15°. It will be attracted and aligned for minimum reluctance position, which is centre line os stator pole (2). So, rotor moves by 15° towards left. In this way excitation sequence changes from (R) , then (Y) and then (B) one by one. Four stator poles 1,4,7,10 are established by phase (R) in a series connected four coils with proper polarity. In this way, poles 2, 5, 8, 11 and 3, 6, 9,12 are established by phases Y and B, respectively. The excited sequence R, Y,B is continued and results in 15° step motion in anti-clockwise direction. In this four pole rotor and eight pole stator arrangement described coil R to R distance is 90°.

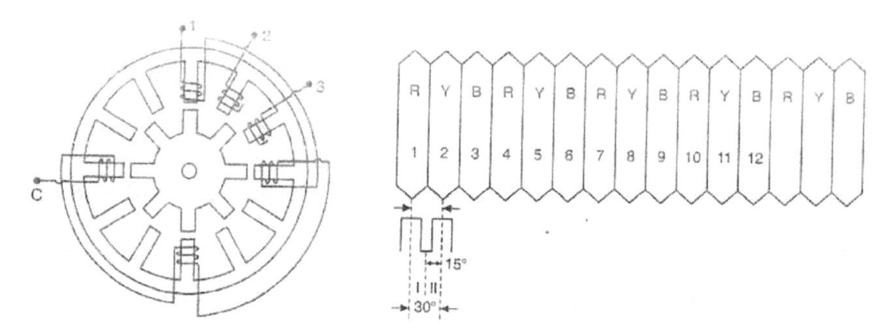

Figure 3.38 15° position

The stepper motor moves in discrete steps when it goes from position (1) to (2) as shown in Figure 3.37. Due to the inertia of the system it behaves as a second order system, subjected to step change. There will be overshooting and rotor will settle down after some oscillations. It happens for every step change. The oscillations should be damped out. In case of permanent magnet stepper motor, the permanent magnets will provide the damping action and in case of variable reluctance stepper motors, damping action will be provided by the eddy currents induced in the rotor. By increasing the frequency of pulsing the settling time can be reduced.

Permanent magnets are used in places where large step size is required and pulse rate is low. Variable reluctance type is used when small step size is required and the pulse rate is high. However, the overshoot of both types of stepper motors is problematic, particularly for precision control systems and servo mechanism applications.

Applications of stepper motors

- Stepper motors are more suitable for digital, numerical, computerised microprocessor and micro-controller based applications.
- In electrical actuators, servomotors and stepper motors are used extensively.
- Due to continuous motion, servo motors have become more popular in servo mechanism and closed-loop control applications.
- In digital control systems with open loop control systems stepper motors are more useful.
- Stepper motors provide one-to-one control since each step is associated with some output. In some applications, it eliminates feedback transducers, amplifiers, integrators, analogue to digital converters.
- The elimination of number of components results in cost savings for components and increased reliability reduces maintenance cost.

- Resolution of stepper motor in open loop control system is number of steps per revolution or step size in degrees.
- Stepper motors are also used in closed loop control systems.
- In instrumentationthe torque requirement is practically very small and stepper motors are very much suited for this type of applications.

4 Instrumentation in textile testing equipments

4.1 Introduction

Conventional method of testing the fibres involves testing the important fibre properties like length, strength and elongation, fineness, maturity, trash in separate instruments. Even though these instruments give reliable estimate on the fibre properties, it takes longer time, laborious, require skill of the operator and also the limitation in number of samples tested. Hence, many modern test equipments have been developed by Uster, Premier and other manufacturers to simplify the testing and also able to test quite a large number of samples required for the spinners.

Cotton fibres have varied fibre properties and require more number of samples to be tested within a short time. Hence, all the required fibre properties are combined in one instrument with separate modules which involve instrumentation for determining the fibre properties. High volume instruments (HVI), advanced fibre information system (AFIS) determine many fibre properties from a sample of cotton which is very useful for the raw cotton purchaser for producing quality yarns.

4.2 Digital fibrograph

Light dependent resistors (LDR) consists of resistors using cadmium sulphide is sensitive to light. The resistance decreases with increase in light intensity. Such resistors are used for the measurement of light intensity. They have been applied for the fibre length distribution measurements in Digital Fibrograph, counting of number of revolutions in rotating shafts etc. It is also known that resistance of textile materials decrease with increase in moisture content. This principle is employed for the measurement of fibres, yarns and fabrics. Measurement of fibre length in textiles measured by Digital Fibrograph is shown in the Figure 4.1.

Fibre length measurement of Digital Fibrograph

Both the high and low volume instruments use an optical principle of determination of fibre length. In this principle, a narrow rectangular beam of light is allowed to fall on the specimen beard. The attenuation of light through the specimen at different areas of the beard is measured and used to obtain the different span length values.

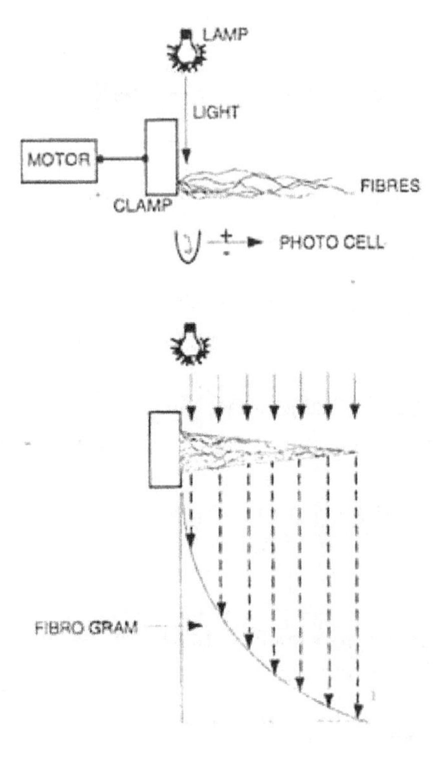

Figure 4.1 Optical measuring principle of length measurement in Digital Fibrograph

4.3 High volume instruments (HVI)

Measurement of different parameters using HVI and low volume instruments (LVI).

Length

Both the high volume instrument and the low volume instrument use an optical principle of determination of fibre length. A narrow rectangular beam of light is allowed to fall on the specimen beard. The attenuation of light through the specimen at different areas of the beard is measured and used to obtain the different span length values. In the HVI, the tip of the beard is scanned first and scanning gradually proceeds towards the clamp while in the LVI, the beard is scanned in the opposite direction. In both the instruments, the span length values are obtained by actual measurement (Figure 4.2).

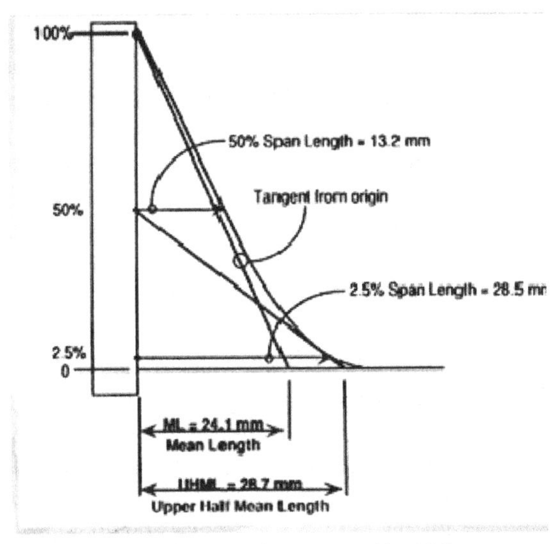

Figure 4.2 Fibre length measured by HVI

HVI instruments provide "Fibrogram" which shows the arrangement of fibres from the shortest to the longest in terms of span lengths (the distances fibres extend from a random catching point). The Fibrogram simulates the way the fibres will behave in yarn making processes. The significance of Fibrogram is recognised in processing of fibres which are caught by drafting rollers or aprons when it is being transferred from one place to another follows a Fibrogram configuration as shown in the Figure 4.3.

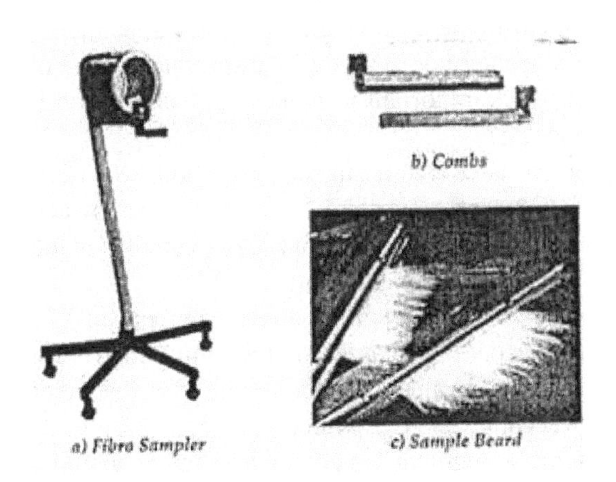

Figure 4.3 HVI instrument

Strength

Fibre strength and elongation determine the toughness of a fibre which has direct bearing on the yarn and fabric strength. It is necessary to test the fibre "bundles" rather than single fibre strength which is representative of the arrangement of fibres in the yarn. Many conventional testing instruments like Pressley Strength tester, Stelometer are available for testing the bundle strength of the fibres. In HVI instruments, constant rate of extension principle is used which is now universal and accepted for testing textile materials. The results of HVI, Stelometer and Pressley should not be expected to match with one another.

In the LVI, the fibro stelo is used. This module uses the pendulum lever principle of loading the specimen to estimate the fibre strength characteristics. A random sample of cotton fibres is prepared, short fibres being removed by combing so that all the fibres in the test specimen extend all the way through the jaws. Uniform sample preparation, proper calibration, control of proper RH% and temperature in the testing lab and proper conditioning of samples would ensure accurate and reliable HVI strength values. Universally, fibre strength is expressed either in breaking tenacity in grams/tex or breaking length is used.

Breaking tenacity in g/tex = (breaking load in kg/sample weight in mg.) × 15 mm

HVI 9000 strength measurement

HVI uses the "Constant rate of elongation" principle while testing the fibre sample. The available conventional methods of strength measurement are slow and are not compatible to be used with the HVI. The main hindering factor is the measurement of weight of the test specimen, which is necessary to estimate the tenacity of the sample. Expression of the breaking strength in terms of tenacity is important to make easy comparison between specimens of varying fineness.

The problem is overcome in the HVl 9000 by positioning the jaws and breaking the fibres at a constant "Amount" location across the beard. By breaking the fibres at a constant amount location, it is made sure that the samples are broken with a constant number of fibres between the jaws.

Therefore, raw data strength is directly proportional to the force to break the fibres. The raw data so obtained are then adjusted to desired levels by testing samples of designated values. In order to make the estimation of the specimen linear density accurate enough, a micronaire correction factor is normally introduced so that the strength values are not affected by variations in micronaire.

Fineness measurement

The micronaire module of HVI 9000 and the low volume fineness tester use the airflow method to estimate the fineness value of cotton. A sample of known weight is compressed in a cylinder to known volume and subjected to an air current at a known pressure. The rate of airflow through this porous plug of fibre is taken to be a measure of the fineness of cotton. The number of fibres in a given weight of cotton will be more in the case of finer fibres than in the case of finer fibres than in the case of coarser fibres. If air is blown through these samples, the plug containing finer fibres will be found to offer a greater resistance than the plug with coarser fibres. This is due to the fact that the total surface area in the case of the former will be greater than the latter and hence the drag on the air flowing past will be more. This differentiating factor is made use of to indirectly measure the fineness of cotton.

The instrument operates as follows. The chamber lid is closed; a piston at the chamber bottom compresses the fibre to a fixed and known volume. A regulated stream of air is then forced through the sample and the pressure drop across the sample is applied to a differential pressure transducer. The transducer outputs an analogue signal voltage proportional to the pressure drop. This analogue voltage is applied to an analogue to digital converter, which outputs a digital signal representing the voltage. Cotton with known fineness values is tested and the voltages obtained are used to obtain the calibration curve, which is used for all subsequent testing to display the cotton fineness.

The fineness is expressed in the form of a parameter called the micronaire value, which is defined as the weight of one inch of the fibre in micrograms. Maturity of cotton also influences the micronaire value.

Moisture

Principle of measurement

Moisture content of the cotton sample at the time of testing, using conductive moisture probe and the main principle involved in the measurement is based on the measurement of the dielectric constant of a material.

Trash content

Principle of measurement

Particle count, % Surface area covered by trash, trash code. Measured optically by utilising a digital camera, and converted to USDA trash grades or customised regional trash standards. The HVI systems measure trash or non-lint content by use of video camera to determine the amount of surface area of the sample that is covered with dark spots. As the camera scans the surface of the sample, the video output drops when a dark spot (presumed to be trash) is encountered. The video signal is processed by a microcomputer to determine the number of dark spots encountered (COUNT) and the per cent of the surface area covered by the dark spots (AREA). The area and count data are used in an equation to predict the amount of visible non-lint content as measured on the Shirley Analyser. The HVI trash data output is a two-digit number which gives the predicted non-lint content for that bale. For example, a trash reading of 28 would mean that the predicted Shirley Analyser visible non-lint content of that bale would be 2.8%.

While the video trash instruments have been around for several years, but the data suggest that the prediction of non-lint content is accurate to about 0.75% non-lint, and that the measurements are repeatable 95% of the time to within 1% non-lint content.

Colour measurement

The HVI colour module utilises optical measuring principles to define colour. The colour module has a photodiode, which collects the reflected light from the sample. The photodiode output is converted into meaningful signals using signal conditioners. The illumination of the sample is done with the help of two lamps connected in parallel. Light from the lamps is reflected from the surface of a cotton sample on the test window. The reflected light is diffused and transmitted to the Rd and +b photodiode. These two signals are conditioned to provide two output voltages, which are proportional to the intensity of light falling on the respective photodiodes. These voltages are converted to digital signals from which the computer derives Rd and +b readings to be displayed on the screen.

Maturity and stickiness

Principle of measurement

Calculated using a sophisticated algorithm based on several HVI™ measurements.

Other information

Near infrared analysis provides a fast, safe and easy means to measure cotton maturity, fineness and sugar content at HVI speed without the need for time consuming sample preparation or fibre blending.

This technology is based on the near infrared reflectance spectroscopy principle in the wavelength range of 750–2500 nm. Differences of maturity in cotton fibres are recognised through distinctly different NIR absorbance spectra. NIR technology also allows for the measurement of sugar content by separating the absorbance characteristics of various sugars from the absorbance of cotton material.

Cotton maturity is the best indicator of potential dyeing problems in cotton products. Immature fibres do not absorb dye as well as mature fibres. This results in a variety of dye-related appearance problems such as barre, reduced colour yield, and white specks. Barre is an unwanted striped appearance in fabric, and is often a result of using yarns containing fibres of different maturity levels. For dyed yarn, colour yield is diminished when immature fibres are used. White specks are small spots in the yarn or fabric which do not dye at all. These specks are usually attributed to neps (tangled clusters of very immature fibres).

NIR maturity and dye uptake in cotton yarns have been shown to correlate highly with maturity as measured by NIR. A correlation of R = 0.96 was obtained for a set of 15 cottons.

HVI calibration principles

HVI uses a unique calibration principle by which routine calibration is performed by presenting samples bf known value to the instrument, which then makes adjustment to the raw instrument values, obtained so that measurements will agree with the calibration cotton designated values. The general principle by which this calibration is done involves a simple, two-point regression analysis (Figure 4.4).

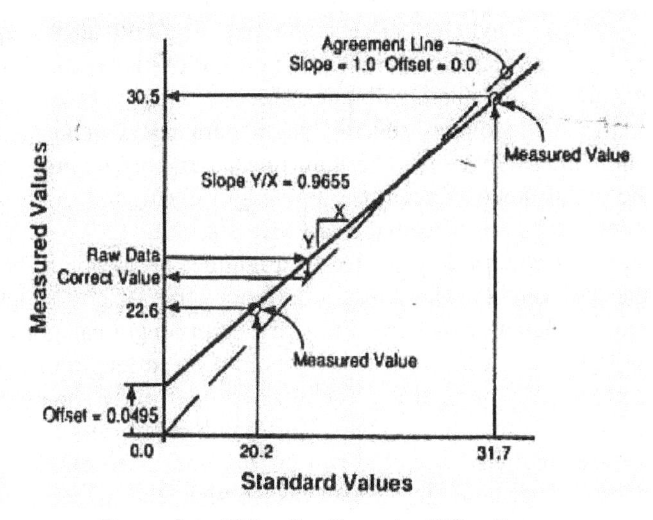

Figure 4.4 HVI calibration using ICC cottons

In actual practice, raw values from the instrument will not match exactly with the designated values due to the variety of reasons including component ageing, atmospheric conditions and other sources of calibration drift. In such cases, the computer software makes adjustments to the raw data by using regression analysis. For the calibration of length, strength and micronaire, ICC cotton can be used while the calibration of colour has to be performed using the standard colour tiles supplied along with the instrument.

4.4 AFIS (Advanced Fibre information System)

Advanced Fibre Information System is based on the single fibre testing. There are two modules, one for testing the number of neps and the size of neps, while the other one is used for testing the length and the diameter. Both modules can be applied separately or together.

Among all physical properties of the cotton, fibre length varies the most within any one sample. There are two sources of variability:

1. Variability that comes from mixing cottons of various lengths.
2. Variability that is biological in nature and exists within a sample of the same origin.

The same variety grown under different conditions, with lower or higher fertilizer doses, irrigation, or pest control, can produce various lengths. This is why fibre length is tested as an average of many fibres. Fibres also break during handling and processing thus, emphasising the need for_measurement of magnitude of the length variation. There are many different measurements of fibre length, including staple length, model length, mean length (average length), 2.5% span length, effective length, upper quartile length, upper-half mean length, length uniformity index, length uniformity ratio, span length, short fibre contents and floating fibre length.

The AFIS test provides several length parameters deduced from individual fibre measurements. The main measurements include: the mean length, the length upper percentiles, the length CV%, and the Short Fibre Content (defined as the percentage of fibres less than 12.7 mm in length). Fibre length information is provided as a number or as weight-based data (by number/by weight). The length distribution by weight is determined by the weight-frequency of fibres in the different length categories, that is the proportion of the total weight of fibres in a given length category. The length distribution by number is given by the proportion of the total number of fibres in different length categories. The length parameters by weight and by number are computed from the two distributions accordingly. Once the AFIS machine determines the length distribution, the machine computes the length distribution by weight assuming that all fibres have the same fineness. Samples do not require any preparation and a result is obtained in 2–3 minutes. The results generally show a good correlation with other methods.

With the introduction of AFIS, it is possible to determine the average properties for a sample, and also the variation from the fibre to fibre. The information content in the AFIS is more. The spinning mill is dependent on the AFIS testing method, to achieve the optimum conditions with the available raw material and processing machinery. The AFIS-N module is dealt here and it is basically used for counting the number of neps and the size of neps. The testing time per sample is 3 minutes in AFIS-N module.

This system is quick, purpose oriented and reproducible counting of neps in raw material and at all process stages of short staple spinning mill. It is thus possible, based on forecasts supervisory measures and early warning information to practically eliminate subsequent complaints with respect to finished product.

AFIS – Working principle

The working principle of AFIS fibre testing instrument is shown in the Figure 4.5. A fibre sample of approximately 500 mg is inserted between the feed roller and the feed plate of the AFIS-N instrument Opening rollers open the fibre assembly and separate off the fibres, neps, trash and dust. The trash particles and dust are suctioned off to extraction. On their way through the transportation and acceleration channels, the fibres and neps pass through the optical sensor, which determines the number and size of the neps The corresponding impulses are converted into electrical signals, which are then transmitted to a microcomputer for evaluation purposes. According to these analyses, a distinction is made between the single fibres and the neps. The statistical data are calculated and printed out through a printer. The measuring process can be controlled through a PC key board and a screen.

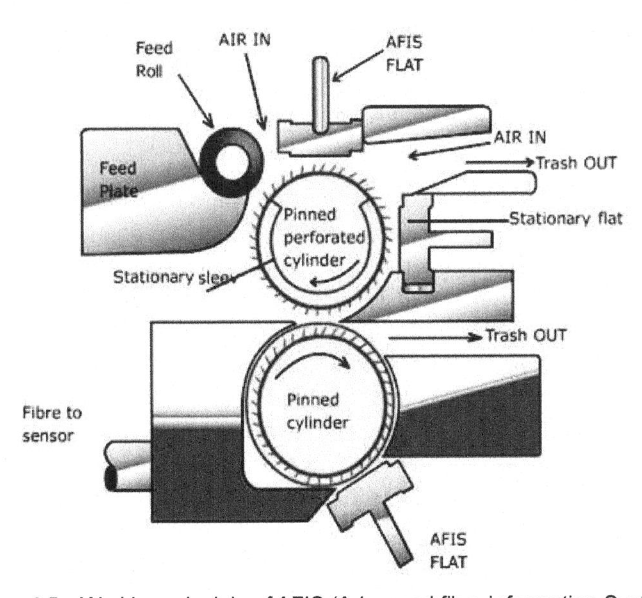

Figure 4.5 Working principle of AFIS (Advanced fibre information System)

4.5 Vibroscope

Vibroscopes have been developed to measure the linear density of the fibres. The measurement of linear density of the fibres by vibroscope method is shown in the Figure 4.6. One end of the fibre is clamped at the top and the lower end of the fibre passes over a knife edge thus providing fixed length of fibre under tension. The tension used is in the range of 0.3cN–0.5cN/tex. The weighted specimen (synthetic fibre)

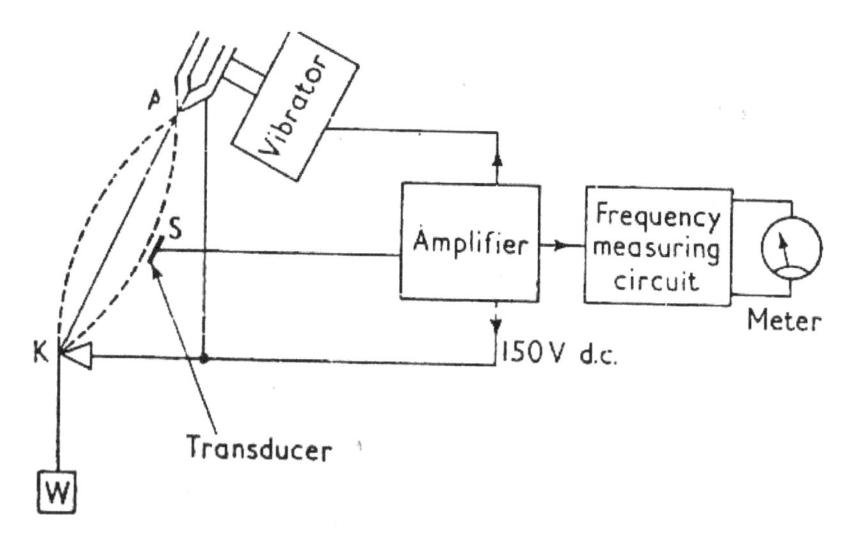

Figure 4.6 Vibroscope

is clamped to the vibrator at (A) as shown in the Figure 4.6 and passes over the knife edge (K). The clamp and the knife edge are connected to a 150 V source and hence the specimen is charged electrically. Transverse vibrations produced on the specimen will induce a charge in a brass screw (S) which is situated midway between the clamp and the knife edge. The brass screw is placed 1 mm away from the specimen. The screw acts as a transducer. The signal from the screw is amplified and fed back to the vibrator; an oscillatory loop is formed, thus causing the specimen to vibrate at its resonant frequency. The voltage developed across the vibrator can then be fed in to the frequency measuring circuit and the frequency of the oscillation is indicated on the meter.

4.6 Variable capacitance type

Principle: The two parallel plates of a capacitor in a capacitive system is separated by a dielectric constant by the relation

$$C \; A / d$$

Where
C = Capacitance
R = Relative permittivity of the dielectric
A = Area of the electrode
d = distance between the electrodes.
From this, variation in the capacitance of the capacitor is obtained by changing

- The distance between the electrodes
- The area of the electrodes and
- The relative permittivity.

Applications

Yarn, slivers or roving materials of relative permittivity (R) is passed between the two parallel plates of a capacitor kept at a distance (d) as shown in the Figure 4.7.

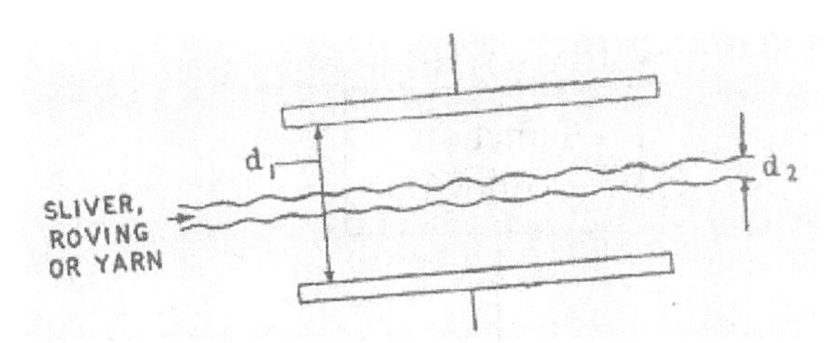

Figure 4.7 Principle of variable capacitance

In this case,
Capacitance C = $1 / (d_1 - d_2) + d_2 / R_1$
R_1 for textile materials is constant for a given moisture content. Variation in d_2 will vary the capacitance which is a measure of evenness. Uster Evenness Tester and other evenness tester works on the capacitor plates are based on this principle.

Textile materials are characterised by variations in mass, twist, strength, elongation, etc. The variation in the fineness of the yarn is the measure of yarn unevenness. Many methods are available to measure the mass variations in the yarn along its length. Cut and weigh method is the simplest way of measuring the mass variation per unit length of a yarn. The method consists of cutting consecutive lengths of yarn and weighing them. However, this method is more tedious and lead to error if the lengths are not cut accurately. Instrumental methods have been developed over the few years for measuring the mass variation per unit length of the yarn. There are two methods available in the market for its determination.

- Capacitance methods
- Optical methods

Both the methods are used around the world. The most commonly used method is capacitance type. In the optical methods, the yarn diameter is measured along the length and the variation in yarn diameter is the measure of yarn unevenness. This method has better correlation with the fabric simulation through the naked eye.

4.6.1 Capacitance method

The working principle of capacitance type is shown in the Figure 4.8. In this method, the mass variation of the yarn to be assessed is passed between two parallel plates of a capacitor whose value is continuously measured electronically.

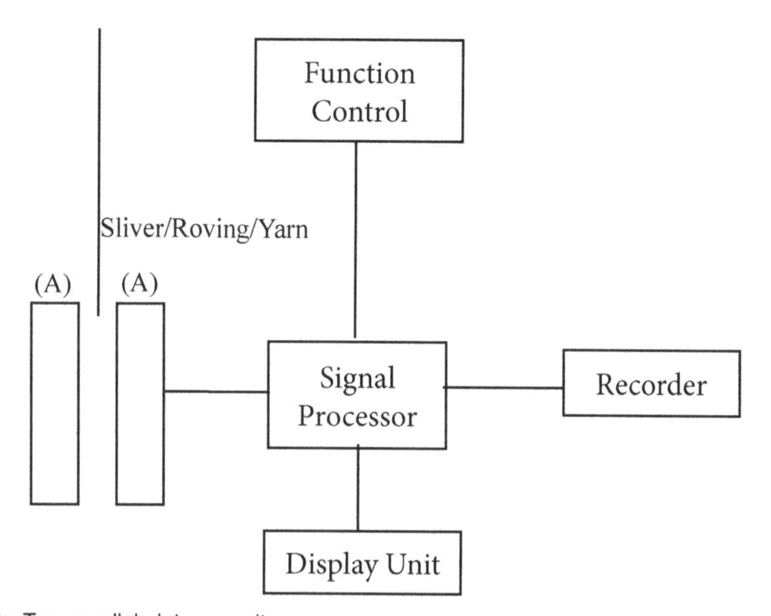

A= Two parallel plate capacitor

Figure 4.8 Capacitance principle evenness tester

The presence of yarn between the plates changes the capacitance of the system which is directly proportional to the mass of the material between the plates and the relative permittivity (dielectric constant). If the relative permittivity remains the same, then the measurements are directly related to the mass of the material between the plates. The measurement is always expressed as between successive lengths and over a total measured length. If the successive lengths are short the value is referred to as short term mass unevenness. The measurement made by the instrument is equivalent to weighing successive 1 cm lengths of the yarn. Uster instruments are based upon the capacitive sensor type concept.

4.6.2 Optical methods

In the optical method of determining yarn unevenness, instead of using capacitance measurements, it uses an optical method of yarn diameter and its variation. In this instrument, an infra red transmitter and two identical receivers are arranged as shown in the Figure 4.9.

The yarn passes at a speed through one of the beams and hence blocks a portion of light to the measuring receiver. The intensity of the beam of light falling on the yarn diameter measuring receiver is compared with that of the reference receiver and from the difference in intensities, yarn diameter is calculated.

The optical method measures only the variation in yarn diameter and not its mass. It depends on the constant level of twist in the yarn. Since the twist level in the yarn is not constant along its length, the imperfections recorded by the instrument differ in nature from those recorded by the other instruments. However, this system is not influenced by the moisture content or fibre blend variations in the yarn. Zweigle G 580 instruments works on optical principle method.

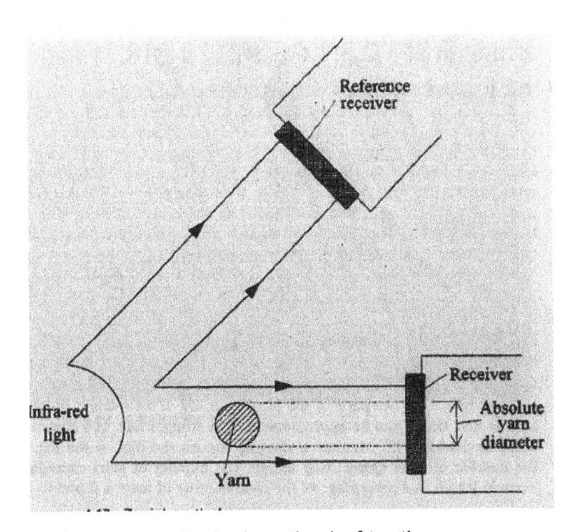

Figure 4.9 Optical method of testing evenness

4.7 Yarn hairiness

Yarn hairiness is defined as the protruding fibres which project from the body of the yarn due to various reasons in material and machinery related factors. In most circumstances it is an undesirable property giving rise to many practical difficulties in downstream processing like weaving, knitting, etc., as shown in the Figure 4.10. It is not possible to represent the hairiness as a single parameter as the number of hairs and length of the hairs both vary independently. In general, a yarn may have a small number of long hairs or a large number of short hairs or a combination of both. Quite a number of instrumental evaluations of hairiness in yarn are available differing in their measurement and expression of results.

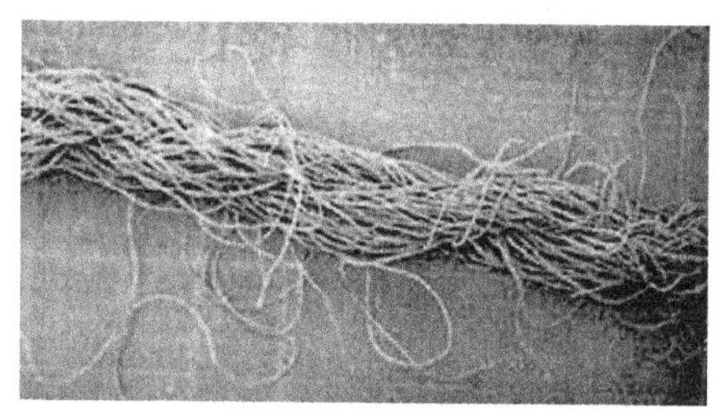

Figure 4.10 Hairiness profile in spun yarns

Zweigle G 565 instrument counts the number of hairs at distances from 1 to 25 mm from the yarn edge. The measurement of yarn hairiness by the instrument is shown in the Figure 4.11.

Working principle

In this apparatus, the hairs are counted simultaneously by a set of photocells which are arranged at 1,2,3,4,6,8,10,12,15,18,21 and 25 mm from the yarn as shown in the Figure 4.11.

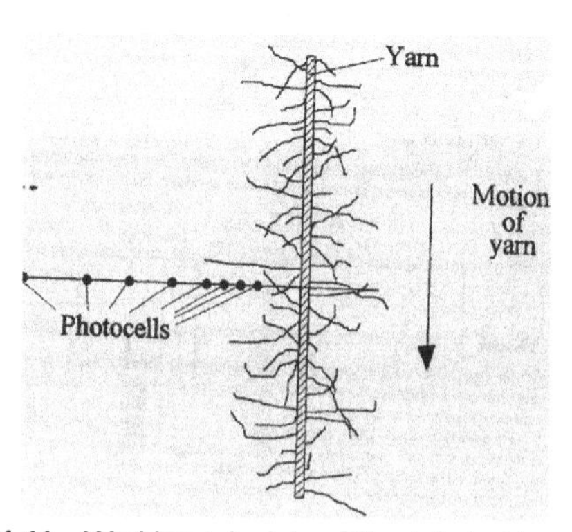

Figure 4.11 Working principle of Zweigle hairiness tester

The yarn is illuminated from the opposite side from the photocells and as the yarn runs past the measuring area the protruding hairs cut interrupt the light off momentarily from the photocells which causes the electrical circuits to count the number of hairs. The instrument measures the total number of hairs in each length category for the set test length. The testing speed is fixed at 50 m/min. The instrument calculates the total number of hairs above 3 mm in length which is the standard referred by the yarn consumers for the yarn quality.

4.8 Determination of contaminants in the yarn

The presence of trash and dust particles in the yarn gives information on the quality of the yarn, appearance of the finished fabric, the machine settings and damage in the machine parts. With the development of Uster 4-SX developed by Zellweger Uster, the determination can be made automatically with the two new opto-electronic sensors. OI is the determination of optical impurities and OM Optical Multi functional. The OI sensor detects trash and dust particles in the yarns and their number and size, whereas the OM sensor determines the diameter of the yarn, the fine structure and roundness of yarns, among other parameters.

Measuring principle of OI sensor

The measuring principle of OI sensor is shown in the Figure 4.12.

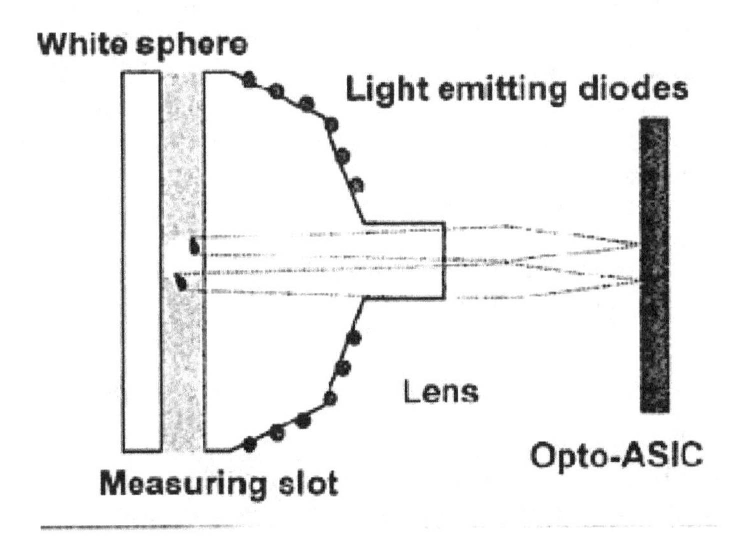

Figure 4.12 OI sensor

The principle is based on the projection of image of the yarn body surface on a linear array opto-ASIC. The yarn passes through a hemispherical, white testing space which is illuminated by blue light-emitting diodes mounted on the hemisphere. Blue light is preferred since it produces a strong contrast between the trash particles and the yarn. The measuring method determines only the trash particles on the yarn surface and will not detect the hidden foreign fibres, spun-in black hairs, etc.

The measuring method determines the mean area of the trash and dust particles. From that, the system calculates the mean particle size in micrometers, detecting particle sizes from 100–1750 μm.

A dimensional comparison of the minimum and maximum sizes of the trash and dust particles to be detected in relation to the yarn diameter of a cotton yarn with count of Ne 30 as well as the limiting size of the trash and dust as shown in the Figure 4.13.

The Uster 4-SX is a fully automatic testing equipment test at a speed of 400 m/min. It provides the following quality data independent of the testing speed.

- Number of trash particles per kilometre and gram.
- Mean trash size in μm.
- Number of dust particles per kilometre and gram.
- Display of the number of trash and dust particles in a histogram.

With the introduction of the opto-electronic OI sensor, it is possible to determine the trash and dust particles in the yarn along with the imperfections. With the existing capacitive sensors, it is not possible to distinguish between neps and trash particles. The OI sensor determines trash and dust particles as separate objects. Seed coat fragments are also counted as trash or dust according to the particle size. The presence of high trash content in the yarn will have an effect on the subsequent production processes. It is possible to have a control on the scouring process which removes the vegetable matters in the yarn by adjusting the concentration of sodium hydroxide and the temperature.

The unit per gram of trash and dust particle is included in the report for immediate comparison of the dust and trash particles measured by AFIS.

4.8.1 Practical importance of dust and trash particles

With the development of OI sensor, it is possible to distinguish between the neps and trash particles where it is not possible with the conventional testing instruments. The OI sensor simultaneous measurement of diameter variation, shape and density by the opto-electronic measuring of the two-dimensional yarn diameter as shown in the Figure 4.13.

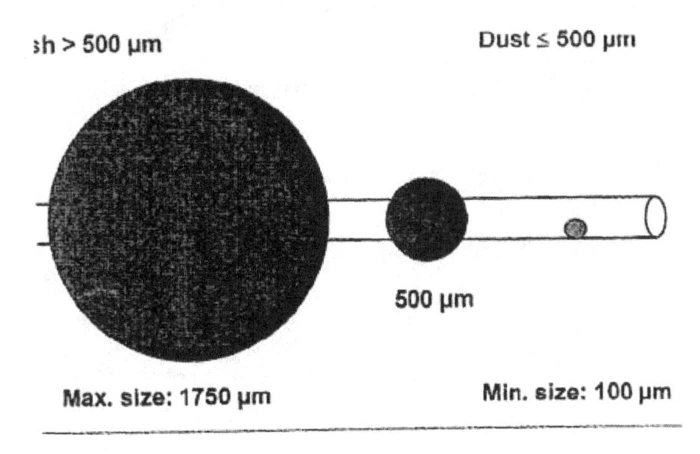

Figure 4.13 Dimensional comparison of trash and dust in a yarn

On the other hand, identifies trash and dust particles as separate objects. Seed coat fragments are also counted as trash or dust depending on the size. An important advantage of testing to determine the objective assessment of the cleaning efficiency. A high trash content in the yarn will also have an adverse effect on the subsequent spinning process especially for carded yarns. It is very helpful in the subsequent scouring process. The information on vegetable contaminants in the yarns or fabrics will be removed in the bleaching process with the help of sodium hydroxide and supporting chemicals at a temperature of 98°C. If the degree of contamination is known, it is possible to adjust the concentration and duration accordingly.

Extensive tests done on Uster Tensojet have shown that the presence of trash particles and seed-coat fragments in the yarn can cause dangerous weak places. The weak places have a mean specific tensile strength of 7.4cN/tex (min. 2.8cN/tex, max 9cN/tex) depends upon the type of yarn and spinning system.

There is also a relationship between the dust content on the wear of the machine parts. At friction and deflection points of the subsequent production process. This would apply in particular to the needle wear in knitting machines and the wear of yarn guide elements at deflection points in the weaving or warping areas. A high dust content will reduce the life of the delivery nozzle in the spin box of the rotor spinning machine. the ring traveller at the ring spinning machine and the yarn guide elements in the winding area.

The pictures (Figures 4.14 and 4.15) show different examples of seed-coat fragments and dust particles measured by OI sensor.

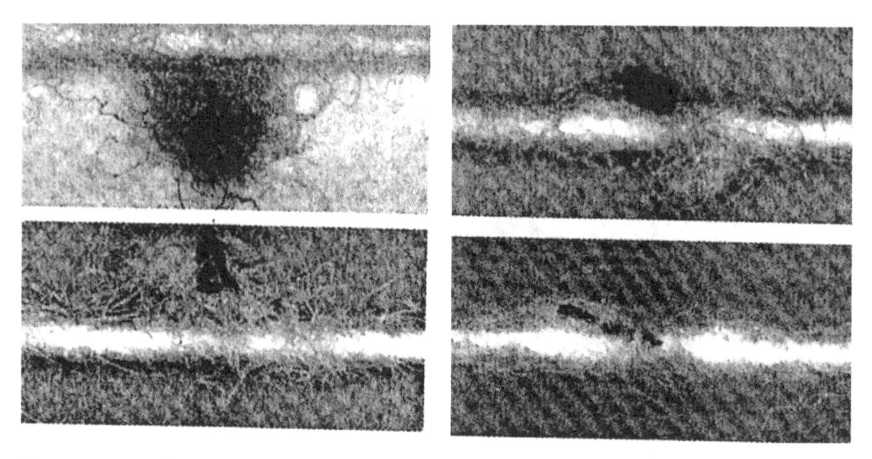

Figure 4.14 Seed-coat fragments

Figure 4.15 Trash particles

Seed coat fragments are always accompanied by more number of fibres whereas trash particles are not attached to fibres

4.9 OM sensor

The problems in dyeing are caused by the variation in shape and density. It is attributed to the spinning methods, raw materials and machinery components which directly affect these quality characteristics. Hence Uster has developed the OM sensor which determines the two-dimensional measurement of the yarn roundness (shape) and subsequently controlling the brilliance and lustre of the fabric. These characteristics are important for the garments. The measuring principle is shown in the Figure 4.16 (A) and (B).

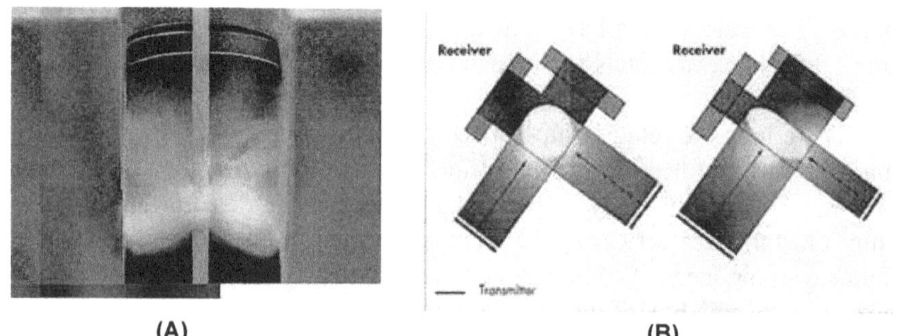

(A) (B)

Figure 4.16 (A) and (B) Measuring principle of OM sensor

4.10 Universal tensile tester

A universal testing machine (UTM), also known as an universal tester, materials testing machine or materials test frame, is used to test the tensile strength and compressive strength of materials. The "universal" part of the name reflects that it can perform many standard tensile and compression tests on materials, components, and structures. The important components in the universal tester is shown in the Figure 4.17 (A) and (B).

- Load frame – Usually consisting of two strong supports for the machine. Some small machines have a single support.
- Load cell – A force transducer or other means of measuring the load is required. Periodic calibration is usually required by governing regulations or quality system.

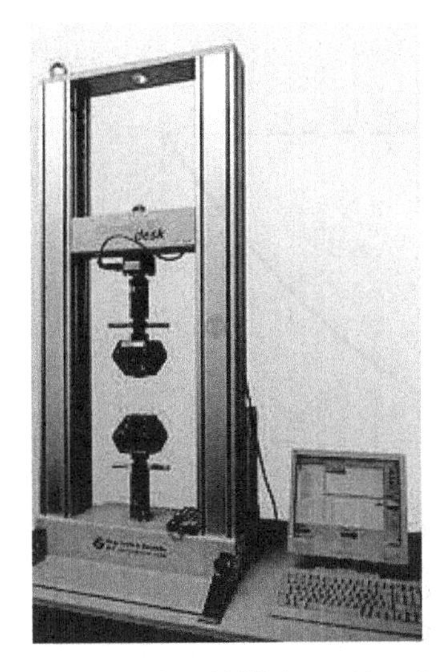

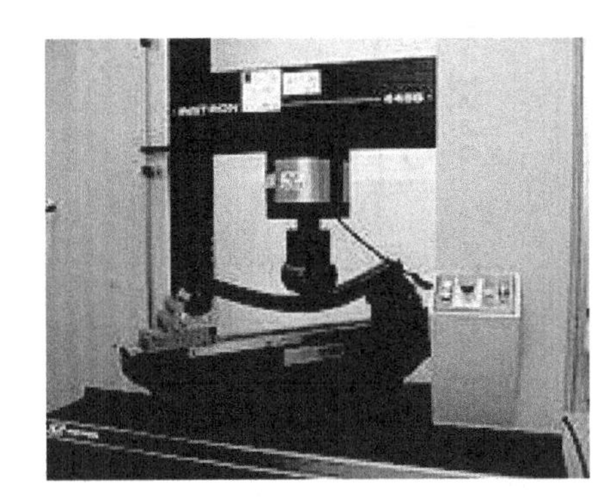

Figure 4.17 (A)Universal tensile tester, (B)Test fixtures for three point flex test

- Cross head – A movable cross head (crosshead) is controlled to move up or down. Usually this is at a constant speed: sometimes called a *constant rate of extension* (CRE) machine. Some machines can program the crosshead speed or conduct cyclical testing, testing at constant force, testing at constant deformation, etc. Electromechanical, servo-hydraulic, linear drive and resonance drive are used.
- Means of measuring extension or deformation – Many tests require a measure of the response of the test specimen to the movement of the cross head. Extensometers are sometimes used.

- Output device – A means of providing the test result is needed. Some older machines have dial or digital displays and chart recorders. Many newer machines have a computer interface for analysis and printing.
- Conditioning – Many tests require controlled conditioning (temperature, humidity, pressure, etc.). The machine can be in a controlled room or a special environmental chamber can be placed around the test specimen for the test.
- Test fixtures, specimen holding jaws, and related sample making equipment are called for in many test methods.

4.11 Moisture measurement

Relative permittivity or dielectric constant of textile materials varies with moisture content. A typical curve of dielectric constant *vs* moisture content for cotton fibres is shown in the Figure 4.18.

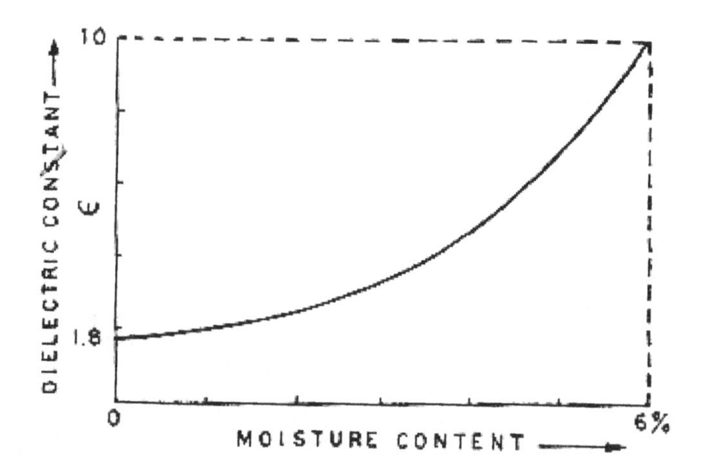

Figure 4.18 Graph of dielectric constant *vs* moisture content for cotton

The graph depicts that the change in capacitance due to variation in (d) would be a measure of moisture content. Fielden Drimeter working on this principle is suitable for moisture measurement in the warp on a sizing machine.

4.12 Measurement of tension

Tension variations in yarn or warp sheet causes the distance between two parallel electrodes to vary and the resulting change in capacitance is a measure of tension. MANTRA Single yarn Tension Transducer and BTRA

Split lease Rod Tension Transducer are some of the examples of tension transducers.

Variable Inductance Type (VI)

Principle

The physical quantity which is being measured can be made to vary the inductance of a coil. Variation in inductance can be achieved by changing:

- By varying the geometry of the coil, i.e. by changing the overall length of the coil (Figure 4.19 (A).

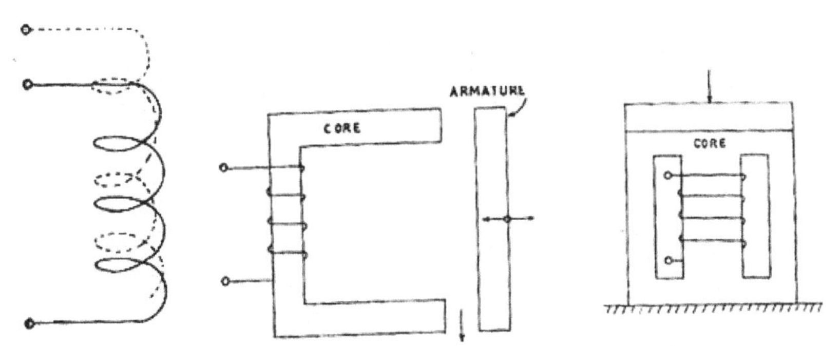

Figure 4.19 (A) Varying the inductance of the coil, (B) Reluctance of magnetic path, (C) Magneto restrictive core material by varying the air gap, (D) Mutual inductance, (E) Transformer type

- Permeability of the magnetic core material by using magneto restrictive core material like Nickel or Cobalt (Figure 4.19 C).
- Coupling of two or more elements e.g. by changing mutual inductance as shown in (Figure 4.19 D) or transformer type operating on variable coupling between two secondary coils with primary coil like linear voltage differential transformer (Figure 4.19 E)

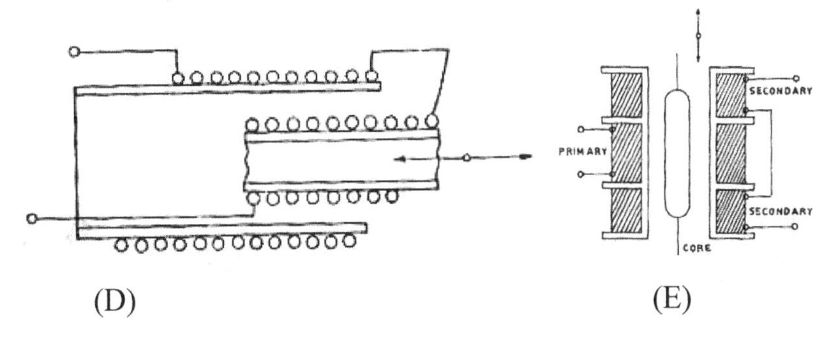

Applications in textiles

- Most widely used variable inductance transducer in textiles machines is the Linear Variable Differential Transformer (LVDT) for the measurement of thickness, displacement, pressure, etc.
- Auto levellers use LVDT to sense and control the thickness variation on card sliver in carding machine.
- Variable air gap type of transducers are also used for the measurement of displacement, pressure, etc.

5
Control systems in ginning

5.1 Introduction

Cotton ginning is an important operation in the cotton production sequence. The cotton producer is interested on optimum yield possible and the spinner requires highest fibre quality possible. The purpose of the ginning operation is the separation of cotton seed from lint and removing sticks, trash and other coarse particles from the input material. The input material for the ginning process usually comes from variety of growing areas. These are processed together and the process acts as a first blending of huge quantities of fibres. To do better ginning, a true knowledge is required since all the cottons are not same. Hence ginning process requires constant monitoring based on incoming raw material, harvest qualities, fibre characteristics, type of picking methods, environmental aspects and cotton growing conditions

5.2 Ginning the seed cotton

Picking of cotton is done by manually by employing labour or by machines. In United States, who is the leading producer of cotton, 90% of cotton is picked by machine picking. Irrespective of the method of picking, the purpose of any ginning machine is to separate the cotton fibres from the seed. There are two types of ginning machine in use, the saw gin and the roller gin. The saw gin is mainly used for short and medium length cotton and roller gin is often preferred for longer fibres. However, in Asian countries like India and Pakistan shorter varieties are roller ginned. Cotton seed is a by-product for the cotton producers and the seeds are normally crushed for oil and sometimes used for animal feed. The quality of the lint produced is very important from the spinner's point of view for best quality yarn with minimum loss in waste. At the same time, if the lint quality is good, it will bring maximum value for the cotton producer. According to the cotton producer, the seeds represent difficulties since the seeds get crushed in the beaters of the opening and carding machines and the presence of excessive seed-coats makes the spinning process a difficult one.

On the other hand, the economics of ginning are affected by the sale of the by-product and thus they have some effect on the cost of the cotton mill. The raw material cost of cotton in the mill roughly represents half of the yarn cost. In addition to the cost of the cotton, the cotton mill has the task of producing good quality yarn which should be accepted by the fabric buyers. Hence the total quality control of the ginning process is a pre-requisite for the ginners and the yarn producer have a strong interest in knowing about the basic fibre properties.

5.2.1 Reasons for monitoring ginning process

It is better to have an understanding about the variables in cotton fibres which affect the ginning process. They are

a. Varieties of cotton.

b. Properties of fibres like length, fineness, etc.

c. Type of picking methods – Spindle, stripper, hand.

d. Growing conditions like rain fed or irrigated.

e. Harvesting conditions.

f. Demands from the market and its fluctuations.

g. Use for domestic market.

h. Use for export markets.

Since all the cottons are not same, the ginning process requires constant monitoring based on incoming harvest qualities, fibre properties, picking methods and growing conditions. To give a practical and effective solution, USTER Intelligin-M monitors and controls the ginning process through a system of online sampling stations located throughout the gin from the module feeder to the bale press. In this system, in formations like fibre moisture, trash and colour are fed to the main console whereas the software analyses the fibre values for optimum dryer temperature and lint cleaning practices. Hence this system provides a total quality control of the ginning process (Figure 5.1). Thus, the ginner has the opportunity to have a complete picture of all variables and manually adjust the process or set the parameters so that the process runs automatically.

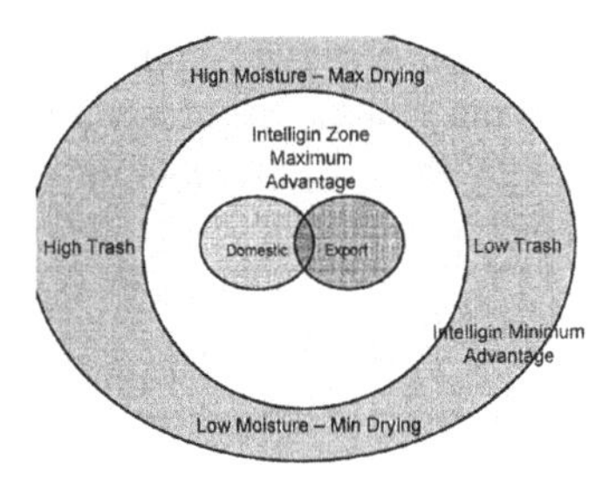

Figure 5.1 Process control measures

5.2.2 Quality control in ginning process

Quality control or quality assurance is the heart of a spinning mill success for a yarn producer. In the present scenario, since the quality requirements are very stringent and every fabric buyers insist on best quality, attention has to be paid for ginning process also. Although, the spinning technology has developed in the machineries from Blow Room to Winding and so on with more on-line and off-line testing equipments which helps the managers to have a control over the quality, the quality of the fibre produced from the cotton buyer is of utmost importance.

The gin must be equipped to remove the foreign matter, moisture and other contaminants that significantly reduce the value of the cotton lint.

5.3 Uster Intellegin – System features

The Uster Intellegin-M system provides information on leaf grade, colour grade, extraneous matter identification and lint moisture at lint flue. The system includes database collection, monitoring system and one measurement unit at lint flue. There are also additional available options or measurements.

a. Determination of incoming module moisture
b. Final bale moisture
c. Energy monitoring
d. Machine efficiency monitoring
e. Louvered grid bar control.

The "Monitoring" system gives the ginners necessary information for processing quality cotton. It also monitors the gin process changes and adjustments. With this information, ginners can use real time information to make decisions for setting dryer temperatures and determining leaf grades. The "Control" system controls the ginning process with set specifications without any constant manual change, thus helping the ginner to focus on other issues like labour and maintenance.

Features of monitor or control system

- Real – time measurement of fibre colour, trash and moisture.
- Patented colour and trash sensor technology to HVI modules.
- Software reporting data for every minute data base analysis.
- User friendly software algorithms with touch screen facility.
- Automatic control of dryer temperature for optimum fibre moisture.
- Automatic control of seed cotton and lint cleaner by-pass valves.

The following diagrams show us the system organisation (Figure 5.2).

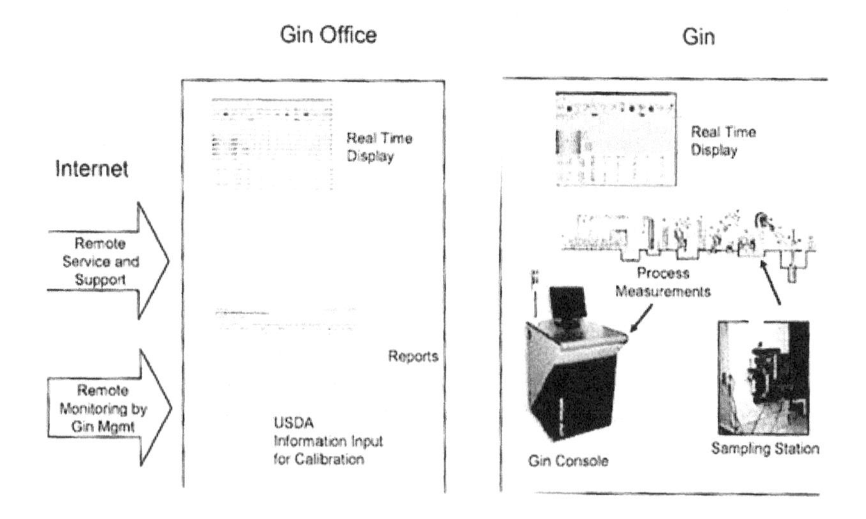

Figure 5.2 Uster Intelligin-M organisation

The system has a sampling station (Figure 5.3) and host console which is positioned just past the lint cleaners. This gives a final reading on fibre moisture, trash and colour to ensure the accuracy of the decisions made before the cotton enters the bale press. This information is useful for the spinners so that they might make adjustments to the ginning process so that the fibre qualities like yield and moisture content can be optimised.

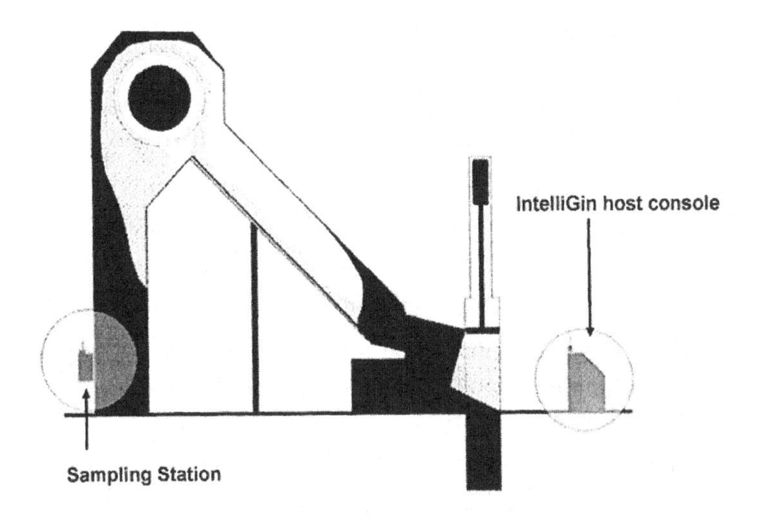

Figure 5.3 Sampling station

The main console gives reliable information to the ginner a complete summary of the process. Using this information, the ginner can make decisions to improve moisture levels and safeguard the fibre properties. In addition, information is also available for contaminations level in cotton which will help ginners to concentrate and remove the same. This will help the ginners and spinners to have special discounts for contamination free cotton (Figure 5.4).

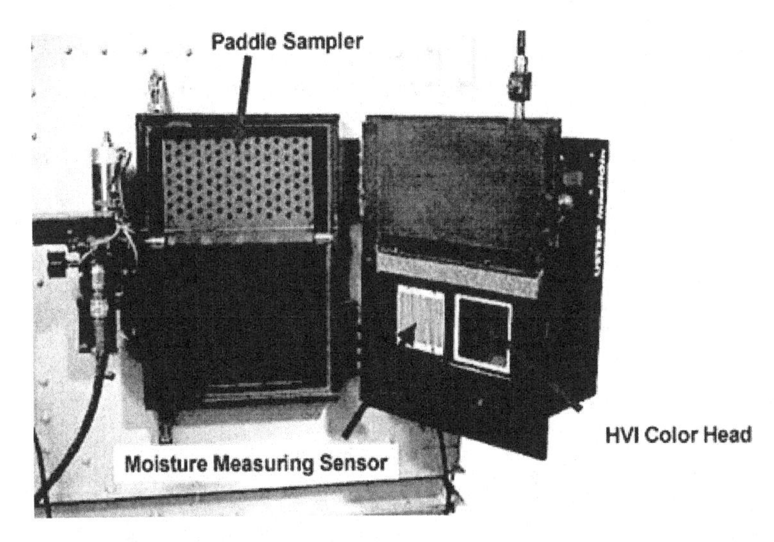

Figure 5.4 Moisture measuring sensor and HVI colour head

5.4 Advantages of Uster Intellegin-M

- The Uster Intellegin-M gives the user easy to read graphical reports and provide summaries as well as in-depth analysis of bale characteristics for the producers.
- The reports also give trend analysis for process improvement for improved gin administration.
- Analytical reports are also available at every level of ginning process.
- Vital reports like colour, leaf grade and moisture provide values to producers while production and shift reports are essential for improving gin profitability.
- Monitoring of the final bale moisture enables the ginner to have an edge over the ginning process as it affects the moisture of cotton.
- Moisture in cotton is critical so as to optimise the dryer settings. This is affected by the final bale moisture sensor.
- Results have shown that the optimum bale moisture level is 7.5% thus matching the USDA conditions for classing cotton. This is maintained by accurate measurements of moisture throughout the ginning process.

5.4.1 Field trials with and without monitoring system in ginning

Field trials have been conducted by Uster Intellegin-M to assess the yield and also the fibre length. This is done by taking the cotton from the same field and under identical conditions; the material is processed with Uster Intellegin first time and without Uster Intellegin-M for the second time. The yield was assessed in both the cases.

These cotton bales were tested by using Uster HVI. From the results, the cotton bales ginned with Uster Intellegin-M were assessed. The fibres shorter than 34 mm are only 1%. On the other hand, the cotton bales without Uster Intellegin-M were assessed. From the results, the fibres have mean length values which are shown in Figure 5.5.

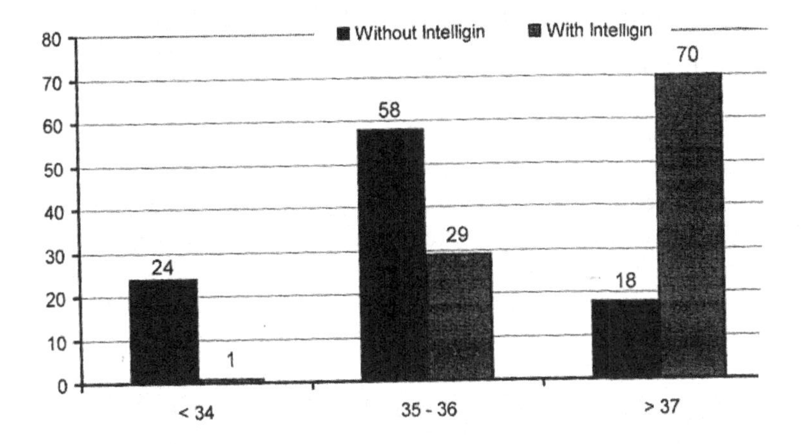

Figure 5.5 Mean length values of cotton fibres with and without Intellegin

5.5 Improvement in strength values of cotton fibres

The strength of the fibres were also assessed for the two types. This is shown in the graph (Figure 5.6). From the results, the fibres with a strength value of 28 or less are only 24% when the cotton bales are ginned with Uster Intellegin-M. The bales which have a strength values of >30 cN/tex or more have a percentage of 52% with Uster Intellegin system. On the contrary, the bales ginned without the monitoring system generally showed lower mean strength values as seen in the graph

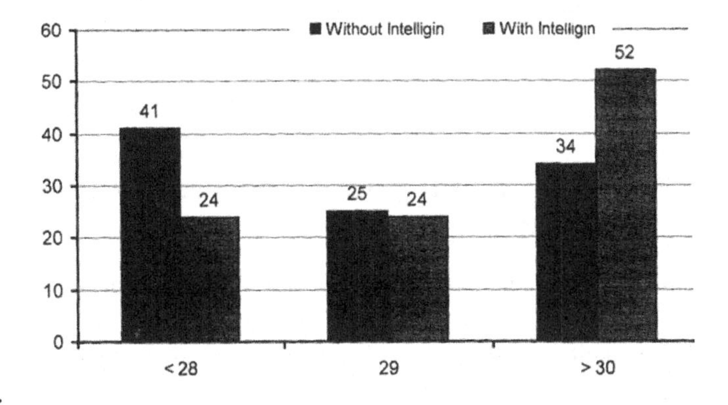

Figure 5.6 Mean strength values of two cotton with Intellegin and without Intellegin

5.6 Improvement in uniformity values

In another test conducted, the improvement in uniformity values of the bales ginned with Uster Intellegin and without the Uster Intellegin. This is shown in Figure 5.7. The fibres ginned with Uster Intellegin have generally better uniformity values. Uniformity index or ration of the cotton fibres is an important characteristic for weaving performance. The fibres which have better uniformity index or ratio will perform well in Ring frames with fewer end breaks and also in weaving with better running performance with fewer warp/weft breaks. The other important parameters like seed coat neps and neps also have considerable influence when the bales are ginned with Uster Intellegin.

Improvement in the cotton fibre properties with Uster Intellegin.

Figure 5.7 shows the improvement in the cotton fibre uniformity characteristics. This property is especially important for weaving operations and also in spinning end breaks. The better the uniformity values, better is the running performance.

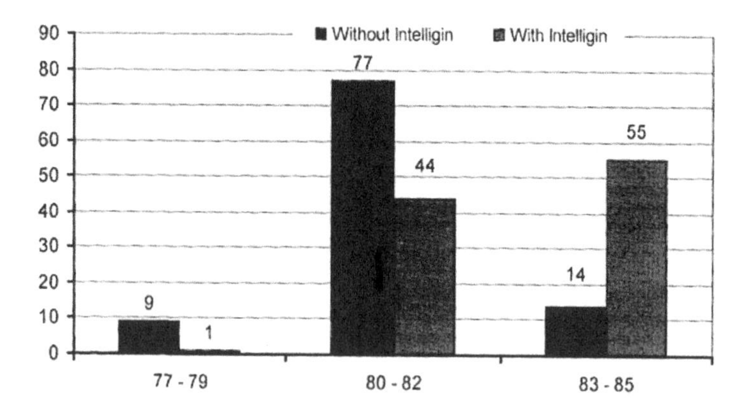

Figure 5.7 Improvement in the uniformity values of cotton

5.7 Improvement in fibre properties like neps, seed coat neps

Figure 5.8 shows the quality characteristics of the cotton fibres like neps, the upper quartile length (UQL) and short fibre content. With the Uster Intellegin, all the results are positive

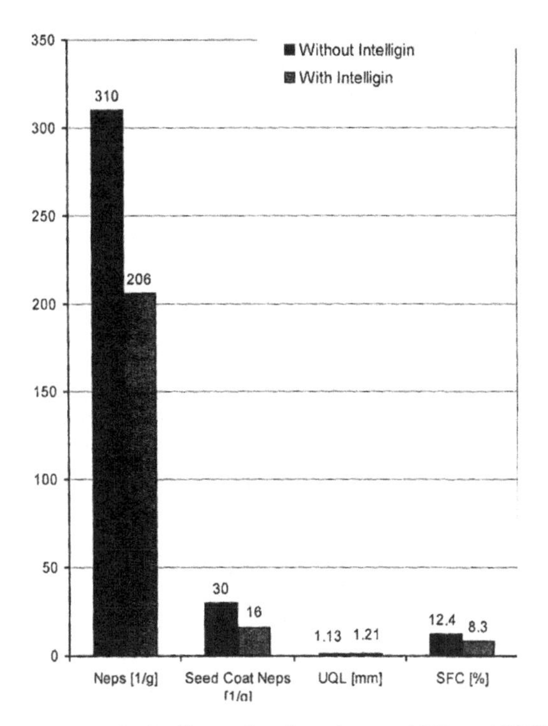

Figure 5.8 Improvement in the Neps, Seed-coat neps, UQL and SFC% with Uster Intelligin

5.8 Benefits for ginners

The Uster Intellegin-M provides the ginner the critical data which enables the ginner so as to process for weight and grade optimisation. The sensor technology of the system monitors the ginning process through a system of on line sampling stations. Online measurements provide information on moisture, colour and trash for optimum processing. A special software analyses the fibre quality for optimum dryer temperature and cleaning practices.

6
Control systems in blow room line

6.1 Introduction

Blow room is the first process stage in spinning line. The contribution of the blow room line is 5%–10% of the production costs in a spinning mill. From the cost accounting point of view, the installation of modern blow room line is not a relevant cost factor. However, the loss of the raw material is a factor to be considered from the yarn realisation point of view. In addition to that, blow room must also eliminate the foreign material like contaminations in par with trash (non-lint) in the cotton. This will only be possible with the simultaneous elimination of some of the good fibres. The lint loss in the blow room line needs attention, since it has significant influence on costs. Even 0.5% increase in waste in blow room line accompanied by at least 50% of good fibres in the eliminated waste would rather end up with reduced yarn realisation.

Nowadays, in the present textile scenario, due to the stringent quality requirements laid down by the European markets, contamination in the raw material plays a major role and a problem for the spinners. Contaminations in the raw material are plastics, feathers, polypropylene fibres, human hair, jute, pieces of fabric present in the raw material cotton have to be removed to the maximum extent. Hence, blow room machines have the additional work of removing the contaminations from the raw material. These contaminations, if not removed properly in the blow room stage will present problems in the yarn since the contaminated material are spun in to the yarn. For this purpose, foreign fibre clearers are installed in cone winding machines also.

6.2 Material flow control

Since, blow room line comprises of sequence of individual machines, each machine must always receive an exact quantity of material per unit time from the preceding machine. The machines receive the material from the preceding machine must pass the same quantity of material to the next succeeding machine. To fulfil the conditions, each machines are adapted in such a way that each machines produce more than the succeeding machine requirements.

Since each machine has excess capacity, a control system is required to ensure the correct delivery of material to the succeeding machines. To adapt this, the basic principles are as follows:

a) Stop-go operation
b) Continuous operation

a) Stop-go operation

The principle of working of stop-go operation in blow room line is shown in Figure 6.1. The example shown here is the Hopper Feeder which is essentially a blending machine in blow room sequence. Here the cotton fibres will have the opportunity to have a thorough and homogenous mixing.

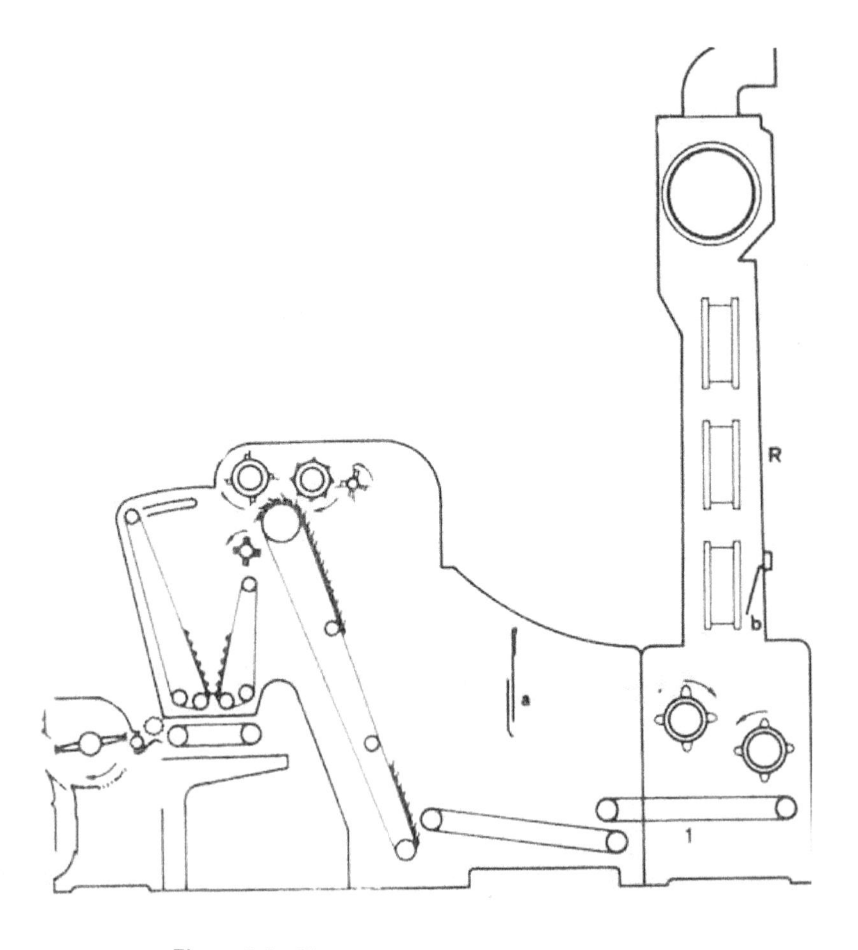

Figure 6.1 Stop-go operation in blow room line

In the Figure 6.1 the conveyor (1) feeds the material in to the hopper. There is a sensing lever (a) provided which is pushed so far to the right. The purpose of this sensing lever is when the material starts filling up in the hopper, it is pushed to the right and when the required quantity of material is filled up, the sensing lever (a) makes a contact with limit switch provided to switch off the drive to the conveyor. Hence, the drive to the conveyor is stopped which means there is no supply of material to the hopper. When the material level decreases in the hopper, the sensing lever (a) is pushed to the left and the drive to the conveyor is started and the conveyor again restarts filling the hopper.

There is also a reserve hopper shown in the Figure 6.1. In exactly the same way, the build up of the material in the reserve chute is sensed by a second sensing lever (b) in the chute. When there is no material in the reserve hopper (R), the pressure inside the reserve hopper is equivalent to atmospheric pressure and the sensing lever (b) will not be depressed. As soon as the material starts filling up in the reserve hopper, the pressure inside the hopper is greater than the atmospheric pressure and the pressure exerted by the column of the material is so great which depresses the sensing lever (b). This causes the material from the preceding machine to be switched-off. When the column of the material has again been removed largely by the conveyor (1), the sensing lever (b) rises and the preceding machine is switched on and the material starts filling up in the reserve chute.

Stop-go operation principle has the advantage that the machines always run with the same speed when they are producing the material is always processed under the same conditions, since there are only two options, namely full on and off. Moreover, the machines are operated only 50% and the remaining periods, the machines are non-productive.

The disadvantage of this principle lies in the material throughput. As the machines do not operate for 50% of the time, in their production periods, say, 300 kg/hr as calculated by the mill personnel's, instead they are actually processing material at the rate of 600 kg/hr. The loading of the machine is also high since it operates on full on and off, the cleaning effect obtained is poor.

b) Continuous operation

In order to overcome the disadvantages of stop-go operation, continuous operation almost operate continuously and without stops. For this, a fine control device serves to maintain material throughput by adjusting the production speeds of the individual machines. In continuous operations, there are always continual delays and accelerations with the possibility of varying treatments to the raw material. This will not have any negative effects where the production rates do not exceed ±20%–±30%.

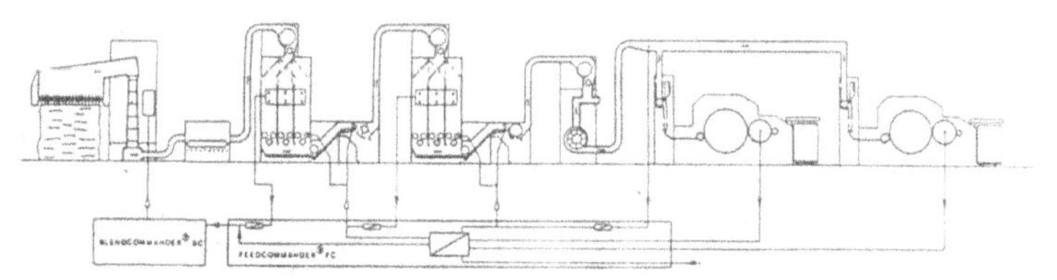

Figure 6.2 Continuous operation

As a context, this continuous operation is not a new concept. This because, in conventional lap feed blow room line, in the scutcher machine, the feed material from the beater may be three bladed beater or Kirschner beater, the material is regulated by sixteen pedal levers which operate continuously and the material is regulated either mechanically or electronically. A similar approach developed by modern machinery manufacturers is shown in Figure 6.2. Feed commander is the central regulating unit in which all the individual machines are connected. This feed commander receives an analogue signal from the tacho generators provided in individual cards. From this analogue signal, the instantaneous demand for material is calculated. Using this data, the micro computer can establish the basic speeds of all drives that determine the throughput and the drives can be correspondingly controlled. A second signal is superimposed on this basic speed signal which is derived from the contents of the succeeding machine. In this way, successive machines are linked via individual looping system. A program has to be made such as speeds, production quantities manually, which represents a fairly substantial initial outlay. When the machine is balanced, the balanced operation is transferred to feed commander and stored there.

6.3 Optical monitoring system

The optical monitoring system for a horizontal cleaner is shown in Figure 6.3.

The optical monitoring devices or sensors (1/2/3) are mounted in the filling chute of the horizontal cleaner machine. In this filling chute, there are three monitoring devices provided (1/2/3). Sensor (1) is an over fill safety monitor to avoid any damage to the machines due to over flow of material. Sensor (2) monitors the level of filling in the chute i.e. optimum filling and sensor (3) monitors the minimum level in the chute.

If the column of the material falls below the light barrier (3), then the preceding machine is switched on and delivery of material takes place. When the material starts filling up in the chute, the material interrupts the light beam of the light barrier or sensor (2) to such an extent that the ma preceding machine is switched-off again. In this way, the material filling in the chute is monitored by the three light barriers or sensors for satisfactory working

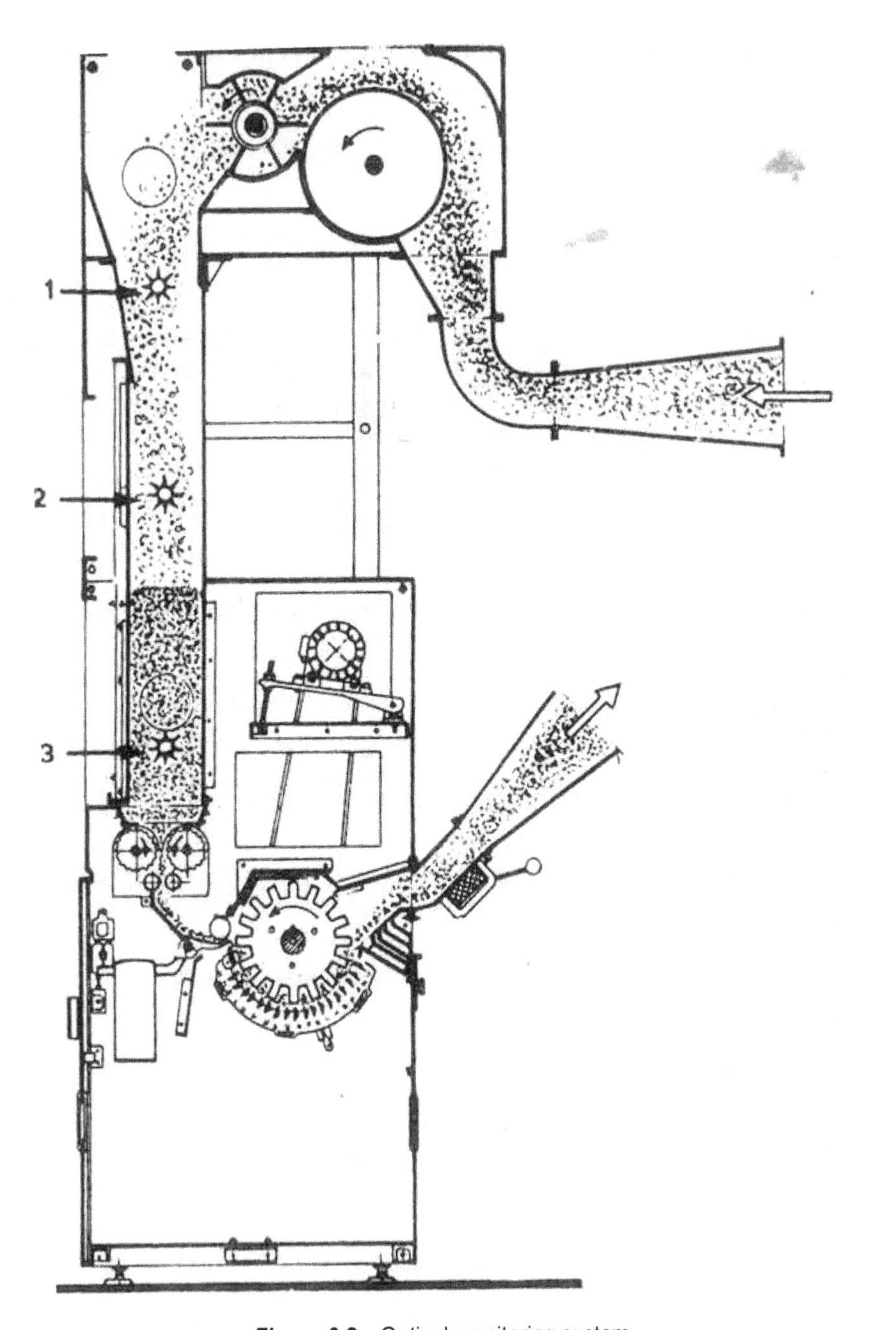

Figure 6.3 Optical monitoring system

6.4 Control systems in feeding unit in blow room line

Feeding Unit BE/BEC

Function: The function of the feeding unit in blow room line is to feed cleaners with a controlled material supply.

Principle

The material is accumulated in the reserve trunk. There is a light barrier provided in the machine which monitors the material level and the material transport of the preceding machine. A pair of delivery rolls forms the base of the reserve trunk. The delivery rolls guide the material to a spiked roll without gripping the fibres. Hence, they provide uniform feed of optimum amounts of material to the next succeeding cleaner in the line. This assists thorough cleaning of fibres without fibre damage. The speed of the delivery roll is adjustable through a variable speed geared motor. The feeding unit BE is an efficient cleaning point in the blow room line.

Control systems in blending chambers in blow room line

In the textile spinning mills, for any given lot of bales, a blend must be produced in which the optimum distribution of the raw material is constantly ensured with respect to the fibre length, fibre linear density, colour maturity, etc. The blend is made not merely on the basis for the production of yarns of uniformly high quality in evenness, tensile strength, behaviour of the yarns during dyeing process, etc., but it should also improve the subsequent processing properties of the material. Furthermore, in modern technology blow room line, many machineries have come up to meet the requirements. Machines like Multi-mixer, Unimix, etc., fulfil the requirements of the spinners for satisfactory yarn appearance and so on. These modern blending chambers produce homogeneous blend and also suitable for homogenisation of coloured and man-made fibre blends produced on a tuft blending installation with weighing hopper feeders. One such blending chamber machine is shown in the Figure 6.4.

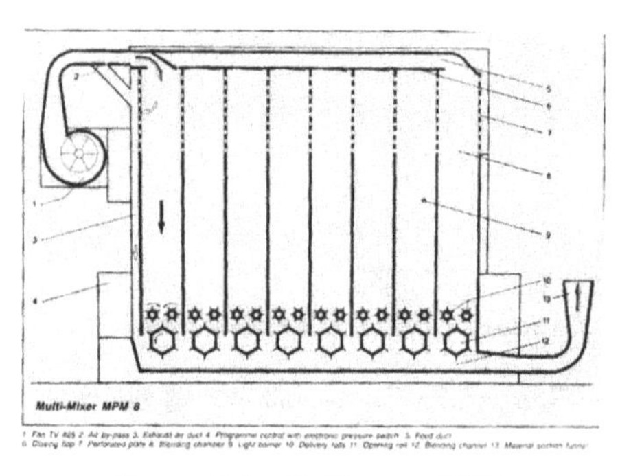

Figure 6.4 Multi-mixer MPM

The charging of the Multi-mixer MPM at the beginning of a blend lot starts with the first chamber at the end of the feed duct. When there is no material in the chamber, the pressure inside the chamber is equal to the atmospheric pressure. It is filled only to just below the light barrier situated in the adjacent chamber which governs the minimum charging level during automatic operation. When the first chamber is filled up half level, the closing flap of the second chamber is opened by push button operation. The second chamber is charged rather higher level than the first and the closing flap of the third chamber opens up. The third chamber is charged to a rather higher level than the second chamber. Likewise, the charging level of the remaining chambers rises uniformly up to the last chamber which is fully charged.

Whilst the last chamber is being charged there is a changeover to automatic operation and the material transport is switched on. The columns of the material in the chambers start to drop. As soon as the last chamber is full, the closing flap shuts automatically and charging restarts with the first chamber as soon as the level of the second chamber has dropped below the light barrier.

As all the chambers are emptied simultaneously, fibres come together in the blending channel which has come from the preceding machine at different times.

Mode of filling of blending chambers

When the charging level rises in each chamber, more and more of the holes in the perforated plate of the upper part of the chambers are blocked with material. This increases the pressure of the conveyor air. The charging level on each chamber is based upon the pre-selected pressure which is sensed by the pressure transducer. Once a pre-selected pressure is reached, an electronic pressure switch closes the flap of the next chamber. The pressure that is set for the changeover is displayed on the manometer. The capacity of the chambers can be varied by altering the change over pressures.

6.5 Foreign matter removal in blow room line

Foreign matter in the cotton material like iron pieces are potential source of danger to the textile spinners since it not only damages the machine elements like spikes, beater pins but also cause fire accidents. Hence, a control mechanism is required to separate from the cotton materials and eliminate them. One such device in blow room line is magnet trap.

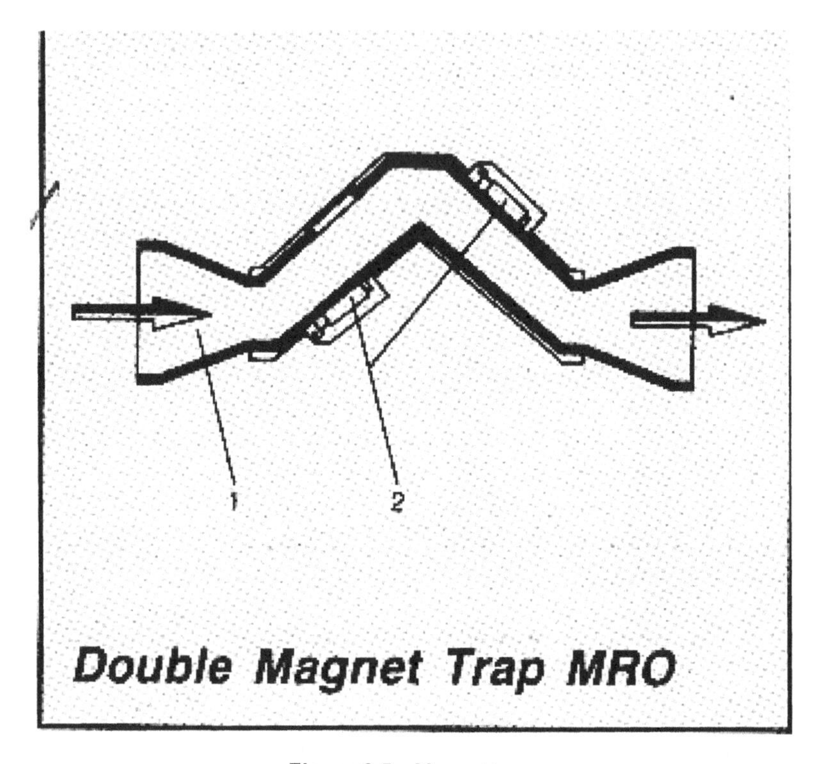

Figure 6.5 Magnet trap

Magnet traps in the ducts or machines capture iron particles which are not entangled with the fibres.

The double magnet trap MRO is fitted in the material transport duct where possible at the start of the line and before the actual opening and cleaning machines like Axi-flo or ERM cleaner. Two sets each of 3 magnets, (as shown in the Figure 6.5) are fitted in an angled section of the duct. With this arrangement, the tufts which carry the iron particles are caught by these magnets and taken out separately.

6.6 Automatic heavy part separator by SEPAROMAT ASTA

Many heavy particles separators in blow room line have been manufactured by various manufacturers and installed in blow room line. The main reason is to eliminate the contamination from the cotton tufts without any damage to the fibres as well as to the machinery elements. One such heavy particle separator is SEPAROMAT ASTA.

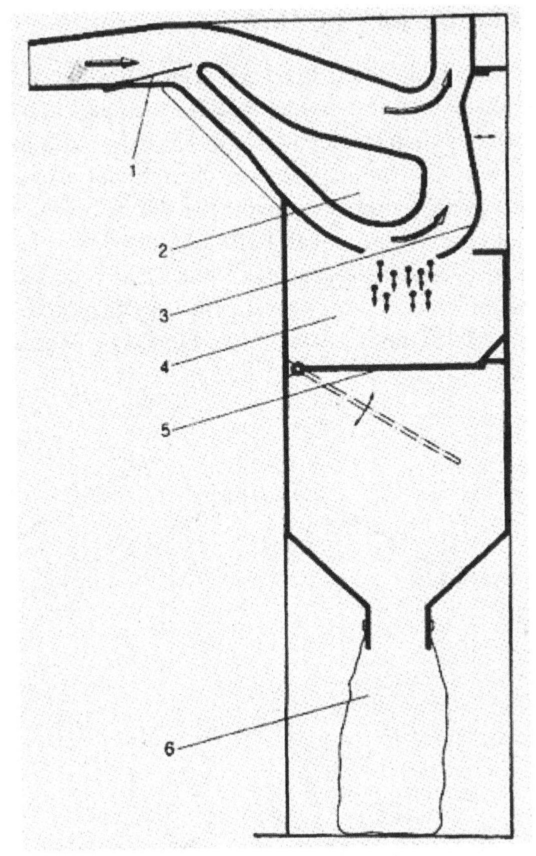

Figure 6.6 SEPAROMAT ASTA

Working principle

The automatic heavy part separator SEPAROMAT ASTA removes all foreign matter which is heavier than the tufts regardless of the material composition. Impurities in the cotton also drop out. Moreover, the loss of cotton is quite small.

In the Figure 6.6, the SEPAROMAT ASTA is fitted in the material transport duct at a point where constant air conditions to the machine exist. The main air stream carrying the tufts is deflected upwards at right angles. A subsidiary air stream passes through a grid (1) which retains the tufts. It is guided back in to the main air stream under an aero dynamically shaped piece. This prevents the normal tufts from dropping down at the deflection point by deflector plate (3). Hence it is unable to carry heavy particles and these may drop in to the collecting box (4). The bottom of the collecting box opens automatically at adjustable time intervals. Then the contents of the collecting box will fall in to a transport sack (6).

6.7 Metal part extractors

Magnetic particles which come from the cotton during cotton picking or in careless ginning operation get packed in to the bales and shipped to the spinning mills. Magnetic particles like iron pieces, sickles and so on cause damages not only to the beating points of beaters but also cause fire accidents due to heat and friction while working the cotton on beaters. Those iron pieces if not carefully removed before laying down the mixing cause troublesome in mechanical processing. Hence magnetic extractors are in general installed in blow room line sequence to safeguard the machinery condition. One such magnetic extractor developed by Marzoli is shown in the Figure 6.7.

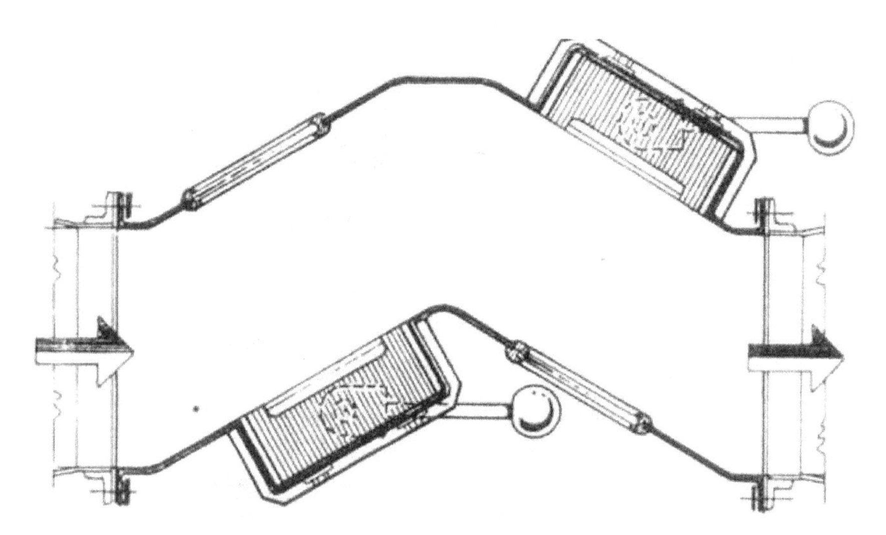

Figure 6.7 Metal part extractor

Magnetic extractors have been used for many years in ducting parts of the blow room line to remove the iron particles from the cotton material. The most effective form of device is a knee-bend having permanent magnets at the two impact surfaces. When flocks of cotton are fed through the ducting, the magnets retain the ferrous particles and allow cotton only to the beaters. These ferrous particles retained by the magnets can be removed from time to time. In this way, damage to the costly machineries are avoided. However, these magnet extractors can eliminate magnetisable metal particles only and let others which are not magnetisable to pass through.

6.8 Electronic metal extractors

In order to overcome the above difficulty, Trutzschler of Germany has developed an electronic metal extractor which is shown in the Figure 6.8.

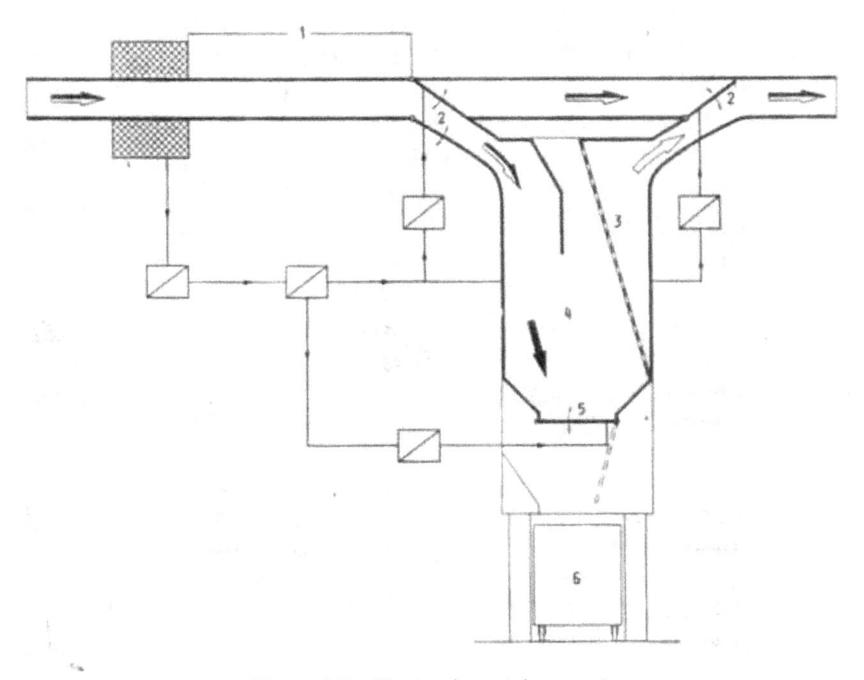

Figure 6.8 Electronic metal separator

In this system, a metal detector (1) is provided in the transport ducting. The detector sensitivity is adjustable. There is an eliminating arrangement for removing the metal particles is located 15 meters after this detector. When a piece of metal of any kind passes through the detector, the entry and exit flaps (2) are set in such a way that the material cannot fly straight through the duct, but must pass through the eliminator. The grid (3) holds the raw material, say cotton, and metal back while the air can flow through. The flaps immediately return to the normal flow condition, and the floor (5) of the collecting chamber opens and the metal contents fall in to a transport container (6).

6.9 Fire control systems for a modern spinning plant

Fire eliminators are necessary in spinning mill blow room line to avoid any fire accidents. One such device developed by Rieter is shown in the Figure 6.9. The equipment consists of a spark detector and an eliminating device. Both are built in to the transport duct. There is a rapidly operating flap which is pivoted by the spark detector. This flap operates immediately as soon as the spark or any burning material is detected by the spark detector. The material passes in to a receiving chamber which is preferably located in the open air. At the same time, a hooting alarm is given and the blow room line is switched-off. The pivoting flap remains in the eliminating position until the line is switched-on.

In the conventional days, in order to extinguish the fire in blow room line which happens due to various reasons, water sprinklers and sand were used. Workers and even personnels were unaware of the fact that for extinguishing fire of various types, which types must be used. Because of this, several water sprinklers were burst and the floor would become wet and also cause damage to the cotton fibres in the mixing area. To overcome this problem, CO_2 gas cylinders were installed around the walls and are manually operated. These wet cylinders were also found to cause damage to the machines and material because of residual formation.

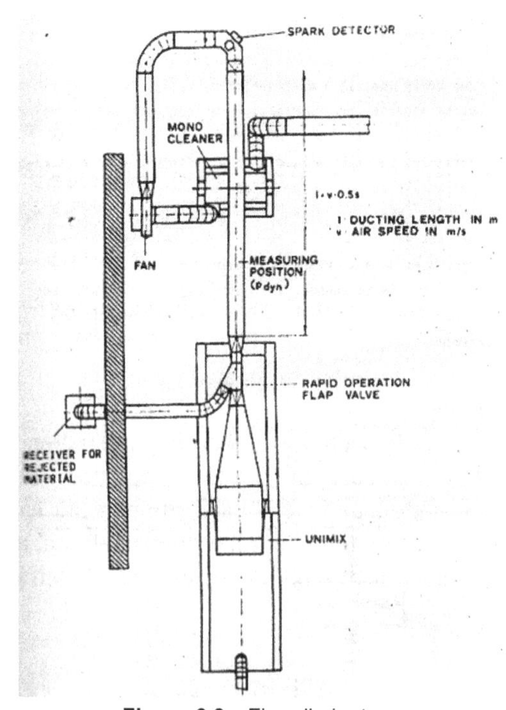

Figure 6.9 Fire eliminators

In general, it is a common experience that the workers become panic when one finds fire takes place and any number of fire extinguishers is freely burst. These dry extinguishers were retained to be used for electrical equipments, but then the experience was far from satisfactory.

With the introduction of modern sophisticated automatic high speed blow room lines and the raw material getting costlier day by day, it was indeed necessary to put in action a highly dependable automatic fire protection system which would take care of the modern costly machines and prevent raw material from getting damaged. With the installations of blow room line with chute feed cards, the insurance premiums went further high. As a matter of fact, these modern fire control systems it is now considered as a primary requirement besides getting substantial reduction in the insurance premiums. At the same time, down time losses on account of the fire had to be positively controlled.

6.10 ARGUS fire control system

A few CO_2 flooding systems were developed to get reductions in premiums, but then these systems were found to operate below the satisfactory levels. Series of false alarms causes constant worry for the mills. The Angus Fire Control Systems was put in to practice by USA. The principle is the detection of smallest ember by electronic infrared spark detector and diverted from the main travel path and suitably extinguished.

Principe of spark diverter

When the beater of a bale plucker strikes metal material or the spindle end of the beater is tangled with fibre, spark may be caused; the spark can mix in the fibre and move in the conducting duct under the action of the wind. When the cotton fibre mixed with sparkle moves through the high sensitive infrared probe detecting area, the controlling program will immediately make the fire alarm with sound and light. Furthermore, while stopping the running of related equipments such as blower fan, etc., the cotton flow with sparkle and combusting scraps are discharged into collect box to assure that the sparkle will not enter the next procedure and eliminate the hidden trouble of fire.

When a fire starts from a spark or a burning ember, it is detected by infrared sensors that are normally positioned on stock transport ducts. Even though the spark or burning ember is moving at a very high speed of 5 m–20 m per second, it is noted by the detector within micro seconds. The detector sends a signal to the control panel. A hooting alarm sounds and the extinguishing system is activated and the machinery containing the fire is put-off. This not only prevents the fire from controlling and also prevents the spreading of fire to other machines. In addition to the extinguishing systems, ARGUS has also developed ARGUS-JOSSI high speed metal and spark diverters. These devices remove particles and embers from the process and divert them to the collection boxes.

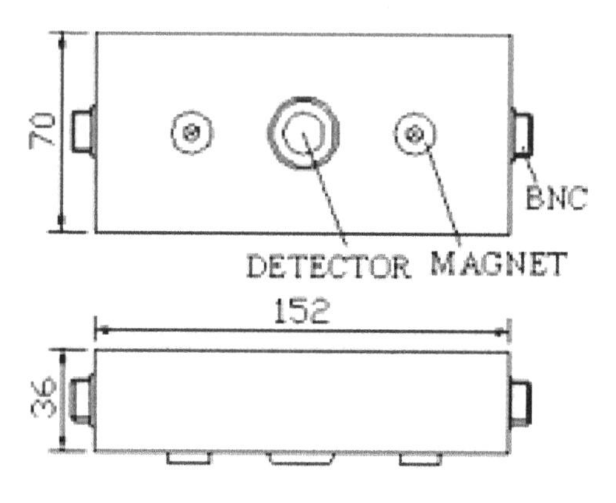

Figure 6.10 Spark detector

Spark detector communicates with the host via the power carrier, and to set a local net in the workshop, what we need are only two lines, which makes the distance available for signal's long transmission, and makes the system more simple and reliable (Figure 6.10).

The host is equipped with LCD. While whenever the alarms are on, the host not only shows alarming area via acousto-optic alarm, but also can record the specific alarming time automatically

The diverters can be installed suitably at the desired points within a length of 3 m in the stock carrying ducts. These diverters act after the detectors sense the ember/spark immediately and divert the same to the collection chamber. The opening and closing operation of these metal and spark diverters is automatically controlled depending upon the size of the ember or size of the metal piece. At the same time, a signal is sent to the control panel to shut the machine off till the troublesome material can be removed.

Advantages

- To control the fire from spreading to other machines.
- To reduce production losses.
- Avoid expensive repair, damages to the machinery and buildings.
- Reduction in the premiums.

7

Foreign fibre detection in blow room

7.1 Introduction

Foreign fibres or more precisely called contaminations have been one of the primary causes for the spinners as it leads to quality claims and rejections from the buyers in the last few years. International Textile Manufacturers Federation (ITMF) publishes data regarding the presence of contaminations in the cottons all over the world for the last few years and also suggests that there is considerable increase in contamination levels for the last few years. This makes it very important to find the most effective solution to combat foreign matter in the cotton.

Many contamination clearers manufacturers are coming out with their advancement in technologies for the maximum possible elimination of contamination from the cotton. It is also more evident that the presence of foreign fibres in textile fabrics cannot be accepted anymore. In general, many foreign fibre particles are detected only after the fabric is finished, and the spinner is finally made responsible for the damage. In order to minimise or overcome the problem of contamination in the fabrics, spinning mills should have a quality management system to fight against the material and reduce the costs due to claims. Experiences made by leading textile equipment manufacturers with the foreign matter removal system at blow room and card will be discussed in the following sections.

7.2 Foreign material in the cotton

Discrimination must be made with trash percentage and foreign fibres in the cotton. Trash indicates the presence of non-lint content in the cotton like leaves, dust, dirt, seeds, seed coats, etc., there are various types of foreign materials in the bales and some of them are shown in the following section, the source taken from ITMF 2003 (Figure 7.1).

Figure 7.1 Contaminants in cotton

Cotton related contaminations
Honey dew or stickiness
Leaf
Stem
Barks, trash, seed-coat fragments
Contaminations of natural and man-made fibres
Fabrics and strings made of
Woven plastics
Plastic films made of polypropylene, polyethylene
Jute
Cotton
Organic matter
- Birds feathers
- Grass
- Paper and carbon papers

- Leather pieces
- Human and animal hairs

In organic matter
- Iron rust due to wrapping of bales with metal straps
- Metal plates/wires

Chemicals
- Oils/greases
- Rubber
- Stamp colour tar

Foreign fibre contamination materials and its count vary depending upon the growth area and harvesting methods. In many spinning mills, the mill employs manual labour to eliminate larger type contaminations present in the bales. However, small foreign fibre materials like human hair, animal hair cannot be eliminated by the above method since the size is too small and cannot be detected.

7.3 Effect of foreign fibre material in textile production chain

The presence of foreign fibre material in textile production process creates three problems as shown in the Figure.

- More ends down in spinning processes.
- Seconds in knit wear or in woven fabrics resulting in quality claims or rejections from the buyers.
- Reputation of the quality image of the yarn or fabric producer is impaired.

Depending on the degree of contamination i.e., number of foreign fibres/ bale the losses per bale can amount from DM 20,000 to DM 10,000.

7.3.1 Degree of contamination in cotton bales

Survey by ITMF in the past has revealed the fact the degree of contamination in cotton bales depend very much on the growth area as shown in the Figure 7.2 (A) and (B). From the Figure 7.2 (B), the machine harvested cotton in the cotton growth areas are less affected by contamination as there was less contact between the workers and the cotton.

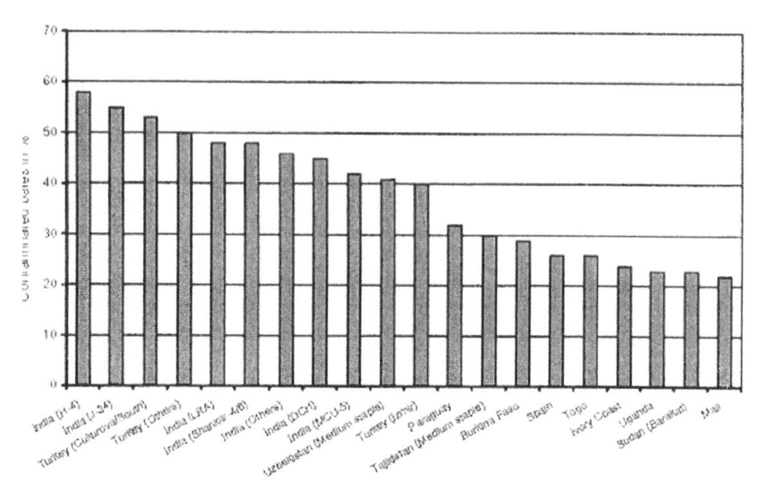

Figure 7.2 (A) Countries with more foreign fibre contamination (Source: ITMF 2003)

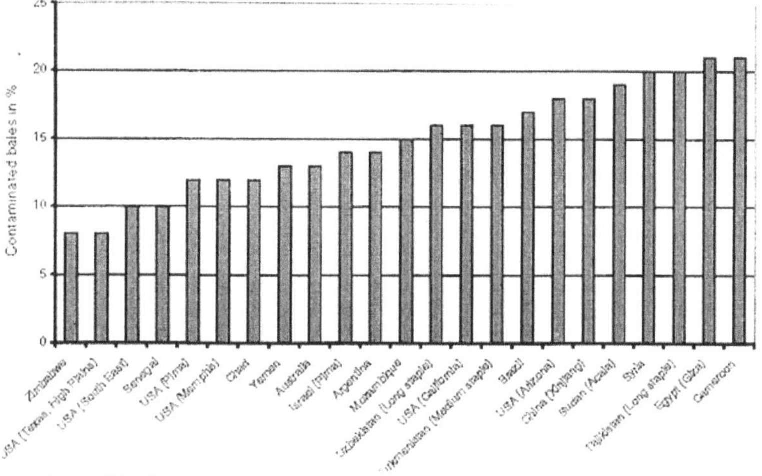

Figure 7.2 (B) Countries with less foreign fibre contamination (Source: ITMF 2003)

7.4 Size and appearance of foreign fibre matter in the textile supply chain

Contaminations of bigger size should be extracted in the blow room process with foreign fibre sensors in the line. If the foreign fibre material cannot be eliminated prior to the card, the foreign material will be cut or torn in to pieces by the card. For example, a piece of plastic can result in a number of individual fibres in the card. As these fibres are mostly coloured in nature, the cluster of foreign fibres can easily be recognised in the output card sliver.

sliver. These clusters of foreign fibres will lead to human interventions, loss in production and consequently labour costs. Moreover, these contaminations also cause problems in electronic yarn clearers installed in auto coners and in OE rotor spinning machines which will trigger an alarm due to the higher amount of foreign fibres within a short interval of time as shown in the Figure 7.3.

In addition to that, often in spinning mills some of the foreign fibres are added accidentally through human ignorance, waste recycling, etc., which usually contaminate the cotton during the spinning process. For removing such defects, clearer in the winding section is the last stage in spinning mills is the only available option to eliminate such fibres.

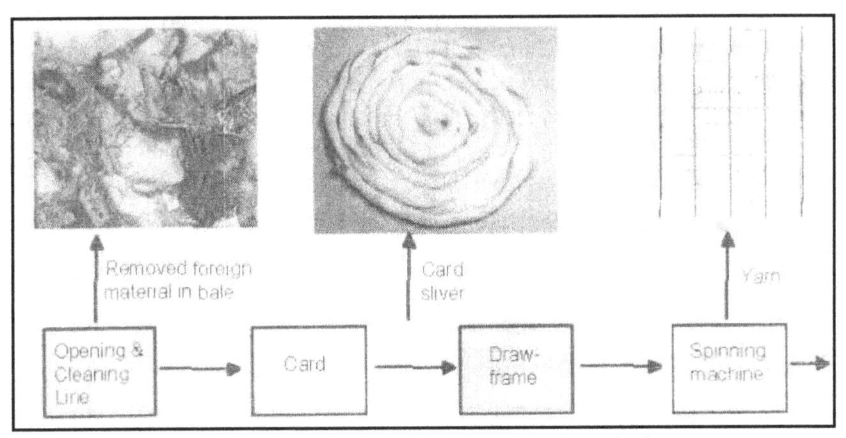

Figure 7.3 Foreign matter at various stages in the spinning process

Cotton contamination sorters (CCS) are optional online quality monitoring devices in blow room. These devices detect and remove the foreign matter from cotton. CCS provides information that the number of foreign fibres detected and removed by the systems and recorded in to system PC as "Ejections /minute or Ejections/hrs". These information are available as hour wise, shift wise, weekly basis, monthly basis and so on depending upon the programming available with the system PC. CCS is an important on-line monitoring device for the removal of contamination. Accuracy of CCS is generally observed by passing the contaminants like paper, foreign fibre, pieces of fabric of 1 cm^2 in the blow room pipeline along with the cotton material just before the CCS and observing after the CCS how many contaminants are ejected by the CCS. This shows the efficiency of the CCS. However, regular maintenance is required for the proper functioning of the CCS by cleaning CCS glasses, cameras and light source, cotton path which affect the performance of CCS significantly. The efficiency of CCS

CCS efficiency % = No. of defects ejected through CCS / No. of defects passed through CCS × 100.

7.5 Appearance of foreign fibres in spinning mills

It is customary to detect the foreign material at the earlier stages in blow room itself. If it is not eliminated, the foreign material will be cut in to many pieces in the carding machine due to the wire points in the card. For example, a piece of plastic film can be disintegrated in to number of small pieces and if it is coloured, the cluster of foreign fibres can easily be recognised in the card sliver as shown in the Figure 7.4.

These clusters of foreign fibres will lead to human interventions, loss of production efficiency and increase in labour costs. The yarn clearers in OE rotor spinning and in winding machines will trigger an alarm due to the higher amount of foreign fibres within a short period. In some occasions, in the spinning mills foreign fibres are added accidently through human errors, waste recycling, etc. For such foreign fibres, yarn clearers at OE rotor or in auto winding machines have to clear the foreign fibres from the yarn up to the permissible limits and if it is not removed from the yarn, it will be clearly seen in the yarn and the fabric thus downgrade the fabric from the customer's point of view.

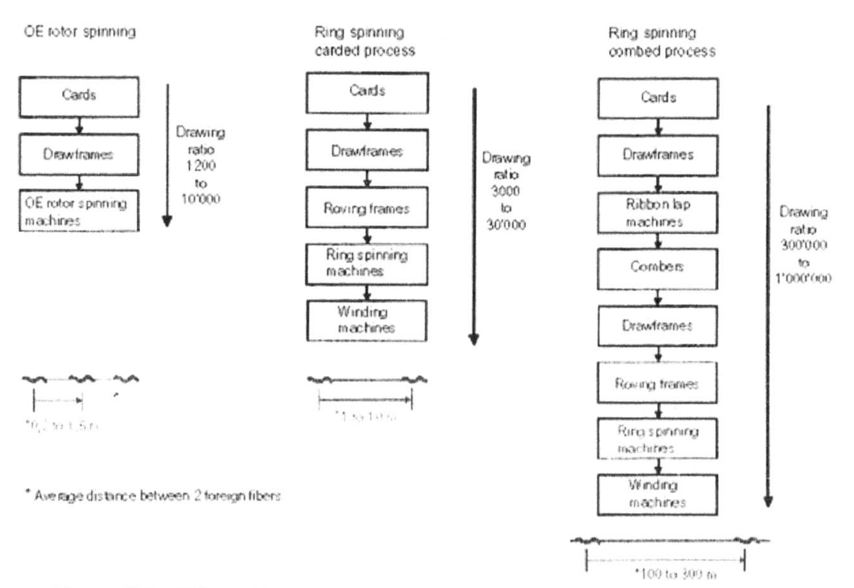

Figure 7.4 Effect of foreign fibres at various stages in the spinning machines

7.6 Frequency of foreign material and elimination methods

The domains of the foreign fibre material removal systems and the frequency of foreign fibre material is shown in the Figure 7.5. It is evident that the frequency of foreign material increases considerably in the area of fine foreign matter due to human and animal hair, plastic fibres, seed coat fragments, etc. For quite a period of time, the spinning mills adopted the following methods to eliminate the disturbing foreign material in order to get their product acceptable by the customers.

- Proper selection of cotton (avoiding most contaminated cotton).
- Manual labour to pick foreign matter in the cotton bales prior to opening process.
- Installation of contamination eliminating devices in the blow room.

Foreign fibres clearers in auto winders or in OE rotor spinning

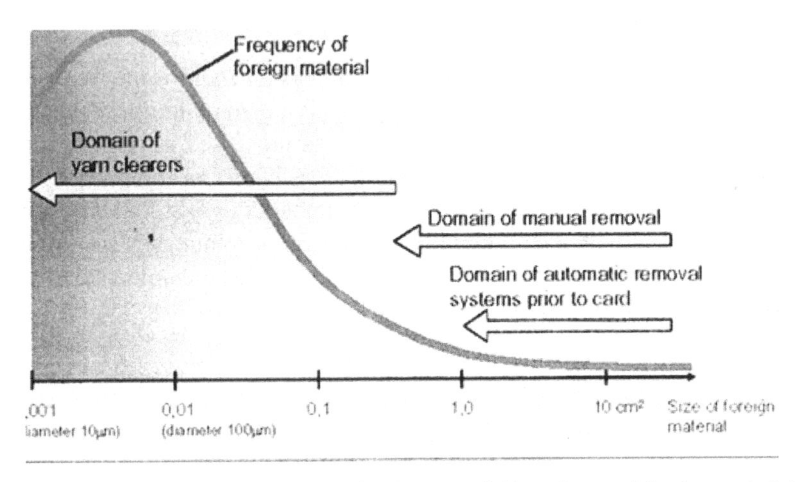

Figure 7.5 Methods to eliminate foreign material in cotton and foreign material frequency

7.6.1 Effect of foreign fibre particles on the spinning process

If the foreign fibre material present in the cotton is not eliminated in the blow room process by the contamination removal system, it will be embedded in to the tufts of cotton and the same will be disintegrated in to many pieces in the carding process. This will produce a large number of individual foreign fibres which form a cluster in the card sliver. After the drawing processes, it will appear in the yarn and its frequency depends upon the number of process stages. The position of foreign material present in the card sliver is shown in the Figure 7.6.

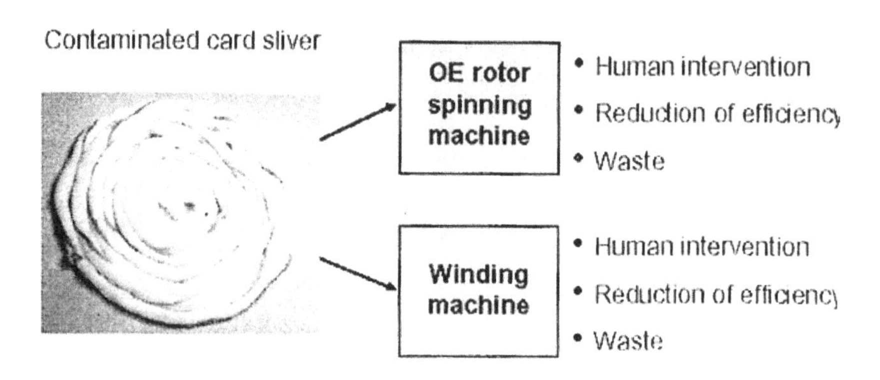

Figure 7.6 Effect of foreign fibre particle in the spinning process

In the OE rotor spinning, the machine triggers an alarm if the frequency of foreign fibre in the incoming material exceeds a certain limit if the clearer is set properly. If a contaminated card sliver is processed in ring spinning system, the distance between two foreign fibres is higher as compared to OE rotor spinning system, but the frequency of foreign fibres is still much higher in rotor yarns than with the regular bobbin in the ring spinning. If the foreign fibre alarm is set properly on the winding machine, the contaminated bobbin will be ejected by the machine.

7.7 Optimum strategies for foreign fibre detection and elimination

The objective is to combine the foreign matter detection and extraction systems in spinning process in such a way that the most suitable technological and economic variants are used. The coupling of different systems will make it possible to use the advantages of the individual systems to the best effect and eliminate the disadvantages which arise with the use of individual solutions. Maximum protection against foreign fibres can be achieved by the simultaneous use of all existent systems. The economic and technological optima will be linked by balancing cost and benefit. Subsequently, existing systems marketed for detecting and eliminating foreign matter in the actual spinning operation have to be taken in to account in different combinations. These will make it possible to produce specific product related requirement profiles so as to ensure the economic elimination of foreign fibres. Furthermore, producing qualities which lie outside the requirements for use is to be avoided in order to remain competitive. On the other hand, specific quality standards must be adhered so as to exclude complaints from customers.

It is essential for the spinning mills to have the most foreign matter detection and extraction machinery available for the trials, equipment to complement the plant already available will be installed at the end. The production stages following spinning will serve to assess the quality of foreign fibre extraction and the type of damage on the basis of size of the particles in the end product.

It is also evident that the detection and extraction of foreign matter require an effective system to overcome the difficulties which arise in the later stages in the processing. Over the years and still in some spinning mills, the following methods are used to eliminate disturbing foreign matter in order to keep the defects in the fabrics within acceptable limits.

- Selection of cotton based upon the presence of foreign matter.
- Employing manual labour to pick the contamination in cotton bales prior to spinning.
- Installation of contamination removal devices in blow room line prior to cars.
- Installation of foreign fibre clearers in winding machines.

In Export Oriented Units(EOU) , spinners take much care on contamination removal by keeping manual labour, say 1–3 bales for one person, to pick manually the contaminations prior to processing in blow room line. In addition to that, persons who pick the contaminations from the bales categorise the various type of contaminations based on their number as well as their weight, this will help the spinners to know about the intensity of the contaminations and also assess the gravity of the situations in the later stages. Moreover, the spinners make documentary evidence and reports on daily basis so that they can inform the cotton ginners and sellers to have an awareness of the contaminations and also the measures to reduce the same.

In vertically integrated textile units, the mending of the fabrics after removing the fine foreign matter is a common practice, but it is not possible to eliminate all the foreign fibres to be extracted.

7.8 Cotton selection

Selection of cotton bales for spinning mills is based upon the most important fibre properties like length, strength, fibre fineness and trash percentage. Nowadays, due to the stringent requirements laid down in the European markets especially on foreign fibre contaminations, due attention must also be paid to foreign fibres also. For this, it makes sense for the spinning mills to know about the growth areas with low foreign material contamination. Further, it must be the aim to order cotton from areas with low number of foreign material contamination in order to keep low the risk of removing foreign fibres low and also improve the efficiency of the contamination removal systems both human and electronically. In addition, the purpose is to keep the number of foreign fibre cuts in the electronic clearers in winding machines on a lower level.

This is true since some customers require "Zero foreign fibres" as their primary requirement. They are also ready to pay significant premium for such a high value addition product. In case, if the spinner does not realise the premium as a significant one and choosing cotton of high contaminations can lead to serious issues like rejections/ compensations thereby affecting his profit margins. Furthermore, the supply of cotton in cotton contracts to the spinners does not include any clause for contamination levels which is a dispute clause. The losses cannot be recovered from the cotton seller but the spinners have to bear the compensation insisted by the fabric buyers.

In addition to that, spinners must also take in to account the other fibre properties like short fibre content, nep content, etc., some spinning mills import cotton on account of low level of contamination in foreign cotton than local cotton. Figure 7.7 shows the other quality characteristics such as the short fibre content of cottons imported from various countries. To balance these issues, effective quality management systems must be followed in spinning mills to reduce the level of foreign fibre contamination in the end product.

7.9 Contamination removal systems in blow room line

There are various contamination removal systems available today prior to the card. In general, such contamination removal systems are important to eliminate the contamination matter of size greater than 1 cm^2. This is because to avoid further disintegration of fibres in carding process which will ultimately increase number of foreign fibre cuts in the winding section. However, even though contamination removal systems are installed in blow room line they do not help to fully meet the quality targets of the end user. It is practically impossible to eliminate the single foreign fibre in blow room line due to their size and number of ejections. The elimination of fine and single foreign fibres is possible to some extent in winding section since this single foreign fibre constitutes the highest amount of disturbing defects in the final yarn or fabrics.

Further, the location of the system and size of the tuft play a decisive role for the efficiency of detection. Hence, in spinning mills, contamination removal systems are installed at two places, one after the first cleaner and at the end before the chute feed since the tuft size progressively get reduced after every opening and cleaning machines. Similar to manual elimination, the electronic removal systems help in reducing major contaminations in blow room line, and finally reducing cuts in winding. This will help to maintain consistency in cuts in winding machine and with less human intervention.

The foreign matter elimination right at the beginning of the process has two major advantages.

- Foreign matter in the form of larger pieces can be more efficiently detected and eliminated than after disintegration in to multiples of single fibres and pieces.
- The disturbances induced by the foreign matter in the complete chain of subsequent processes are effectively reduced. Particularly the excessive number of end breaks and splices.

7.10 Contamination detection and elimination systems

Various manufacturers have come out with their products to combat the foreign matter contamination in cotton material. Some of the systems are discussed in the following sections.

a) Uster Opti Scan 1 Figure 7.7 shows the components of Uster Opti Scan in a spinning preparation machine. The installation dimensions are it is 4.4 m high, 2.2 m long and 1.8 m wide.

The key modules are as follows:

- Condenser at the top
- Internal transportation belt
- The optical sensor system
- The sectional extraction nozzles.

Principle of working

Uster Opti Scan 1 works on the principle of separation of incoming fibre tufts from the transporting air stream by means of a condenser. A rapidly running white transportation belt then accelerates the fibre tufts. Due to the difference in speed, the fibre tufts are distributed over the belt before reaching the optical sensor. The thickness of the tufts at this stage is around 8 mm. Uster Opti Scan 1 eliminates the most disturbing foreign matters by means of three extraction functions.

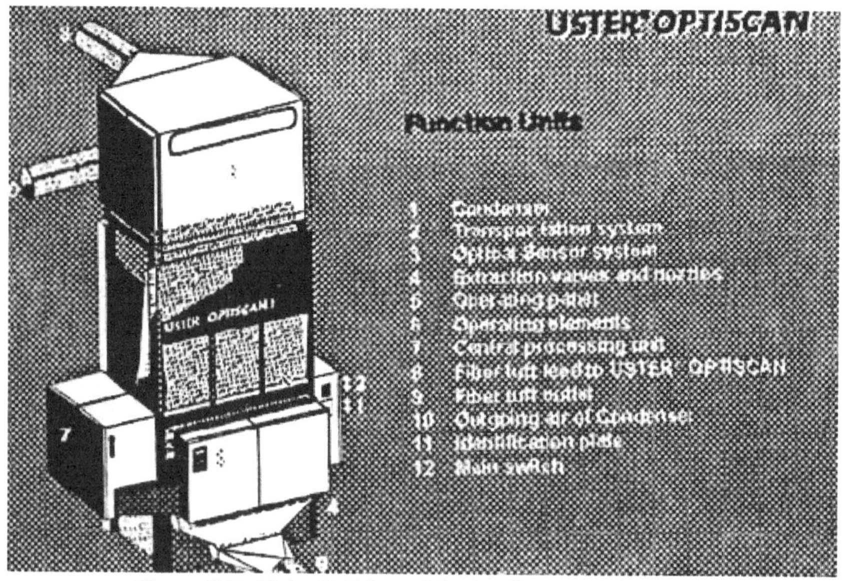

Figure 7.7 Uster Opti Scan 1 foreign fibre removal system

- The optical sensor system monitors the variations in colour and brightness of the cotton tufts. When a pre-selected threshold is exceeded, the pneumatic nozzles are actuated which extract the foreign matter in a collection box.
- In addition, an optical metal detector can be included which activates the extraction nozzles as the optical system.
- Finally due to the change of direction of the flow of tufts, heavy particles such as stones are extracted and get collected in the collection box.

7.10.1 Arrangement of optical sensors

The optical sensor arrangement consists of 64 optical colour sensors which are distributed across the complete width of the machine. In each sensor, there is a processor integrated. This processor evaluates the intensity of the colours like red, blue, green and also their brightness values. As soon as the threshold values are overstepped, the optical sensor actuates the respective blowing nozzles and the foreign matter is collected in collection box provided.

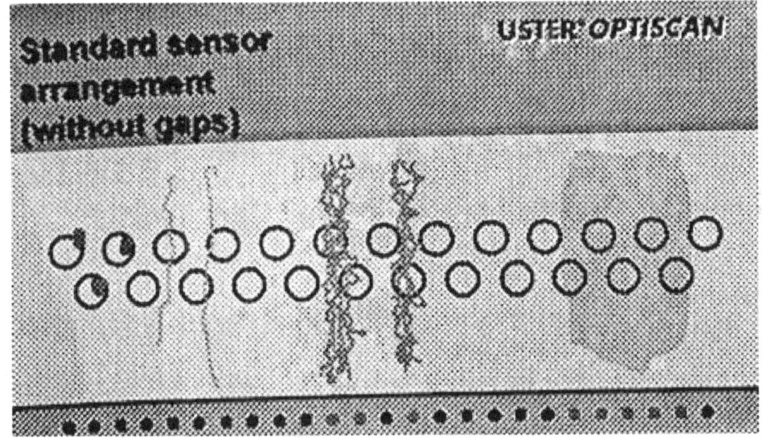

Figure 7.8 Sensor arrangement in the Uster Opti Scan 1

The arrangement of the circular colour sensors shown in the Figure 7.8 are in two displaced rows. It means, there should not be any gap and the complete stream of tufts is checked. In the figure, the dots are shown in front of the colour sensors on the left side which refers to the trash particles. These trash particles need not be extracted because it will be removed by the opening and cleaning machines.

7.10.2 Positioning of Uster Opti Scan in the blow room line

It is necessary to extract the foreign matters from the cotton material at an early stage instead of breaking in to many pieces after the openers and beaters. It means more tasks will be given to electronic yarn clearers in winding to extract very tiny particles. This also adds to the loss in efficiency of the auto coners. Hence the Opti Scan has to be placed at a position right at the beginning of the blow room line sequence. The arrangement is shown in the Figure 7.9 (A) and (B).

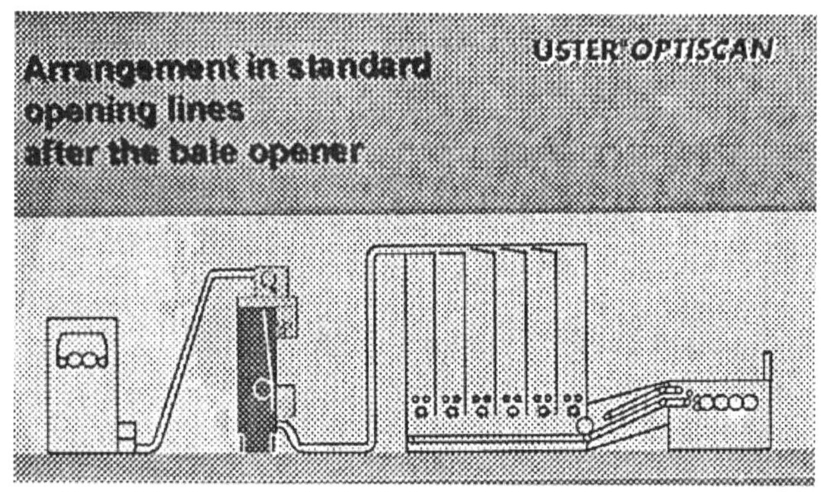

Figure 7.9 Arrangement of the Opti Scan 1 after the bale opener in blow room line

Figure 7.9 Arrangement of the Opti Scan after the mono cleaner in blow room line

With this arrangement the detection of foreign matter is more effective when the cotton tufts are presented to the sensors have been opened to a certain level. The ideal position of this sensor will mostly found suitable after the pre-cleaner or after the blending machine.

A study has been conducted by feeding 25 pieces of woven fabric of 10 × 10, 20 × 20 and 30 × 30 mm in to the automatic bale opening machine. The waste material at each processing stage was collected and examined for any fabric pieces. In the opening and cleaning machines, only the edges of the fabric were extracted and the fabric remain intact. It means up to the card, there is no actual disintegration of the foreign material. Only in the carding section, the actual disintegration of the fabric pieces takes place. It is to be stated that smaller the pieces, better will be the elimination 10 × 10 mm fabric pieces were extracted to more than 95%, whereas in the other sizes, the efficiency of extraction is only 50%. Both the blenders and lickerin in the card extract the small pieces effectively. Hence it is customary to eliminate the large pieces of contamination in the bale itself before processing in to the opening lines.

The Uster Opti Scan 1 extracts foreign material like woven fabrics, plastic films, paper, card board, larger trash and stalk particles and also oily, dirty brown and yellow cotton.

7.10.3 Contamination removal system in Jossi vision shield

Many machine manufactures have come out with innovative ideas to eliminate or minimise the foreign fibre contamination in cotton. One such contamination detector and eliminating device in the blow room line is Vision Shield and Magic Eye 1 from Switzerland. This ensures that the contaminations are eliminated from the cotton before the yarn building process starts. Contaminations which have been eliminated at an earlier stage will not interfere with the production process. The disadvantage associated with low efficiency in winding due to foreign fibres in the yarn is drastically reduced. This ensures the right quality for highly efficient down-stream operation in all textile processes.

Working principle

The vision shield system is the most effective system to detect and eliminate foreign material in cotton. It consists of two ultra fast colour cameras which scan the cotton tufts from both the sides. Due to the high resolution and photo realistic colours, the cameras recognise the size and shape of each particle. This ensures a good differentiation between good material and cotton contamination. The vision shield direct detects the colour of the cotton automatically and the dynamic process corrects drifts of colour shades constantly. Contaminations are ejected with high reliability with a minimum loss of spinnable fibres. The ejected material is disposed in to the waste collecting system.

Another development made in the Vision Shield is the Magic Eye1. The Magic Eye1 is attached to the vision Shield system and allows the detection and elimination of transparent, semi-transparent and white polypropylene from the cotton material. The Magic Eye 1 detects these contaminations by a newly developed optical system. This technology measures and differentiates clearly the characteristic between synthetic and cotton material. This system is better than the camera based system as it immediately recognises the necessary parameters for best results of the detection of foreign fibres.

Advantages

- High productivity and efficiency in weaving and knitting.
- Production of second grade quality fabric is reduced.
- Less machine stoppages in weaving.
- Better warp beam preparation for satisfactory running in high speed weaving machines.
- Less yarn breaks in shifts is the goal in many knitting mills. The cross bar for first quality in knitting is getting higher and higher with faults only 3 faults/100 m is allowed.

7.10.4 Positioning of contamination removal system in blow room sequence

In general, the contamination removal systems have to be placed at a position right at the beginning of the spinning process. The reason is to avoid tearing of foreign matter in to smaller pieces by cleaning machines and separate individual fibres.

On the other hand, the detection of foreign matter is more effective to the optical sensors when the cotton tufts presented achieved a certain degree of opening. The ideal position of the contamination removal system with regard to availability of space and also the efficiency of the foreign matter extraction system will mostly be found after the pre-cleaner or blender machines. In order to study about the efficiency of the system, a case study has been conducted with Uster Opti Scan 1.

Case study

To ascertain the efficacy of the system in the extraction of foreign matter in the blow room line, 50 pieces of woven fabric with sizes of 10 × 10, 20 × 20 and 30 × 30 mm are fed in to the automatic bale opener. The waste material at each processing stage was examined for the presence of foreign material. The foreign material extracted by the opening machines did not only provide an insight to the level of disintegration of woven fabric in to pieces, but also gave reference to the ability of the particular opening machine to extract foreign matter.

The study shows that up until the card, no actual disintegration takes place. Certainly, some pieces of yarn are extracted out of the edges of the woven fabric, but, in most cases, the woven fabric remains intact. The edge yarns are mainly extracted in the blow room. Only at the card and more particularly between cylinder and flats, does an actual integration of the pieces of woven fabric take place.

Now it is the question that which amounts and which types of foreign matter in a modern blow room line able to extract. The solution is that as a general survey, it would seem that the smaller the size of the woven fabric pieces, the more the possibility that they will be extracted. In the above case study, of the 10 × 10 mm pieces, about 95% of the pieces were extracted, whereas in 30 × 30 mm pieces of woven fabric, only approximately 50% were extracted. For this reason, it is particularly important that the larger pieces of foreign matter should be detected and extracted. Both the cleaning machines in the blow room line and the licker-in of the card are able to extract foreign matter. However, a fine cleaner does not extract much foreign matter. Hence, spinning preparation machines itself is able to extract large amounts of foreign matter.

Advantages of Uster Opti Scan1

- Better running conditions.
- Better end product.
- Higher efficiencies.
- Fewer customer complaints.

7.11 Elimination of foreign fibres of polypropylene type

Foreign fibres made of poly propylene (white or colourless) are still a massive problem to the spinners which lead to rejections from the buyers. Enormous sums are spent every year at all levels in spinning, weaving , knitting and finishing compensating for damages caused by polypropylene contamination. To combat these problems with polypropylene fibres, Trutzschler's SECUROPOP-SP-FP is a ground-breaking innovation in the field of foreign fibre detection. Besides, the tried and tested technology and the excellent performance of SECUROPOP-SP-FP can detect coloured foreign parts and also can separate white, colourless and transparent polypropylene in the blow room with the expense of minimum loss of good fibres in the blow room.

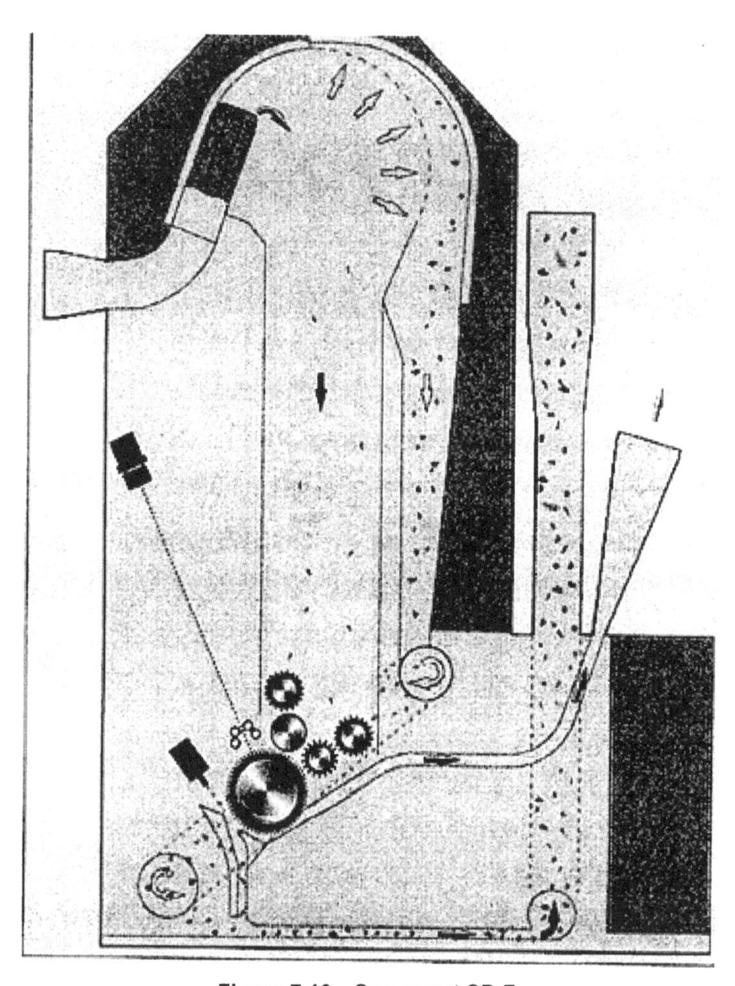

Figure 7.10 Securomat SP-F

Trutschler's uses an entirely new patented technology to detect these foreign fibres – polarised light and CCD colour line cameras as shown in the Figure 7.10.

Working principle

Fibre tufts and foreign matters located in the rectangular duct are illuminated with fluorescent tubes. The light emitted from the fluorescent tubes is first channelled through a polarised filter. A second polarisation filter is located in front of the colour camera's lens shown in the Figure 7.11. With polarised light, foreign matters like translucent, bright or multi-coloured objects are illuminated to a certain extent, while cotton tufts remain unnoticeable.

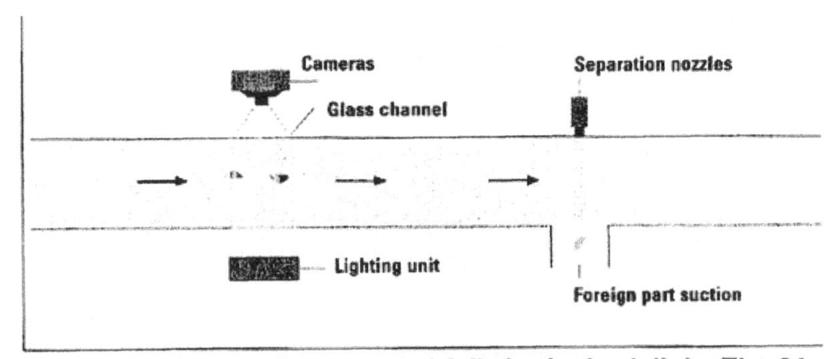

Figure 7.11 Detection of polypropylene by cameras

As soon as the foreign fibres like polypropylene or polyethylene are detected, they are selectively separated via one or two of a total of 64 compressed air nozzle groups which are arranged over the whole working width, thus reducing the loss of good fibres to a minimum.

A similar system for detection and separation module for polypropylene is arranged behind the opening roller in the doffer zone. Here the transmitted light method is applied, while the conventional system uses the reflected light method to detect coloured foreign parts by monitoring the fibre tufts and surface of the rotating roller.

7.11.1 Benefits of SECUROPOP-SP-FP

- Separation of fibres and transport air current and complete dust extraction.
- Storage of material and continuous feeding to card (CONTIFEED) with variable speed drive of the feed unit.

7.12 Securomat SCFO

This Securomat foreign matter extraction system is positioned immediately before the card. It simultaneously performs the function of a de-dusting machine. After de-dusting, the fibre material is fed by delivery rollers to a porcupine roller via a feed chute. A CCD camera senses the porcupine roller zone. Unique colour and brightness values which differ from those of foreign fibres can be allocated to the fibre material and porcupine roller. If any foreign fibre is detected, the signal is transmitted to the blower control unit. The foreign particle is blown out of the fibre stream by the compressed air jets and gets separated from the material flow (Figure 7.12). Contaminations like feathers, film fragments, yarn, thread and fabric are few examples which are eliminated.

SECUROPROP-SP-FPU

The latest technological advances in the foreign fibre detection in blow room include the detection of transparent materials like the most problematic polypropylene contaminants. This is achieved by specially designed sensors. Trutzschler GMbH introduced a foreign matter separator that removes transparent and semi-transparent contaminants by applying polarised transmitted light and 3-CCD colour line scan technologies. This sensor is able to detect the polypropylene, polyethylene of individual strings and large pieces of fabric. Coupled with CCD line scan cameras which operates in the visible spectral range of visible light it clearly distinguish between the contaminants and cotton by the differences in color contrast and or its brightness. The cross-sectional view of the SECUROPROP-SP-FPU is shown in the Figure 7.12.

The cotton material in the form of opened tufts enters the machine from the top through a rectangular duct. This ensures uniform distribution of fibrous tufts across the entire working width of 1,200 mm. In order to visualise the contaminants clearly without getting it concealed by bulky cotton tufts, a high degree of opening is a pre-requisite. Hence this machine has to be located after an opening or cleaning machine in the blow room line. The maximum throughput of the unit is 1,000 kg/hr

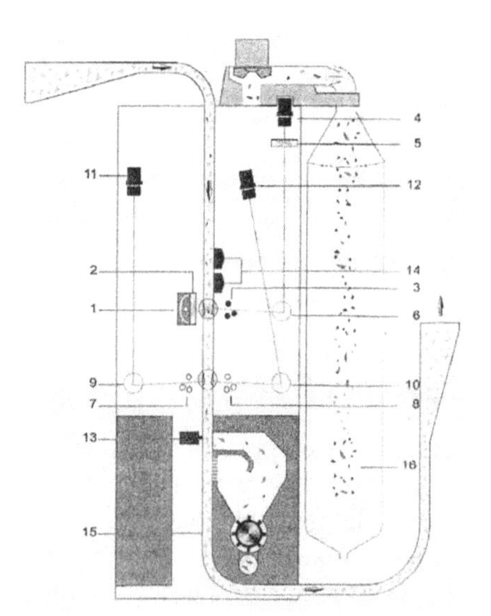

Figure 7.12 Cross-sectional view of SECUROPROP-SP-FPU

The first inspection unit consists of a glass duct segment and an illumination unit (1) which accommodates a fluorescent tube and a polarisation filter

(2). On the other side of the glass duct, 3 ultraviolet (UV) fluorescent tubes (3) are utilised to illuminate the fibre flow. The tufts entering the duct are monitored by a 3-CCD colour line scan camera (4) with the second polarisation filter (5) attached to it. A tilted mirror (5) detects the polarised transmitted light from the illumination unit (1) and the reflected light from the UV tubes (3) to the camera (4). The first inspection unit serves to detect transparent and semi-transparent objects such as polypropylene as well as any other fluorescent contaminants. These modules are called PP and UV modules.

The second inspection unit consists of fluorescent tubes (7,8) on both sides of the glass duct. Subsequently there are two tilted mirrors (9,10) direct the light to two 3-CCD colour line scan cameras (11,12) which are mounted on the top section of the machine. By this arrangement scanning the flow of tufts from both sides increases the detection efficiency and the contaminants cannot escape behind solid cotton tufts. The second inspection unit identifies any contaminants that deviate from the cotton tufts in terms of colour/brightness values. This is module 3 – the colour module.

These cameras provide 2,048 effective pixels per line. Uniform detection sensitivity across the width of the machine is accomplished by using high quality lenses. With a working width of 1,200 mm, a total of 2,048 pixels per line, 10 m/s velocity of the tufts passing through and a scanning frequency of 10 kHz, a resolution of 0.6 mm across the working width and 1.0 mm in the direction of material flow can be achieved. A powerful real time computing device transforms and analyses the unprocessed camera data which serves an integral part of the machine control unit. The actual colour components and the brightness values of the specific material processed are established as reference values. During the normal operation, the signals received from the cameras are compared with the reference values and if it exceeds a certain threshold values the objects are identified as contaminants. This can be adjusted manually via a full touch screen.

Contamination removal

The detected contaminants from the tuft flow are removed by a total of 3 × 48 compressed air nozzles across the working width of the machine (13). Along with 48 pneumatic solenoid valves and a compressed air reservoir which is integrated in to a single compact nozzle beam. The compressed air impulse is confined to three adjacent nozzles covering the actual horizontal position of the foreign material in the rectangular duct. The nozzles fire precisely at the very moment when a foreign object passes the nozzle beam. This is achieved by two specially designed optical tuft velocity sensors (14) which are located before the first inspection unit. The compressed air flow is directed towards a collection container perpendicular to the tuft flow and its duration lasts for 30 minutes. Hence compressed air consumption is minimal even at high production rates. The loss of usable lint accounts for only 0.3 g of fibre per 100 kg/hr throughput per ejection, which corresponds to a total waste amount approximately 1% under practical conditions. The foreign matter in the collection container is discharged continuously by means of a rotary air lock (15) and blown in to a filtering bag (16) with a fan. The cleaned flow of cotton then proceeds to the next machine in the sequence of the blow room line.

Polypropylene contamination is more prevalent in most of the Indian cottons. Chinese cottons contain fluorescent plastic films and polypropylene strings and coloured foreign material is evident in all the cotton varieties in all countries.

8

Monitoring system in carding machines

8.1 Introduction

Carding is an important process step in the spinning mills. If carding sliver quality is maintained according to the subsequent process, it is possible to achieve the best quality yarn. There are many quality factors need to be ascertained in the carding process like neps, seed coat neps, CV% of the sliver. Auto levellers monitor the sliver quality in the carding machine similar to the auto levellers in draw frame machine. Both short-term and long-term auto levellers can be used in carding machine depending upon the type of fibre, process, etc.

8.2 Auto levellers in carding

The aim of the auto levellers is not to correct errors but rather to avoid them particularly at the start of the process in spinning mills. Many mill personnel have an argument whether auto levellers have to be installed at carding or at finisher draw frame stage. In general, card is the first process where the intermediate product sliver is produced where relatively high degree of evenness is required. For various reasons, the card cannot operate absolutely, evenly, for example, owing to uneven material feed. However, spinning mills are forced to use auto levelling equipment under highly varying circumstances. Different principles can be selected according to the quality requirements and the operating conditions of the individual mills.

8.3 Classification of auto levellers

Irregularities in the material can be compensated for in the material supply system either at feed or at delivery. Furthermore, the material supply system should operate with greatest possible accuracy, since this has a direct effect on sliver evenness. Many manufacturers of carding machine offer double chute system with a certain degree of coarse regulation in the lower chute system.

However, the main regulating system is in the feed. Auto levelling is usually performed by adjusting the feed roller speed. Adjustment of delivery speed is hardly ever used. A distinction should also be drawn between different types of levelling systems:

- Short-term auto levelling systems, which regulate the lengths of production from 10 to 12 cm.
- Medium-term auto levelling systems which correct lengths above 20 m.
- Long-term auto levelling systems which correct variations of length over 20 m.

8.4 Principle of short-term auto levelling

This principle of auto levelling is regulation at the delivery. This requires a drafting rolls at the delivery end before the coiler. The maximum draft required is 1.2. A system of open loop control is illustrated in Figure 8.1.

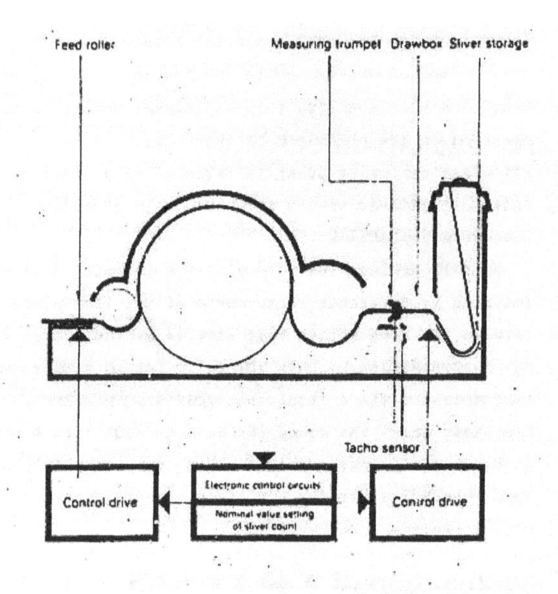

Figure 8.1 Short-term auto levelling in card

In this system, the measuring point is provided upstream from the drafting arrangement. It senses the volume of the incoming material (sliver) and transmits the signals to the control unit. The control unit generates the control signals and these signals are passed to the regulating unit. The regulating unit is of various designs which adapt the speed of the delivery roller according to the measured sliver volume.

If the measuring point is provided downstream from the drafting arrangement or the delivery roller itself acts as a measuring system, then the system is said to be a closed-loop principle. If an open loop principle is used in a short-term auto leveller, certainly short length variations can be corrected but there is no guarantee that the average sliver fineness is held constant. On the other hand, closed-loop principle is not suited to correct short-term variations because of the time delay or dead zone which is inherent in the system.

Moreover, the low draft in the drafting arrangement is also another source of problem, since the low draft always cause stick-slip effect on the fibres. Finally, the drive to the delivery speed must be continually varied. There are only two possibilities of this auto leveller, one where the card sliver is directly fed to the rotor spinning system and the other one is in processing of comber noils.

8.5 Auto levelling at the feed zone

Rieter has introduced auto levelling at the feed zone in addition to the combination with the normal long-term auto leveller. This additional device operates on the principle of short-term auto levelling. However, the operating device is followed by a draft of over 100, it cannot correct short-term mass variations, it can correct only medium to long-term mass variations.

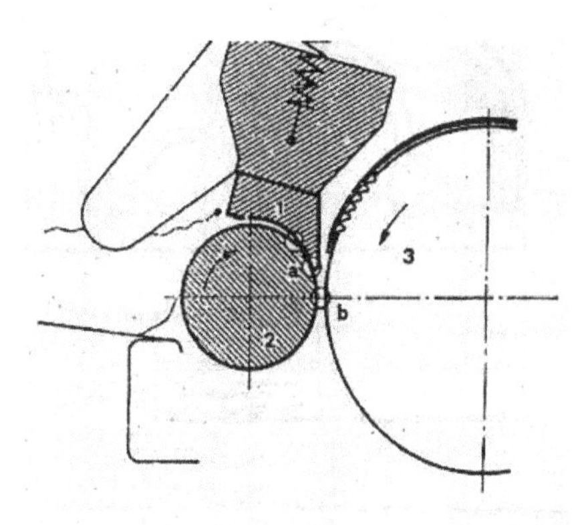

Figure 8.2 Auto leveller at the feed zone of the carding machine

Operating principle

In this device (Figure 8.2), the volume of the incoming feed material is detected by the movement of the feed plate relative to the feed roller which lies below the feed plate. Corresponding to the volume of the material a signal is fed to the micro computer which operates through a regulating gear transmission, continuously adapt the speed of the feed roller according to a pre-determined set in feed volume of the material. This system is an open loop control and such systems can be useful if it works in conjunction with long-term auto leveller.

8.6 Principle of medium-term auto levelling

The working principle of medium auto levelling in the card is shown in the Figure 8.3. Rieter levelling system corrects medium-term variations by means of proportional – integral controller which is assisted by a micro computer. At the in feed, a measuring device detects the cross-sectional variations in the feed batt. The electronics then vary the speed of the feed roll so that these variations are corrected. A pair of stepped rolls senses the variation in the fibre cross-section at the delivery. The electronics compare this measured value with the selected nominal value and correct any deviations again by varying the feed roll speed.

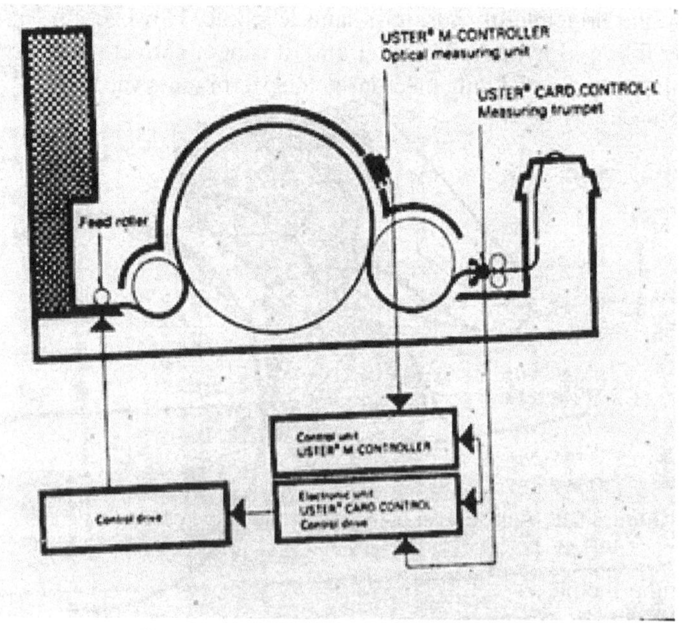

Figure 8.3 Medium-term auto levelling

The Rieter levelling system operates from a delivery speed of 30 m/min up. The auto levelling remains switched on during acceleration and braking of the Doffer provided that the delivery speed is over 30 m/min and the light scanner of the sliver monitor at the step rollers is uncovered. Medium-term auto levellers can be used in places where flock feeder does not supply even feed material to cards or for rotor spinning machines in case if only one draw frame passage is used.

8.7 Principle of long-term auto levelling

The principle and working of long-term auto levelling is shown in the Figure 8.4.

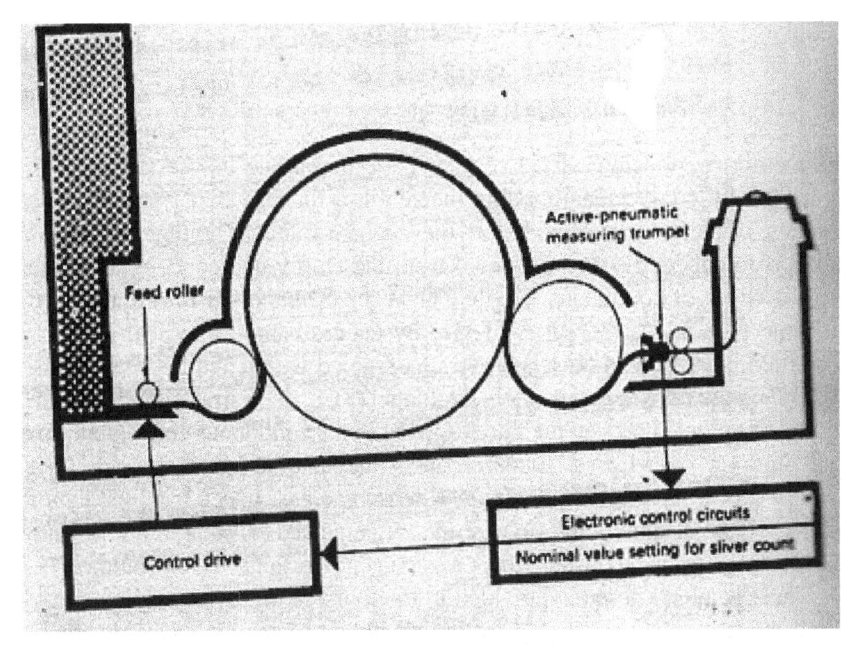

Figure 8.4 Long-term auto levelling

The most commonly used auto levellers in carding machines of a spinning mill is long-term auto leveller. The principle of working is illustrated in Figure 8.4. In this system, the measurement of sliver weight is performed at the delivery side. The electronic pulses generated according to the variation in this way are processed electronically so that the speed of the input feed roller can be adapted according to the required delivery sliver weight through mechanical or electronic regulating devices. Long-term auto levelling is primarily useful in production of carded yarns in the rotor spinning mill.

8.8 Measuring device in card

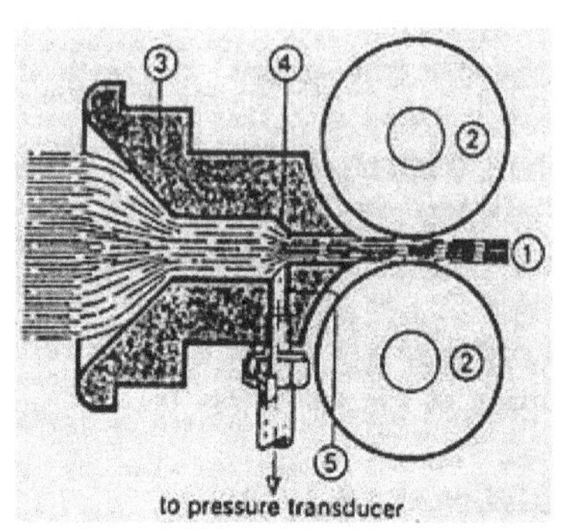

Figure 8.5 Active pneumatic pressure transducer

The measuring principle of active pneumatic measuring device is shown in Figure 8.5. In conventional carding machines, a funnel which may be made of plastic or metal is provided before the calendar rollers. The function of this funnel is to collect the web delivered from the Doffer and condense it in the form of sliver of continuous length. Zellweger technologies have developed the same funnel as a measuring device by using a simple physical concept. In textile fibres, there is always air space occupied within and between the fibres. When the fibre material enters the funnel (3), it also carries along quite an amount of air held between the fibres. Owing to the continuous convergence of fibres in to the funnel, air is squeezed out as the material passes through. As a result of this, a super-atmospheric pressure is generated which is a function of the sliver cross-section, if the sliver speed is maintained constant. If all the fibre characteristics remain constant, this pressure is proportional to the volume.

There is lateral bore (5) provided in the funnel with corresponding leads, which transmits the pressure in to the chamber of pneumatic-electric transducer. Using electrical induction the pressure is converted in to electrical signals. The electrical signals developed are compared with the set values which enable the control of the electronic units in the regulation equipment. The benefit of this active pneumatic measuring device is simplicity in operation and does not require additional or sensitive moving parts. The disadvantage of this system is the measuring principle is affected by fibre fineness and variation in fibre fineness will affect the result. However, in textile since the raw material like cotton fibres are variable in nature, this effect can largely be ignored.

Measuring device at the input

The card sliver regularity is also dependent upon the input feed material weight. Unless a system of regulating the feed material supply, it is not possible to achieve consistent delivered sliver weight and CV%. For this purpose, the in fed material sensed mechanically for its cross-section by the feed trough. The measuring unit registers the movement of the feed trough and gives signal to the electronics which controls the amount of feed material according to the cross-section.

Principle of working

A variation in the feed cross-section is detected before it get passed in to the machine. The microprocessor is able to vary the feed roller delivery speed according to the signal strength so that the material in feed rate is kept as constant as possible. At the start, the electronics measures the mean value and keep this value in memory. When the material is fed, the momentary in feed signal is compared with the already memorised mean value, and the difference signal is passed to the variable speed motor which then alters the delivery speed of the feed roller.

8.9 Measuring device at the delivery

Principle of working

This unit senses the delivered sliver and gives signal an electrical variable according to the proportional sliver weight. The card sliver leaving the apron delivery enters in to the trumpet closer to the nip of the two stepped rolls (Figure 8.6). The bottom of the two is driven and the top roll is able to move in vertical direction through a spring loaded lever. This roll is made to turn by the emerging sliver.

The actual value of the card sliver is measured by the electronics regardless of the delivery speed. The collected actual values are compared with the sliver weight nominal value. Through a non-contact displacement transducer, the position of the top moving stepped roll in relation to the bottom fixed roll is sensed. The distance between the rolls is the sliver cross-section measured at the nip. If there is any deviation, it is registered and compared with the nominal sliver weight and the feed roll speed is altered through the control.

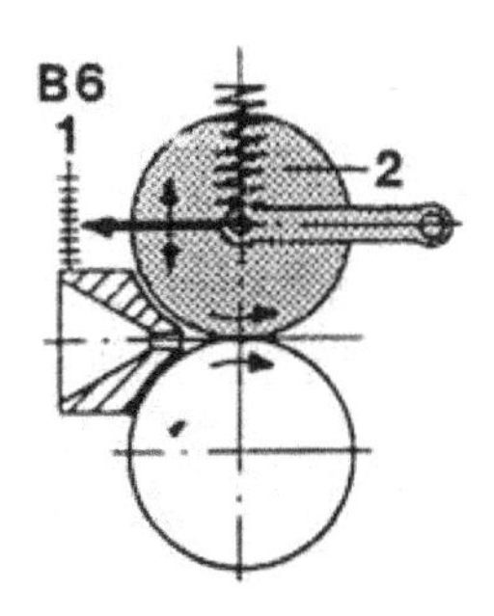

Figure 8.6 Measuring device at the delivery of the carding machine

8.10 Delivery speed monitoring

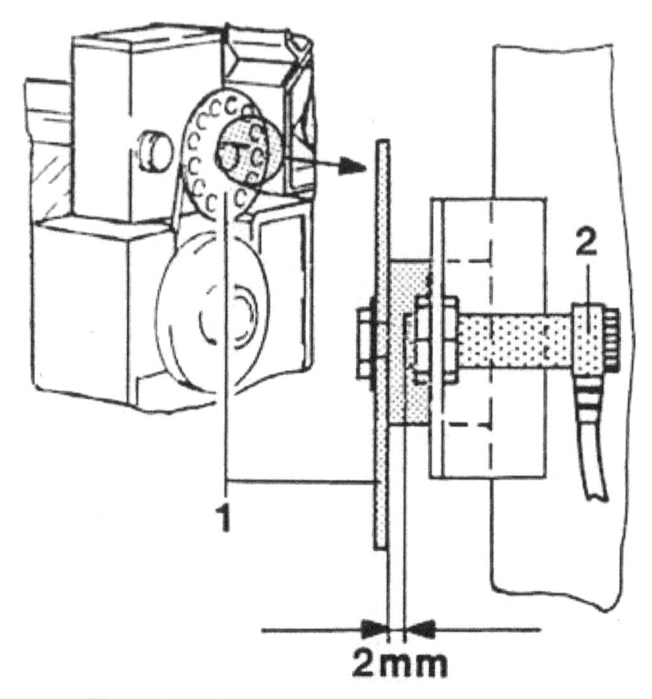

Figure 8.7 Delivery speeds monitoring in carding

The delivery speed is monitored in the following way as shown in the Figure 8.7.

The perforated disk (1) produces pulses in the proximity switch (2). The pulses are then utilised for calculation of the delivery speed.

8.11 Cylinder speed monitoring

The cylinder speed in the carding machine is shown in the Figure 8.8.

Working

The perforated disk (1) produces pulses in the proximity switch (2). The information reaches the computer and the speed is displayed. The proximity switch monitors the minimal cylinder speed. If the cylinder reaches a speed of 250 rpm, the Doffer can be accelerated to the pre-programmed speed. At the same time, during the switching on if no cylinder motion is registered the main motor is shut off. In the same way the cylinder speed is monitored during braking also.

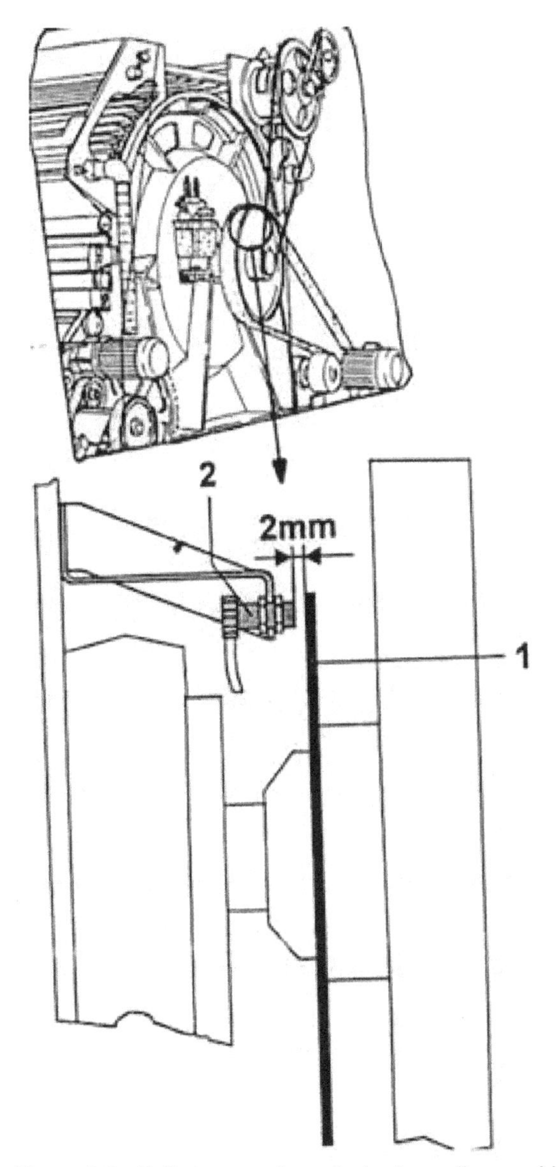

Figure 8.8 Cylinder speed monitoring in carding machine

8.12 Flats monitoring

In the flats monitoring, the movement of the flats is registered by the proximity switch (1). If there is no motion of the flats, the machine shuts off the flats (2) (Figure 8.9).

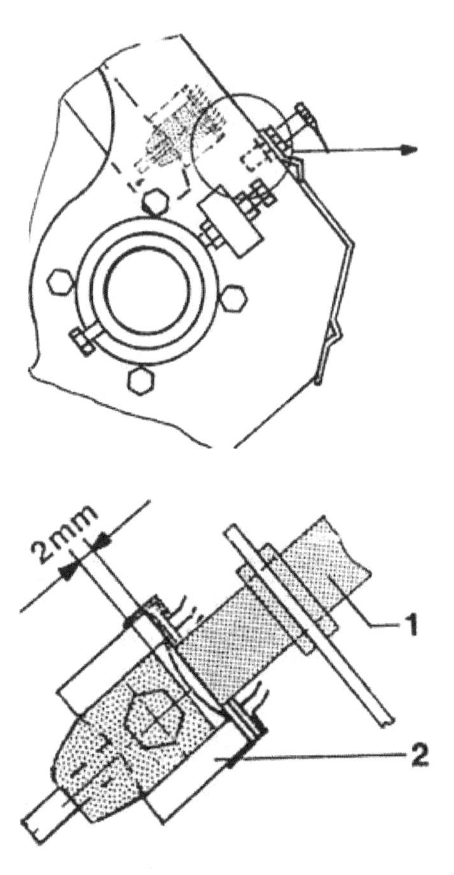

Figure 8.9 Revolving flats speed monitoring in carding machine

8.13 Doffer regulating system

The delivery speed is checked by an electronic control circuit. The function of this control unit is to automatically register the draft in the delivery and correct the speed of the delivery motor (5) accordingly.

Working

The speed monitoring of the Doffer in the carding machine is shown in the Figure 8.10. The proximity switch (1) delivers pulses to the computer (2). These pulses are converted in to the speed values and compared with the preset values. The resulting value is sent to the computer. The regulator type is PI (i.e. proportional – integral) and is active only for 30 second after it has been activated. After that it remains inactive until the next nominal value change.

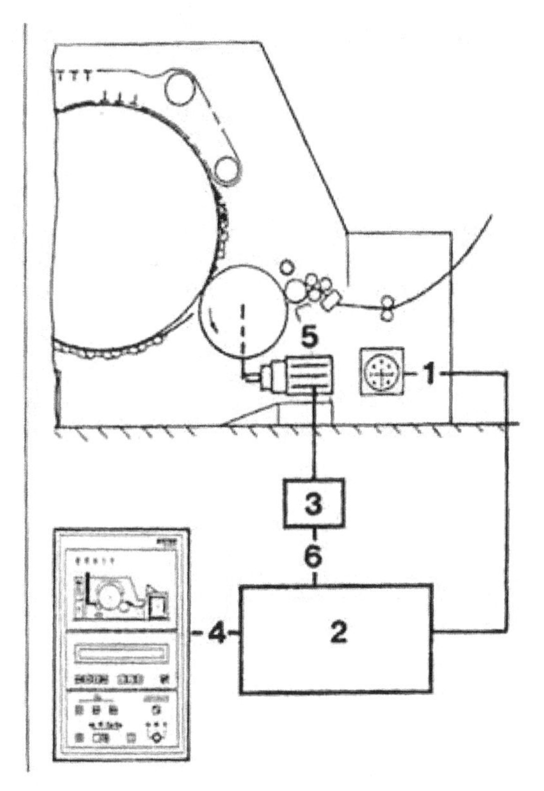

Figure 8.10 Doffer speed monitoring

1 = Perforated disk with proximity switch for calculation of the delivery speed.
2 = RMC
3 = Inverter
4 = Control unit
5 = Delivery motor
6 = Serial communication line to the inverter (3).

8.14 Quality data on the carding machine

The following are the quality data in the UQM.

- A% – Count variation or the deviation of the sliver volume from a set value. It is generally expressed in percentage.
- CV% – Unevenness calculated over a length of 100 m (based on 12,800 pieces of 7.8 mm each)

- CV1m – Unevenness calculated on the basis of 100 pieces of 1 m each.
- CV3m – Unevenness calculated on the basis of 33 pieces of 3 m each.
- CV10m – Unevenness calculated on the basis of 1o pieces of 10 m each.
- The CV% values recorded on-line by the measuring device are with comparable with the CV% values of the Uster Statistics.

9

Control system in draw frames

9.1 Introduction

A system is a combination of different physical components which set together to achieve desired output. A system can also be defined as a set of elements and functional blocks interconnected to produce a desired output. The process is considered as a system in industrial and day-to-day applications like electrical, mechanical, thermal, hydraulic, pneumatic, chemical and textile engineering. Electrical system has voltage, current, power, energy, etc. Mechanical system involves linear and angular displacements, velocity, acceleration, etc. Hydraulic system involves pressure, flow, temperature, viscosity, etc.

A control engineer has to decide what to control, how to control and with what accuracy? Each control system has a set of parameters describing its condition. Depending on the application, accuracy and complexity may be fixed to limit the cost effectiveness. In air conditioning system used for control of air, usually control is limited to room air temperature and humidity. While controlling electric motor, usually speed and torque is controlled.

Suppose if X is the input signal given to any system as shown in Figure 9.1 then the output will be Y also known as response.

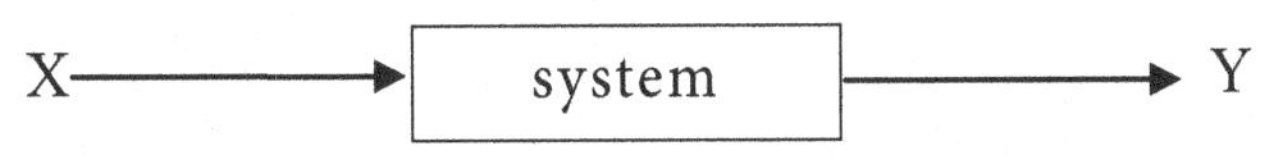

Figure 9.1 Fundamentals of control system

9.2 Control system

Control system is very important in all engineering applications and in machineries. The first significant control device was invented by James Watt in his fly ball governor. This was invented in 1767 to keep the speed of the engine constant by regulating the supply of steam to the engine.

The control systems can be classified in to three main types in textile applications.

a) Open loop system
b) Closed loop system

c) Combined loop system

All the above systems will be discussed in detail with its practical applications in textile machines especially in carding and in draw frame machines.

a) Open loop system

The open loop control system is also known as control system without feedback. In open loop systems, the control action is independent of the desired output. In this system, the output is not compared with the reference input. This is shown in Figure 9.2.

Figure 9.2 Open loop system

The component of an open loop systems are controller and controlled process. The controller may be amplifier, filter, etc., depends upon the system. An input is applied to the controller and the output of the controller and fed to the controller which gives the necessary output.

Few examples of domestic nature are:

1. Automatic washing machine.
   In automatic washing machines, we immerse the clothes in water along with the washing powder. Then the machine is set manually for a time limit of 10 minutes. After completion of the process, we do not know whether the clothes are washed properly or in other words, dirt is removed to the maximum extent. It means there is no feedback to the washing machine. This is an example for open loop system.
2. Immersion rod
   Immersion rod heats the water for a specified period of time say 45 minutes. After 45 minutes, it is not sure that how much water is heated. It means there is no feedback to the immersion rod. This is another example for open loop system.

Merits of open loop control system

a) Open loop control systems are simple to operate and suitable for high speeds.
b) They are economical and less maintenance is required.
c) Calibration of open loop system is not difficult.
d) They are faster than closed loop system.

Demerits of open loop control system

a) Open loop systems are inaccurate which meansthere is no feedback that the defect is corrected or not.

b) These are not reliable.

c) In textile applications in draw frames, combinedloop system (open and closed loop) is followed which gives the reliability in results.

b) Closed loop control system

Closed loop control systems are also called as feedback control systems. In closed control system, the control action is dependent on the desired output. In closed loop system, the output is compared with the reference input and the deviation is detected and the error signal is produced. The error signal is fed to the controller to reduce the error and the desired output is obtained. The closed loop control system is shown in Figure 9.3.

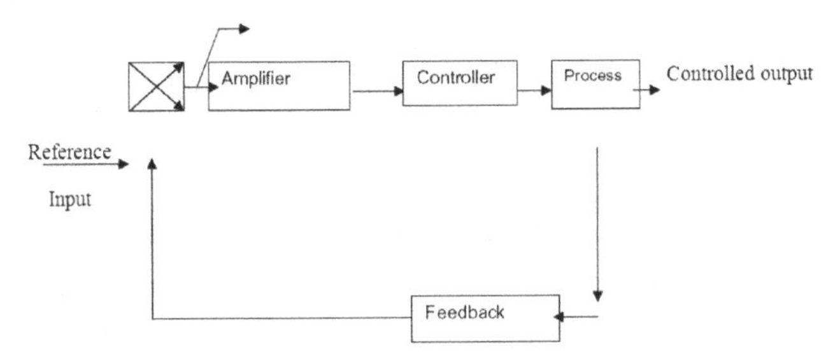

Figure 9.3 Closed loop control system

Example: In an air conditioned room, the temperature and humidity for comfort is regulated by a thermostat. Thermostat measures the actual room temperature and compares with the desired temperature. As soon as the required temperature is reached, an error signal is produced, the thermostat turns ON or OFF the compressor. The same is shown in Figure 9.4.

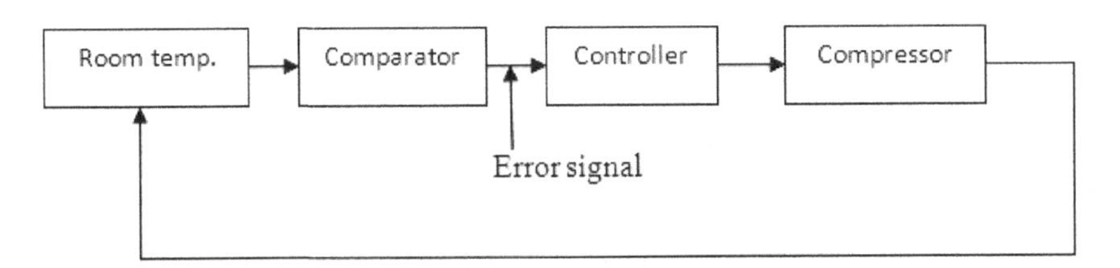

Figure 9.4 Working of thermostat

Merits of closed loop system

a) Closed loop systems are more reliable.

b) Number of variables can be handled simultaneously.

c) Optimisation is possible.

Demerits of closed loop system

a) They are slow because the time taken for sensing and correction is slow and hence some of the material delivered will not have the chance to get corrected.

b) It is not suitable for high speeds.

c) Combined loop system.

In order to overcome the deficiencies in both the open loop and closed loop control systems, combined loop systems was developed. In this combined loop system, the material is scanned or sensed for its input quantity and the same is fed to the comparator unit. In the comparator unit, the sensed values at the input are compared with the reference values for any deviation. If the deviation is detected, the same is corrected before the material is delivered to the output. In the output, the material is again sensed whether the correction is made properly or not. In this way, the output material is sensed at the feed as well as in the output also. In this way, the deviations are corrected and the quality is kept under control. Such type of combined loop systems is incorporated in draw frame machine in spinning mills.

Combined loop system is shown in Figure 9.5.

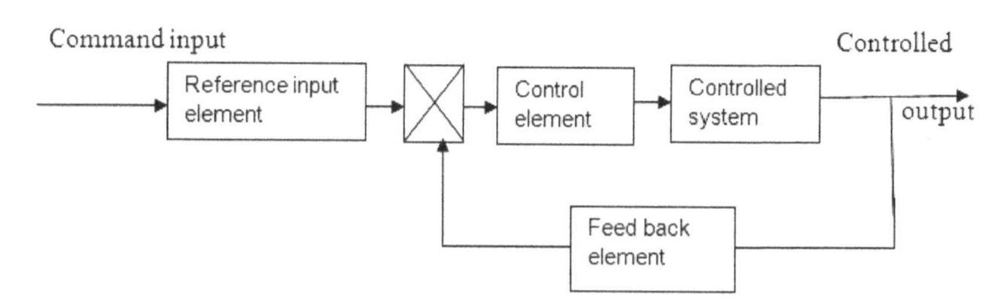

Figure 9.5 Combined loop system

Command input

The command is the externally produced input and independent of the feed-back control system.

Reference input element

Reference input is the standard values or signals proportional to the command input.

Error detector

Error detector receives the measured signal and compares it with the reference input. The difference of two signals produces the error signal.

Control element

This regulates the output according to the signal obtained from the detector.

Controlled system

This represents the control system controlled by the feedback loop.

Feedback element

Feedback elements will feedback the output signals to the error detector for comparison with the reference input.

9.3 Auto levellers

Auto leveller is an additional device which is an additional device for correcting the linear density variations in the delivered material of card slivers and draw frame slivers. The control of coefficient of variation in count of yarn (CV) is very important in many aspects. The control of count variation is important for yarns meant for knitted fabrics. Before the invent of auto levellers, there is no mechanism to control the count variation. The count or the linear density of the material varies from time to time. In spinning mills, the quality assurance department often used to check the linear density of the slivers in carding and draw frames by wrapping 6 yards of material and if there is any variation, the mill personnel has to adjust the wheel to get the required sliver count.

This is tedious process, since the variations introduced, if not corrected in the earlier stages cannot be corrected in the later stages. Normally, without auto levellers in spinning mills, a clear understanding is required about the wheel changes in the department. In spinning mills, the wheel changes are made only in draw frame machine, since the number of teeth in the change wheel of draw frame is higher when compared to the change wheels in carding, speed frames and in ring frames. The effect of change in the delivered sliver material by changing the change wheel in draw frame is less as compared with the effects on other process. Hence, mill personnel change the wheel in draw frame process only. Another concept is the delivery speed of the draw frame is many times greater than any other processes. If a defective count material is left unnoticed in draw frame, it will end up with many more meters of defective count in successive processes.

The main objective of modern draw frame is to improve evenness over short-term variations. It can be achieved by doubling of slivers, say 6 or 8 and then by levelling or equalising.

9.3.1 Concept of auto levelling

The object of any auto levelling equipment is to measure the sliver thickness variation and then continuously alter the draft so that more draft is applied when thick place in the sliver comes and less draft if the sliver is thin. The result is that the delivered sliver thickness is more regular than it otherwise would have been. Theoretically there are two distinctly different ways in which auto levelling can be done:

a) Radical auto levelling
b) Conservative auto levelling

This will be explained in the following sections:

a) Radical auto levelling
The main count of the product is determined by the setting of the auto leveller unit. Radical auto levelling is at present is a radical process it controls the irregularity of the product produced and also the main count of the product.

b) Conservative auto levelling
Conservative auto levelling is defined as the process which means local thickness variation are corrected but the main count of the product is beyond the control of the auto leveller as in this case the mean count is already pre-determined by the conservative auto leveller.

History of auto leveller

Auto levelling principle was applied to the textile spinning industry in the 19th century. A partial regulator was utilised to improve the irregularity of the material. Then it was followed by the commercially successful raper auto leveller used in the woollen industry for the first top finisher gill box which was invented by G.F. Raper.

9.3.2 Classification of auto levelling

Auto levelling can be classified in to three major groups according to the basic principle of operation.

a) Open loop systems
b) Closed loop systems
c) Combined loop systems.

a. Open loop system

Block diagram of open loop auto leveller is shown in the Figure 9.6 (A) and (B). Principle: In this system (shown in the Figure) indicates material flow and the broken line indicate the information flow to the auto leveller unit. The control unit compares the measured signal with the reference signal according to the comparative measurement. The output from the regulator unit increases, decreases or remains unaltered which in turn provides variable speed to the back roller of the drafting system to give the required draft. The popularity of the open loop system is if the directions of arrows are followed from any starting point it always leads "out in to the open" and hence this system is known as open loop auto levelling system.

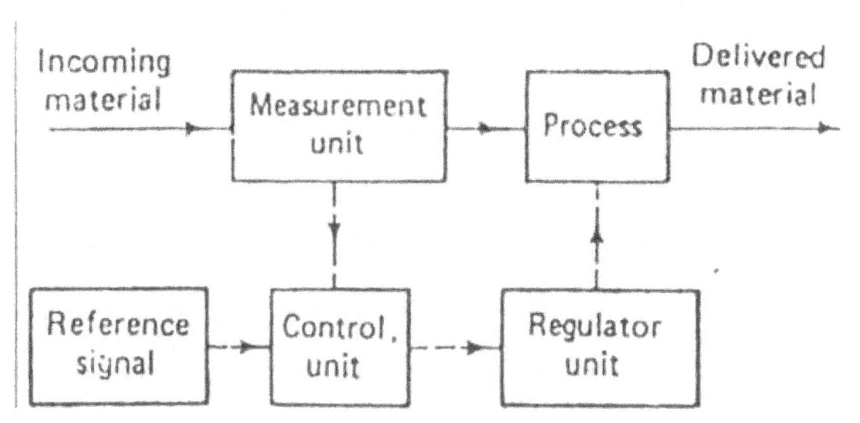

Figure 9.6 (A) Block diagram of open loop system

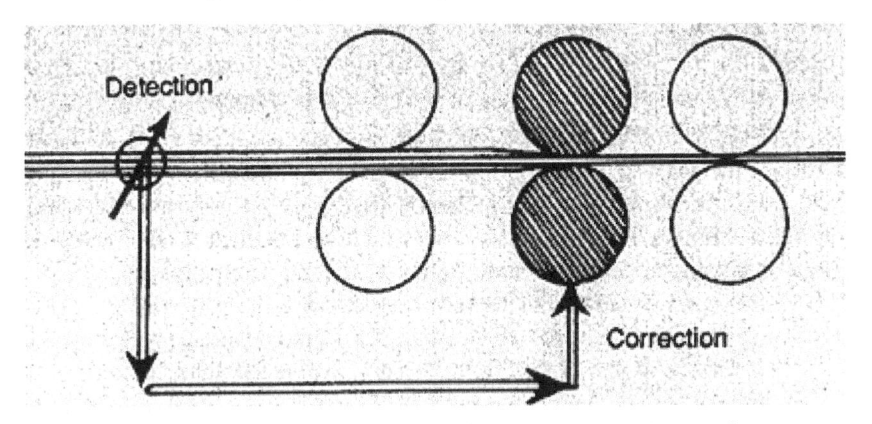

Figure 9.6 (B) Schematic diagram of open loop system in draw frame

The open loop control principle which can be used for the correction of short-term variation (1–10) times of the fibre length. Most of the draw frames are of open loop type. In open loop auto levellers, the sensing of the feed material (card sliver) is done at the feed and the correction is done by changing the break draft or main draft of the drafting system before the material is delivered to the can. The control unit compares the measurement signal with the reference signal according to the comparative measurement. The output of the regulator unit increases, decreases or leaves unaltered which in turn provides the variable speed to back roller of the process to give the required draft.

Advantages

- The system can correct short-term mass variations.
- As the detection is done at the entry itself, it is possible to correct the fault at the right location.
- Problem of hunting can be avoided.

Disadvantages

- The main drawback of the open loop system is the sensing is made at the feed and the correction is done before the material is delivered, there is no checking made at the delivery whether the correction is made properly or not. However, open loop system is very effective and the correction length is many times lower than the closed loop system. Open loop system is suitable for draw frame, since the sensing is made at the feed itself.

Application: In modern draw frame spinning preparatory machines.

b. Closed loop system

Block diagram of closed loop auto levelling principle is shown in the Figure 9.7 (A) and (B).

Principle

In this system also, the incoming sliver material is scanned by the scanning roller. The measured values are sent to the measurement unit. The measured signal is compared with the reference signal by the control units which then determine the output of the regulator unit. The output of the regulator unit provides the required variable output speed to the back roller of the drafting system to give the required draft.

In this system, the measurement always takes place at the point where corrective action is applied. Thus, if the measurement is made at the output, the corrections must be applied at the back roll. It is apparent that the control unit continuously checks the result of its own action. With reference to the diagram, if the direction of arrows is followed by any staring point except the delivery point it always leads to never ending circuit of the loop which link the process and control unit together and hence named as "Closed Loop Auto leveller".

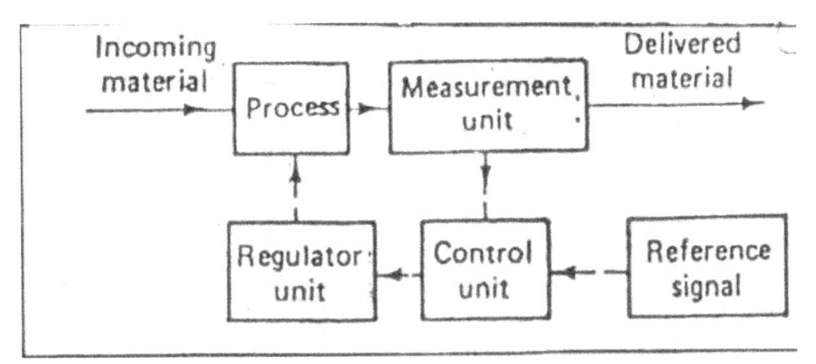

Figure 9.7 (A) Block diagram of closed loop system

Closed loop system is suitable to correct effectively medium and long-term variation (10–100) and 100 % above times of the fibre length. In this closed loop system also, the measured signal is compared with the reference signal by the control units which then determines the output of the regulator units which provides the variable output speed to the process to give the required draft.

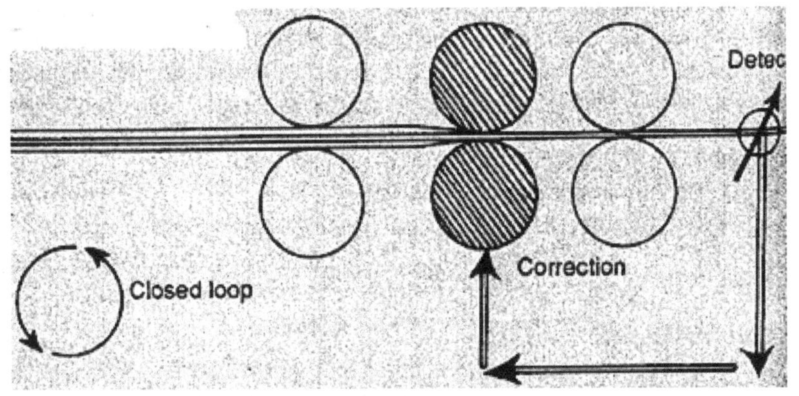

Figure 9.7 (B) Schematic diagram of closed loop system

Measurement always takes place sensing is made at the delivery side and the correction is done by changing break draft or control unit continuously checks the result of its own action, after the point where the corrective action is applied.

Thus, if measurement is made at the output, corrections must be applied at the back roll, it is apparent in draft zone. Most of the carding machines of latest developments are closed loop systems. In the latest generation cards, cards have sensing at the feed rollers and as well as at the delivery calendar rollers. Hence both open and closed type auto levellers are in use in carding machines.

Advantages

- There would be continuous monitoring of the slivers after the corrections have been made.
- Disadvantages
- The measured portion will pass the adjusting point before the adjusting signal arrives from the control unit. Hence closed loop system is not suitable for correcting the irregularity over short lengths.
- The system is more prone for unwanted oscillation in the output which is known as hunting. Hence system must be designed to avoid hunting.
- Closed loop systems are suitable for carding machines to correct medium and long-term mass variations.

In most of the present day draw frames, Sliver Monitor Plus is provided to confirm that the delivered material has required linear density. In some conditions, Evener Draw frames with closed loop systems are mostly used in the first passage. This is considered to be necessary since larger, sudden faults are not fully eliminated. In addition to the dead-time, the control detects a fault only after a time delay. Moreover, the system cannot adjust immediately to the set value but has to adapt gradually. Thus, part of the fault passes to the material which can be compensated by doubling in second passage finisher draw frames. Since both open loop and closed systems are having merits and demerits, several manufacturers incorporate both these systems in the draw frames for better control of sliver count. With this, compensation is usually effected in the range of ±25%.

c. Combined loop system

A combined loop auto leveller as shown in the Figure 9.8 corrects long, medium and short-term variations. Various loop arrangements can be made, for example, combination of open and closed loop or alternatively two separate closed loop arrangements. Although this system gives marked advantages, it is bound to have extra price and complexity.

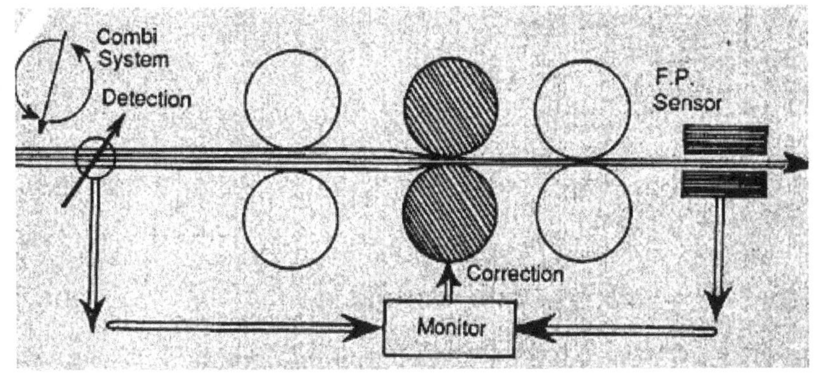

Figure 9.8 Combined loop system

9.3.3 Levelling system in draw frame

The RSB 951 drawframe is equipped with an electronic levelling system shown in the Figure 9.9. The slivers (6 or 8) doublings fed to the creel zone of the machine pass through a pair of scanning rollers. One of the two rolls is equipped with a movable bearing. This roll moves according to the sliver variation. These mechanical movements are transformed by a signal converter in to corresponding voltages and forwarded to an electronic data storage.

The data storage ensures that the draft change takes place exactly in the very moment in which the particular piece of sliver is in the main draft zone. The electronic memory passes the measured voltage on to the set point stage with a certain time delay. The set point stage calculates the required target speed for the servo drive from the measured values and the machine speed.

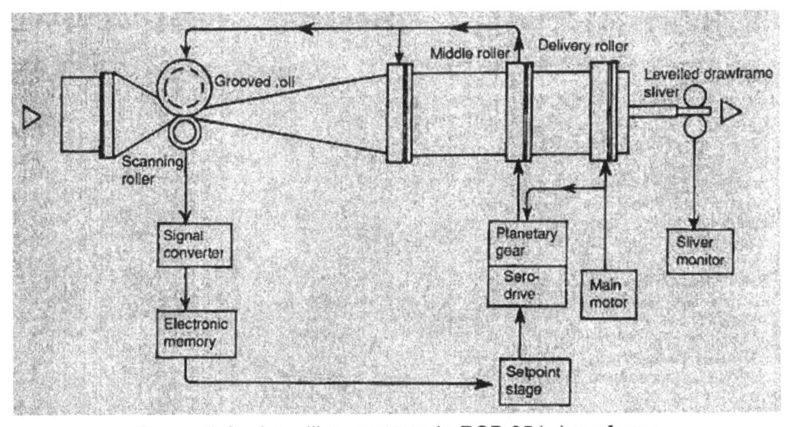

Figure 9.9 Levelling systems in RSB 851 draw frame

The servo drive forwards these target speeds through the planetary gear as additional speed to the middle roll of the drafting system. The required change in draft is hence accomplished.

The speed of the feed roller, scanning roller and the break draft roller are varied analogically with the speed of the middle roller. This system ensures constant delivery speed which is independent of levelling action and thus exact production rates can be calculated.

9.4 Electronic data storage

This electronic data storage stores the signals provided by the scanning rolls (i.e. voltage values) and passes those signals on to the set point stage at the exact right moment. The right moment is the point in time in which the particular piece of sliver arrives in the main drafting zone, the so called levelling action point (E) as shown in the Figure 9.10.

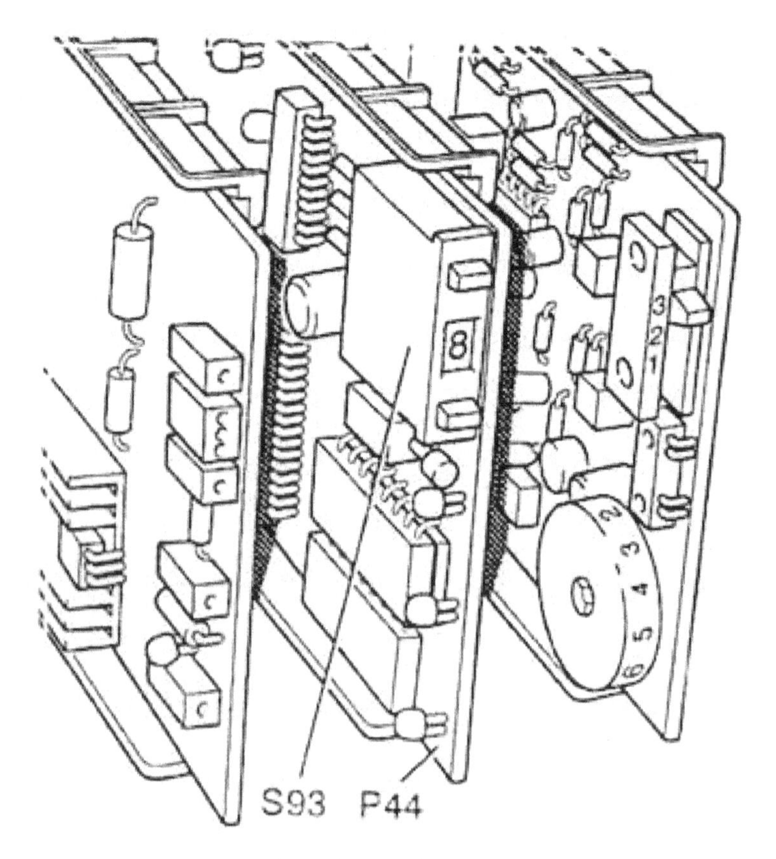

Figure 9.10 Electronic data storage

9.5 LED display of levelling motion

The LED display of levelling motion is shown in the Figure 9.11. The levelling display consists of one green LED with 5 red LEDs on each side. The purpose of these LEDs display is to show the weight variation of the incoming slivers. The green LED represents the mean value, (±, zero). Staring from the mean value, each red LED represents a variation of 5%. The total range for the levelling is calculated as ±25%. In some cases, if the scanning rolls sense a variation of more than 25% on either plus or minus side, the machine is stopped immediately and signal lamp indicates the reason for the stop.

The LED display points out the weight variations whether it is on the "+" or on the "-" side.

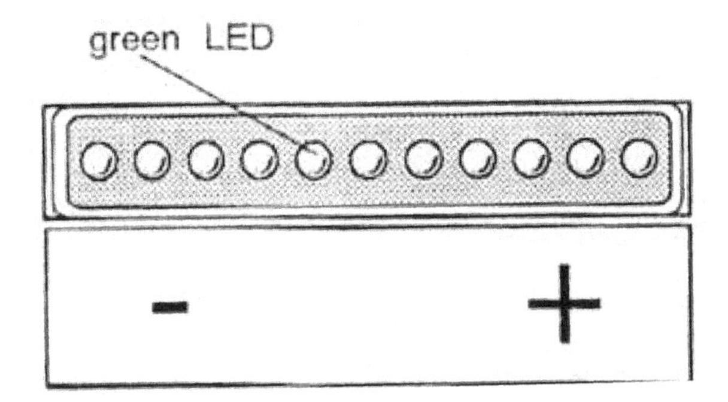

Figure 9.11 LED display in draw frame auto leveller

Auto leveller systems which is in use in most draw frames like RSB 951 is shown in the Figure 9.12.

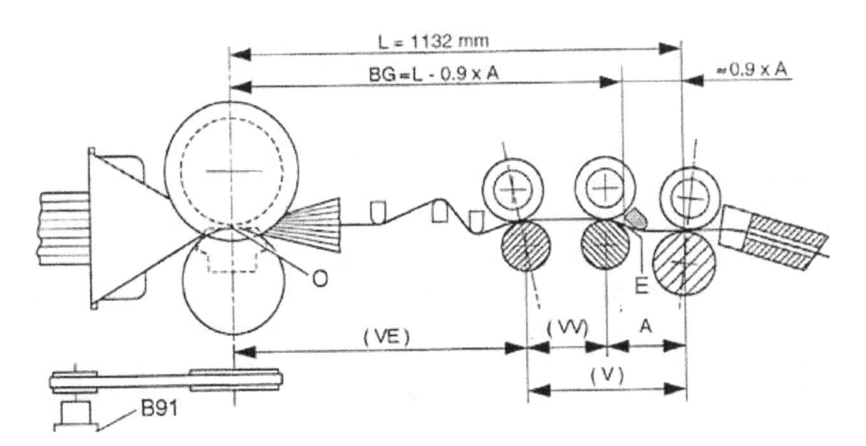

Figure 9.12

In the Figure 9.12 the levelling action point (E), the exact point in time when the levelling starts to act depends on

- Tension of the sliver entering the drafting system VE.
- Draft roll setting (A) and the position of pressure bar.
- Break draft (VV).
- Setting of sliver guides.
- Delivery speed of the machine.
- Fibre characteristics.

- B 91 – Impulse generator which makes 8 impulses per revolution.
- S 93 – switch with numbers located in the electric panel.
- E – Levelling action point, which is approx. 0.9 × A.
- BG – Delivered length of sliver from (O) to (E), during "I" number of impulses.
- L = number of impulses for distance BG.

The optimum levelling action point (E) can be obtained by setting the sliver tests.

Auto leveller is meant for correcting continuous long-term variation, medium-term variation, seldom occurring abnormal variations in the fed sliver due to deviations in the card or comber sliver and also short-term mass variations in the comber sliver due to piecing.

This system is an electronic levelling system The major components in this system are scanning roller, signal converter, levelling CPU, servo drive (servo motor and servo leveller), Differential gear box (Planetary gear box). The function of the scanning roller which consists of tongue and grooved rollers is to measure the variations in the feed material. The most widely used method of measuring the feed sliver arrangement is tongue and grooved rollers. This system has some criticism that the feed material is compressed at the sensing which may cause problems in drafting zone. However, such problem does not arise in practical situations since the material is spread sufficiently by using sliver guides in the sliver path on its entry to the drafting zone. The influence of moisture also has negligible effect on the compressibility of the material. But different materials have different compressibility or different orientation may register different thickness for the same count. However, such differences are meted out by changing the scanning rollers when a lot is changed. The pressure on tongue and groove rollers and the width of these rollers are changed in accordance with the material being processed.

9.6 Sensing mechanism

In the Figure 9.13 (A) and (B), the groove roller (B) is fixed one on axis B'. The tongue roller (A) is spring loaded on a movable axis A'. The clearance between the two is about 0.1 mm on both the sides. This clearance of 0.1 mm should always be maintained. Whenever these rollers are changed, these rollers are changed in pair. The loading on roller (A) is normally set in accordance with the material and sliver hank. The load on these rollers is changed manually or through control panel when the feed hank or feed material changes. One of the scanning roller is movable and these scanning rollers are loaded by spring or pneumatic weighting system. Pneumatic system is always better because of consistency in loading system irrespective of the sliver feed system. In case of spring loading system, the pressure on scanning rollers varies depending upon the feed variation.

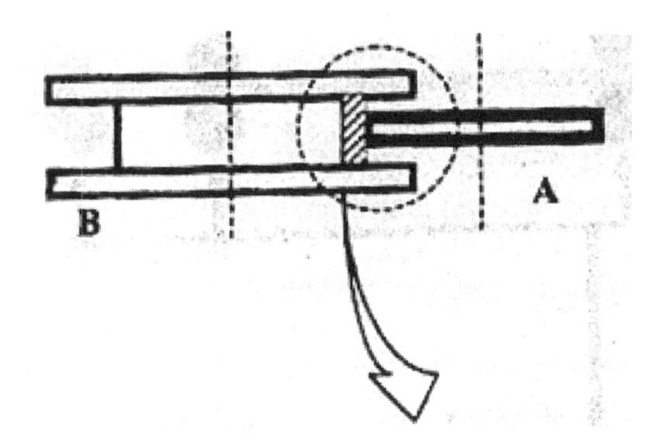

Figure 9.13 (A) Sensing mechanism in auto levellers

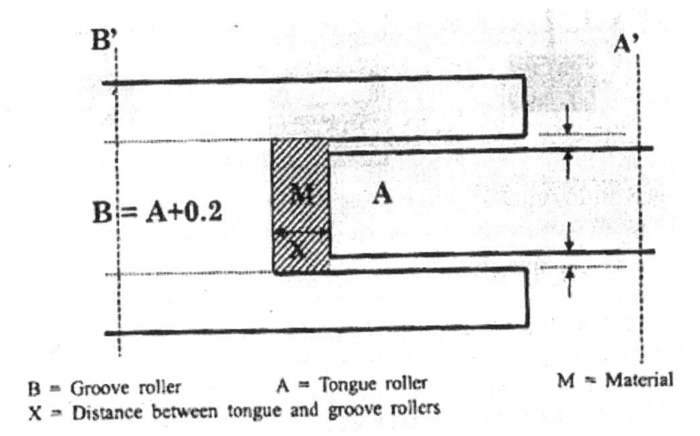

Figure 9.13 (B) Settings in the feed zone in auto leveller draw frame

The variation in sliver mass of incoming slivers displace the scanning rollers. The distance moved by the scanning is proportional to the sliver mass fed. The displacement of the canning rollers is transformed in to voltage by a signal converter and is fed to an electronic levelling processor. In case of analogue system, electronic levelling processor is a servo amplifier, but in the case of digital system, it is CPU (Central Processing Unit). This CPU furnishes the correct target value to the servo drive (Servo motor and servo amplifier).

Working principle

The roller (A) is mechanically attached to a sensor which measures the deflection and converts it in to appropriate voltage. The deflection is done by means of a transducer or a photo sensor. The deflection of the roller (A) from the mean value is converted linearly in to voltage signals and is sent to the comparator or regulation unit of the auto leveller. These signals are stored here about the information about the magnitude of the fault and the direction of deviation from the mean value. These signals are fed to the servo motor which through planetary gearing varies the break draft or main draft of the drafting zone and the linear density of the material delivered is made correct according to the required value.

For example, let us assume a mean value distance of 4 mm between the tongue and groove rollers for a particular levelling system. The range of linearity is 2–5.5 mm, it means that the auto levelling is performed effectively for this range of distance between these rollers. This means that when the distance between these rollers is 4 mm apart in the running condition, there will be no levelling action required i.e. X = 4 mm. Hence the material passing between these rollers will have a space equal to the shaded area as shown in the Figure 9.14. In other words, the mechanical draft set at the draw frame will exactly produce the required hank by drafting the fed mass of the material. When the feed material becomes thicker, it will occupy more than the shaded area. This will make the tongue roller to move outwards. The displacement of the tongue roller will be measured and converted in to electrical signals The magnitude of the signal depends upon the displacement of the tongue roller. Depending upon the magnitude and direction of the voltage signal, the draft will be changed in the main draft zone (Draft increase will take place as the feed material is deflecting the tongue roller outwards). The change of draft will take place after some time delay so that the fault reaches the main draft zone by that time.

Delivery speed of the machine and the electric signal values arrived at the slivers fed are the two important signals for correction. Servo drive takes the information and converts it in such a way that the servomotor RPM and direction is decided for appropriate correction. The purpose of the Planetary gearing (Differential gearing) with its controlled output speed drives the middle and back rollers. Since the servo motor RPM and direction varies according to the feed material variation, the servo motor and servo leveller generates the control speed of the planetary gearing, the required change in draft is accomplished for the weight variation of the sliver fed.

- If the slivers fed are too heavy, the entry speed is reduce, i.e. draft is increased.
- If the slivers fed are too light, the entry speed is increased. i.e. draft is reduced.

9.7 Auto levelling of RSB draw frames

In order to ascertain the efficacy of the Auto levellers in draw frames, two functions are required to be checked. They are:

a) Levelling magnitude setting, and
b) Timing of the levelling action.

Before starting, important points have to be remembered. Mechanical draft or nominal draft should be selected properly. Before switching on the auto levellers, gears should be selected properly such that the rapping average 9 linear density of the sliver) should be less than plus or minus 3%. If the feed variation indicated in the A% display of sliver fed is continuously showing more than -5% or +5% then the mechanical draft selected is not correct and need to be adjusted. If the mechanical draft selected is correct, then the indication in A% display of sliver fed lie between -5% red lamp and 0% green lamp or +5% red lamp and 0% green lamp. More precisely saying that green lamp (0% variation indication) should be on at least 80% of the running time.

Selection of scanning rollers is very important. In other makes of draw frames like DX7-LT or DXA7-LT, the scanning roller is same for all sliver weights and all types of material. However, in RSB draw frames, there are different sizes of scanning rollers. It depends on sliver weight fed and the type of material processed. The pressure on the scanning roller is not constant and depends on the material being used. The reason is to get minimum A% in the sliver. For the same material, if the scanning roller pressure is changed the linear density of the material will also change. Hence enough care has to be taken and whenever the pressure is changed, wrappings have to be checked and adjusted accordingly.

There are two important points for quality levelling. They are

- Levelling action point(time of correction)
- Levelling intensity.

Both sensing of feed variation and correction of the material are done while the machine is continuously running at two different places. Sensing is done at the feed and the correction is done at the delivery side. Hence, the calculated correction should be done on the corresponding defective material. This is decided by the Levelling Action Point. This is defined as the time required for the defective material to reach the correction point should be known and the correction should be done at the right time.

Levelling Action Point depends upon three factors.

- Break draft
- Main draft roller setting and
- Delivery speed of the machine.

Levelling Intensity

Levelling intensity is to decide the amount of draft change required to correct variation in the feed material. There is no good correlation between the mass and volume of different fibres since fibres have different specific gravity. Therefore, the levelling intensity may be different for different fibres.

There are two quick methods to check the levelling intensity.

1. Along with the normal feeding of slivers (6 or 8) introduce an extra sliver. The sensing roller senses the constant increase in feed material weight. The draft will increase constantly for the time the extra sliver is running. If the regulator is under correcting or over correcting the "Quality Monitor display" will show a constant deviation from the mean value. This deviation over a long period can be seen clearly on the sliver quality monitor display.
2. Another method is to take "USTER DIAGRAM". In this method, an extra sliver of one meter is fed along with the normal doublings. The delivered material is tested in Uster tester with a material speed of 8 m/min and chart speed of 10 cm/min. A deviation towards lower side of the diagram indicates over correction and deviation over upper side of the diagram indicates under correction.
3. Wrapping of the delivered sliver is checked for "n", "n+1" and "n-1" sliver at the feeding side. The sliver weight of the material for all the three samples should be checked by checking the wrapping of the delivered material for a minimum of 100 m. In all the three combinations, the sliver weight difference should be minimum. If levelling correction point and levelling intensity is selected properly, then the cut length CV% of 1 m will be less than 0.5%. If the A% deviation is deviating on either side more than 0.5%, then auto leveller has to be adjusted to get the correct value.

a) Timing of the levelling action

Levelling action takes place in the main draft zone of the drafting system. The magnitude of the fault (Thick or Thin) in the material is sensed by the sensing rollers is stored in the electronic memory of the auto levelling system. This variation is corrected before the material reaches the main draft zone. This is continuous process. The variation in the incoming material is of any nature probably of any length at any distance apart. Thus, the function of the levelling system is to correct these variations in exact time when the particular variation is in the main draft zone.

Correction takes place in the main draft zone, just as the material emerges out from the middle rollers. Therefore, the break draft also plays an important role in determining the time taken by the material to reach the font zone. Higher the break draft more the time taken by the material to reach the main draft zone. In addition to that, the front zone setting is also important as the levelling should start when the fault has crossed 10% of the distance of the front zone. The time of travel of the fault in the material from sensing roller to the main zone is calculated from the distance between the two and the amount of break draft applied.

Time is calculated by the tachometer which generates the number of pulses during the travel of variation from the sensing roller to main draft zone. This time is not affected by the speed of the draw frame. When the speed of the machine increases, the speed of the tachometer also increases proportionally in the same manner. Hence the number of pulses required for the feed material to cover the distance from the sensing rollers to the main draft zone remain same. The distance from sensing rollers to the main draft zone vary depending upon the setting of the sliver guides. If by any chance the time calculation is misjudged in terms of number of pulses, then the correction will be at different place where is should not be. Thus instead of correcting the variation, the levelling system will introduce variations. This can be explained by an example. Let us assume that the feed material consists of a fault combination of thick followed by a thin place of 1 cm each. When the thick place reaches the main draft zone, the draft should increase and as soon as it is neutralised the draft should reduce as now the thin place will be in this position. This time of switch over of the draft from higher side to lower side is very critical as even a minute delay will produce a thin place and an early switch over will cause a thick place in a sliver. This can be adjusted by controlling S93 switch in RSB draw frames.

b) Levelling magnitude setting
Sliver test is normally performed to assess the optimum levelling intensity. When a fault in the shape of thick or thin place is sensed by the tongue and groove rollers, the draft in the main drafting zone should be varied according to the magnitude of the fault i.e., if the deviation in the incoming sliver is of the magnitude of 5% towards thicker side, the correction should be (-) an increase in draft by 5%. If the variation in the draft actually taking place is more or less than that what is required, the auto leveller will not totally eliminate the variation. In other words, it may increase the mass variation of the final sliver by over drafting or under drafting. In RSB draw frames shown in the Figure 9.14 R 38 will regulate the magnitude.

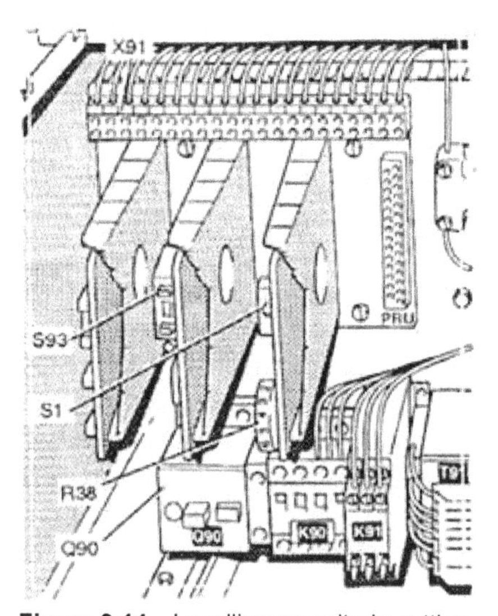

Figure 9.14 Levelling magnitude setting

9.8 Adaptation “C”

With the machine producing the correct sliver weight, the setting of the electronic adaptation (Figure 9.15) should show the incoming sliver weight variations around the green LED as centre of the levelling action. The following procedures are required to produce a correct sliver weight.

- Produce a sliver.
- Adjust the adaptation “C” until the weight variations of the incoming sliver bounce around the centre of the levelling range (green LED display).

- The adaptation "C" ranges from 1 to 999 values. This represents approximately ±30% from the centre of the levelling range. Therefore 1% deviation represents a different setting of approximately 16 points on the adaptation scale.
- The change of the adaptation "C" to a higher number indicates the sliver weight delivered is heavier and reduces the draft.
- The change of the adaptation "C" to a lower number indicates the delivered sliver weight is lighter and increases the draft.
- If the weight variation of the incoming sliver consistently favours on one side, it is advisable to change the draft wheels and the new setting in adaptation "C" must be entered.

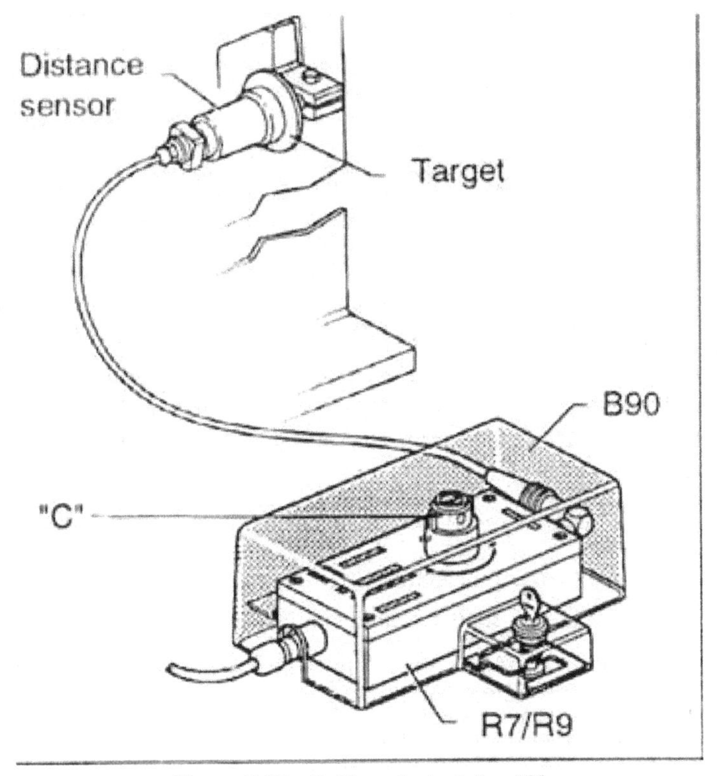

Figure 9.15 Setting of adaptation "C"

9.9 Assessment of auto levellers performance

The efficiency of any auto leveller is judged by three main parameters. They are

- Correction length
- Correction time and
- Correction capability.

Shorter the correction length and time and higher the correction capability the higher the efficiency of the auto leveller as shown in the Figure 9.16 (A).

Correction length

It is defined as the partially corrected sliver portion and the time the system takes to bring back to its desired set value is known as "Correction Time". This can be explained as follows:

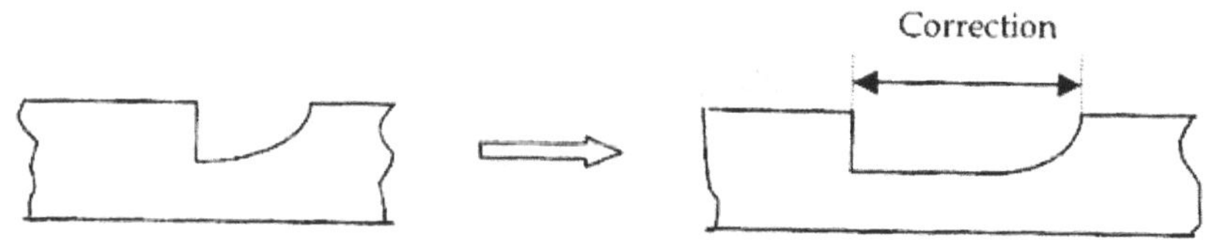

Figure 9.16 (A) Correction length

If there is a certain deviation from the set volume as the material passes, a corresponding signal is sent to regulating device so as to correct the fault. Owing to the mass inertia of the system, compensation cannot be effected immediately but must be carried out by gradual adaptation. A certain time elapses before the delivered sliver has returned to the set volume. During the time, faulty sliver is still produced although the deviation is steadily reduced. The total length that departs from the set value is referred to as correction length.

Correction Length depends on the following factors:

a) Delivery speed of the machine
b) Inertia of the regulating system and its design
c) Extent of mass variation in the feed sliver cross-section whether it is ±10% or 25% range or more.
d) Weight per unit length of the sliver whether it is from normal to lighter side or normal to heavier side or lighter to normal.

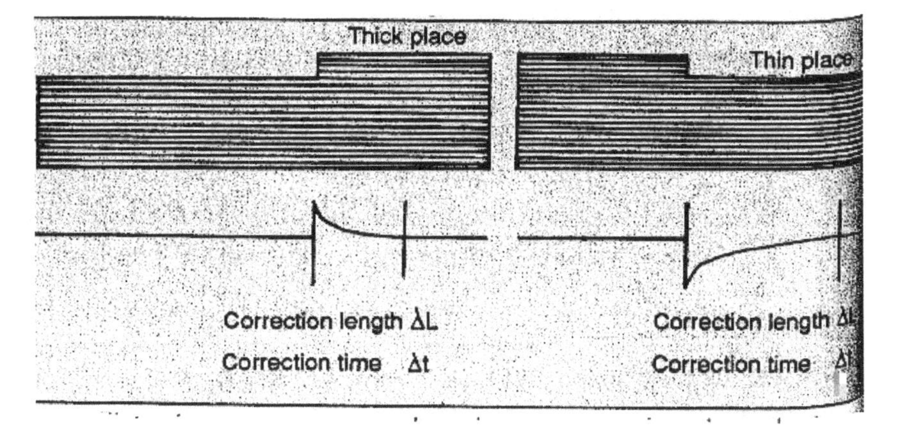

Figure 9.16 (B) Correction time and correction length

Correction time

It is the time taken to correct thick or thin places at the feed point. Time taken to accelerate is more than the time taken to de accelerates. Hence, it takes more time to correct thin places than thick places.

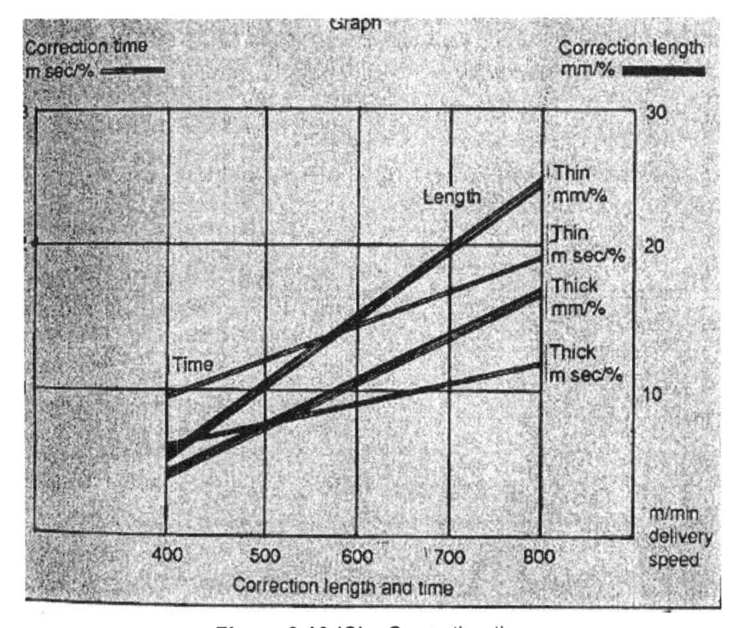

Figure 9.16 (C) Correction time

For example, if an auto leveller takes t sec to level a certain percent increase in mass variation of the sliver delivered at v cm/min, then correction length (l) would be

$$l = vt \text{ cm}$$

Assuming a delivery speed of 400 m/min and the system takes 0.02 sec to level 10% increase in mass variation, the correction length would be

$$l = 400 \times 0.02 \times 100/60 = 13.33 \text{ cm.}$$

If the delivery speed increases, correction length also increases proportionally.

Shorter the correction length, better is the CV%. In RSB draw frames, the correction takes place between the feed and the delivery rollers as explained below:

a) At constant delivery speeds, the feed roller speed is
   - Reduced for thick places and
   - Increased for thin places.

It is the time taken to correct thick or thin places.

The time taken to accelerate is more than the time taken to decelerate. Therefore, it takes more time to correct thin places (feed roller speed to be accelerated), and It takes less time to correct thick places (feed roller speed to be decelerated).

It can be seen from the Table 9.1 and the graph that the minimum correction length is 17 mm and the minimum correction time is 1.3 m sec.

Table 9.1 Correction length and correction time taken by the auto leveller

Speed (mpm)	Thin places length/mm	Thin places time/msec	Thick places length/mm	Thick places time/m sec
400	8	0.9	6	0.7
500	10	1.2	8	0.8
600	14	1.3	10	0.9
700	18	1.5	12	1.1
800	22	1.8	17	1.3

Correction capability

In modern draw frames, during the mechanical processing, the sliver is well-guided and completely compressed (almost to the cross-section of the slivers)

and also according to its volume and weight. The fastest change in "fibre cross-section" occurs between the feed and delivery rollers. The minimum distance between the feed and delivery rollers is 36 mm. Since the correction length is only 17 mm the correction capability of the auto leveller draw frame is twice that the fastest change in fibre cross-section occurring in the mills. At a delivery speed of 800 m/min. the fibres take 2.7 m sec to traverse the distance of 36 mm but the correction time needed is 1.3 m sec. Thus, every sliver deviation is corrected at a 100% safety margin.

9.10 Setting the auto leveller

Procedure for setting the auto leveller.

1. For a given number of doublings (say 6 or 8), material and hank of the sliver required select the appropriate sensing rollers, pressure to be applied on the sensing rollers, web tube condenser and required drafts.
2. Keeping the auto leveller OFF, run the material (Count adjust to 500).
3. If the sensing rollers show a deviation of more than 5% in the material thickness or if more than one red LED is glowing, then there is a need to change the sensing rollers in accordance to the material.
4. If not, change count adjust so that the green LED glows.
5. Keeping the auto leveller OFF run the material and set the draft to get the required draft and also the required sliver hank.
6. As soon as the required hank is achieved or come close to the required hank (recommended within ±2%) turn ON the auto leveller and take wrapping.
7. The deviation of up to ±3% in the final sliver hank can be controlled by sliver adaptation unit.
8. Repeat the procedure for fine tuning the hank.

9.11 Servo drive

Servo drive motor used in the RSB draw frame is shown in the Figure 9.17. The servo drive receives the target values, calculated by the set point stage, and transforms these values in to corresponding revolutions per minute and also the rotational direction. The servo drive controls thereby the planetary gearing, which determines with its output speed, the speed of the middle bottom roller of the drafting system. The speeds of the creel rolls, scanning rolls and the speed of the back bottom roll of the drafting system are also changed accordingly.

The weight variations of the slivers fed determine the speed of the servo drive. If the sliver weight of the incoming slivers is right on target (LED display green is on) the speed of the servo drive is “0”. The levelling action is accomplished by varying the speed of the entry, which means

- If the slivers fed are heavy, the entry speed is reduced which means the draft is increased.
- If the slivers fed are too light, the entry speed is increased which means the draft is reduced.
- The delivery speed of the draw frame remains constant and so does the production rate.

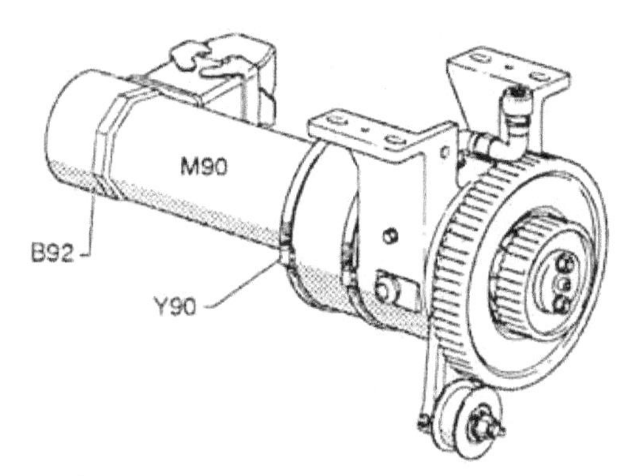

Figure 9.17 Servo drive in draw frame

9.12 Sliver test

The sliver test is generally done while the electronic levelling is active. There are certain procedures to be followed in sliver test (Figure 9.18).

1. Step 1: Turn the switch S1 to position “3” and push Q 90, “I” button.
2. Step 2: Set R38 to intensity level “5” for the first test.
3. Step 3: Set the time of levelling with S 93 in accordance with the table provided. It shows the recommendations for levelling delay. (E).

The sliver test is then performed with

- n + 1
- n (normal number of doublings)
- n – 1

The sliver test serves as an indicator for the correct levelling action. In general, the test should be conducted at normal production speed and normal operating conditions. The adjustment of the weight variations of the incoming slivers with adaptation "C" should cause the variations to be in the middle of the levelling range (i.e. around the green LED).

The sliver test is recommended for the following:

- Initial start of the draw frame.
- Changes in the processes like number of drawing processes, type of material, sliver weight fed, sliver weight delivered, number of doublings and drastic change in speed say from 400 m/min to 600 m/min or so.

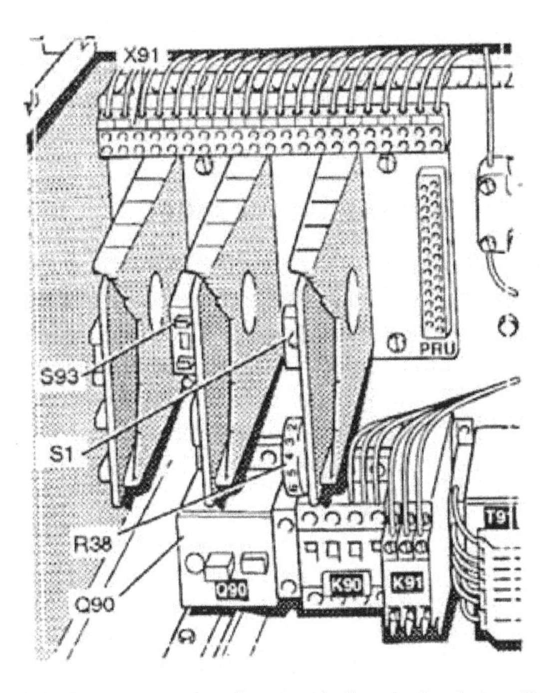

Figure 9.18 Adjustments for sliver test in the electronic levelling device

The following procedures are required to carry out the sliver test.

- Produce 3 sets of slivers each approximately 100 m in length with

a) Normal number of doublings (n)

b) With (n-1) slivers and

c) With n + 1 slivers.

Cut each of the slivers in to 10 × 10 meters pieces and calculate A% according to the formula mentioned below:

$A\% = (k\,tex_{(n-1)}) - ktex_{n}) \times 100 / ktex_{n}$

A% (+) means Over levelling

A% (-) means Under levelling

$A\% = (ktex_{n+1} - ktex_{n}) \times 100 / ktex_{n}$

A% (+) means Under levelling

A% (-) means Over levelling

The output of these calculations will show in most cases either an over compensation or an under compensation expressed in percentage (%). The setting of the levelling device is sufficient if this error is 0.5% or less. The intensity of the levelling device is set with the potentiometer R38. The change of the setting from one point on the scale to the next point on the scale causes a change in value of approx. 0.25%.

The sliver test should be repeated until the error is insignificant. In majority of the cases, an error of 0.5 % or less can be achieved. After the intensity is set, the sliver weight must be checked in normal production conditions. The sliver weight could be fine tuned with the setting of adaptation "C".

9.13 Quick test on auto leveller

In order to assess the working of auto leveller for its effectiveness, quick test can be conducted with the evenness tester diagram. While the machine is in production, a piece of sliver of approx. 1 m length is added to the normal number of doublings. From the delivered sliver, a diagram is made. This diagram from the sliver with (n) doublings, n+1 doublings, and (n) doublings again, will have one of these diagram appearances. The three mass diagrams derived from Evenness tester is shown in the Figures 9.19 (A), (B) and (C).

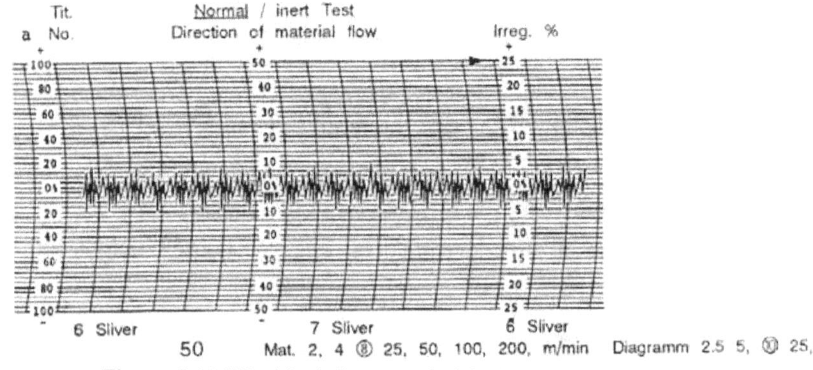

Figure 9.19 (A) Ideal diagram of 'n' doublings in a draw frame

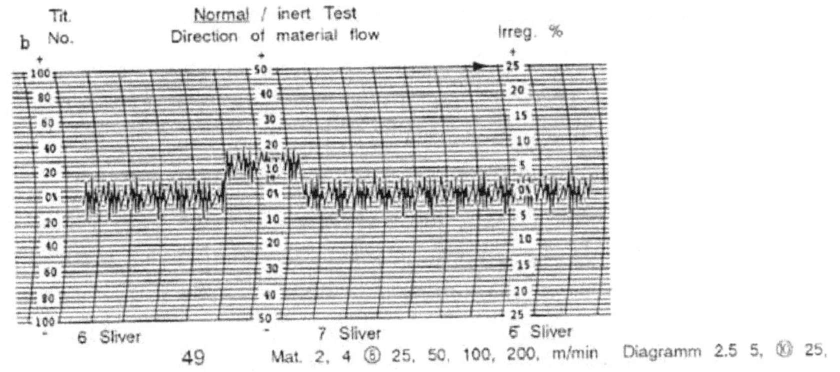

Figure 9.19 (B) n+ 1.doublings, under compensation

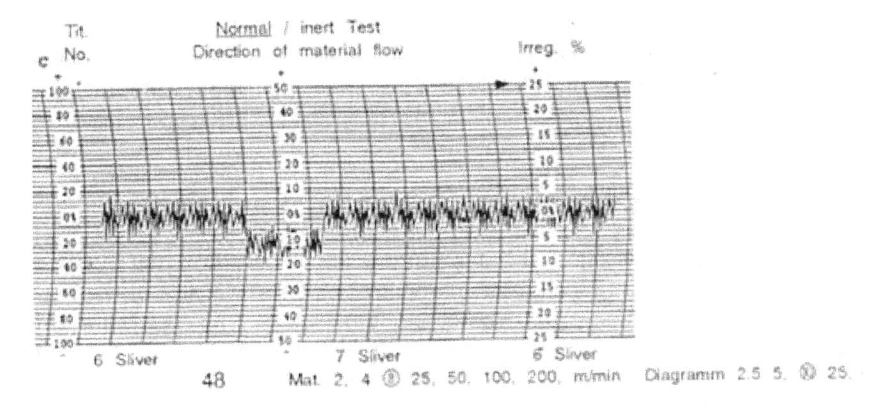

Figure 9.19 (C) n-1 doublings over compensation

From the three diagrams, (A, B and C) the following interpretations can be done.

- Diagram (A) is an ideal diagram. There might be a short peak in the diagram at the point where the extra 1 m sliver is fed (n + 1).
- Diagram (B) represents under compensation of levelling. Levelling action is too late. The potentiometer (R 38) must be adjusted to a higher scale mark towards 10.
- Diagram (C) represents over compensation of levelling. Levelling action is too early which indicates the potentiometer (R 38) must be adjusted to a lower scale mark towards 0.

9.14 Economics of auto leveller

The usage of auto levellers in draw frame has brought many advantages in textile spinning mills. The following are the observations made towards the effectiveness of the auto levelling systems.

- CV% of count of yarn is reduced significantly below 1.5.
- The variations introduced by the speed frame process get reduced by 25%.
- B-L curves clearly demonstrates that the variations are reduced in almost all wavelengths while using auto levellers in draw frames.
- For combed counts, the use of auto levellers in post combing draw frame has reduced the piecing waves fault considerably.
- CV% of strength gets improved by 4–5%.
- End breaks in spinning per 100 spindle hours usually comes down from 12.8/100 to 9.3/100 in Ne 40s combed yarns.
  Considering the above facts and the quality demanded by the international markets, the choice is to use open loop auto leveller in draw frame and closed loop auto leveller in carding machines.

9.15 Online monitoring system at draw frames

After the carding process, quality monitoring devices at draw frames function for better control of evenness and control of linear density of the material delivered. On line quality monitoring systems provides the following information:

- % Feed variation.
- CV% with different cut lengths.
- A% deviation or actual sliver weight.
- On-line mass spectrogram.
- Slub/Thick places per kilometre in sliver with respect to different diameter.

According to quality requirements, warning and stop limits may be set for different quality parameters. Machine warning light blinks and the machine stops immediately if any quality parameter is beyond the set warning limit.

Setting of QMS at RSB Draw frames

Draw frame machines are very sensitive machine for quality aspects. Hence the setting of on-line quality parameters is a judicious one. It should be very perfect and accurate. Warning limits and stop limit systems should be treated as “Action limit” and “Control Limit”.

Warning limit = Action Limit
Stop Limit = Control Limit.

For example, if in a spinning mill allowable Finisher draw frame sliver hank limit is ±2.0%. Beyond this limit is treated as "Non-confirming product", then

- Warning limit should be set on ± 1.5% where warning lights will blink for A% deviation.
- Stop limit should be set at ±2.0, where no production will allow beyond this limit and the machine will stop immediately.
- Likewise, warning and stop limits may be decided for CV%, thick places and Spectrogram.

9.16 Sliver watch

Sliver watch system is another foreign fibre monitoring in the draw frame machine. The Sliver watch system makes it possible to detect foreign fibres in slivers. The quality of the sliver determines to a great extent the quality of the yarn manufactured. Irregularity present in the sliver cause yarn faults and also more ends breaks in the ring spinning and more clearer cuts in auto winders. In weaving and knitting, it leads to more warp breakages or holes in the knitting and ultimately leads to loss in the machine efficiency due to frequent stops. The final fabric quality can have visible faults which are due to the defects in the spinning process. The aim of the system adapted to the draw frame is to stop the draw frame as soon as the foreign matter is detected so that it can be removed. With this system, the sliver is passed through a tube on the inside of which light sources or luminous diodes are located. At the other end, receivers measure the quantity of light in the sensor. Foreign fibres which have a different colour from the sliver, absorb part of the light, changing the signal, and making detection possible. The light is absorbed over the entire sliver surface, so that the foreign fibres can be detected over the entire circumference of the sliver. Standard sliver stop motions can be dispensed with if this system is used.

9.17 Thick places in sliver

Thick places in the sliver shown in the Figure 9.20 cause potential quality problems in downstream processing. They can lead to roving and yarn breaks in speed frames and in ring spinning machines. In addition, it will increase clearer cuts in winding. Thick places are short pieces of silver with fibres getting stuck together or fly getting drawn-in. In order to detect this short thick place, a measuring unit with a resolution down to 1–2 cm is required to detect such thick places in the sliver.

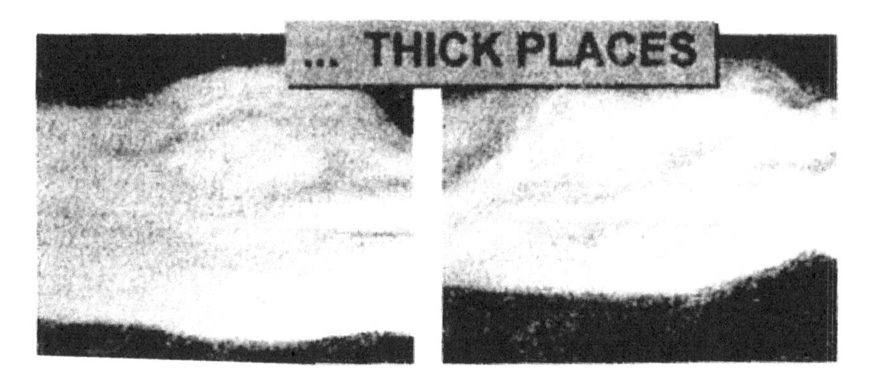

Figure 9.20 Thick places in the sliver

These thick places arise due to accumulation of fly, sliver or fibre parts which get caught on coarse surfaces and insufficient suctioning-off or inadequate cleaning. This defect can be overcome by systematic maintenance and adequate cleaning.

9.18 Long-term analysis of thick places

All the quality parameters are constantly monitored with the given limiting values. A warning can be given by means of a lamp which alerts the operator in case of the thick place values exceeds the nominal values. The production machine is stopped and analysed for the reasons by the USTER SLIVER DATA. This is shown in the Figure 9.21 if there are two deliveries with unacceptable thick place frequencies and the acceptable level for a draw frame running at 300 m/min should be 1.1–1.5 thick places per hour.

Quality of the sliver produced on a draw frame can be monitored over a longer periods of time. A sample report is shown in the Figure 9.21.

The graphical representation shows the thick place behaviour over the past month and the past 24 hours. The rounded area shows that the machine produced unacceptable values for 15 days and it was due to the wrong setting of the machine. Similarly the CV% (m) and the count variation (A%) can also be shown in the same way as the thick places.

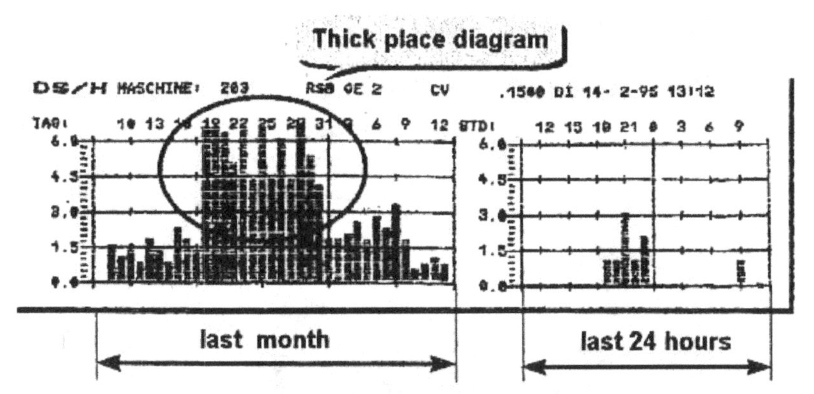

Figure 9.21 Long-term analysis of thick places in the draw frame sliver

9.19 Quality parameters checked by Uster sliver data

Uster sliver data checks four quality parameters.

- Count variation (A%).

  Deviation in count of the sliver produces costly defects in the subsequent processes. It may be due to incorrect draft zone settings, incorrect sliver weight in feed and improper functioning of auto levellers.
- Sliver evenness (CV%) which is frequently caused by mass variations due to faulty drafts. It is due to the incorrect settings, poor functioning of auto levellers and unfavourable fibre blends.
- Periodic faults are detected through spectrogram analysis. These mass variations are not random like the CV% but due to mechanical defects in the machinery and can be from few centimetres to meters.
- Thick places in the sliver which is due to accumulation of fly, insufficient suctioning-off or poor cleaning of the machines.

Uster sliver alarm

Uster sliver alarm is an online monitoring device which can be installed directly by the machine manufacturers or as retrofit on an installed machine. It is essential for a quality management system and can be installed on all preparatory machines like carding, combers and in draw frames. As shown in the Figure 9.22 it works as an independent unit on the draw frame and monitors the same quality parameters monitored by sliver data. It has its own panel and display and can stop the machine in the event of overrun of any one of the quality parameters.

USTER® *SLIVERALARM* -
UQM solution on a finisher draw frame

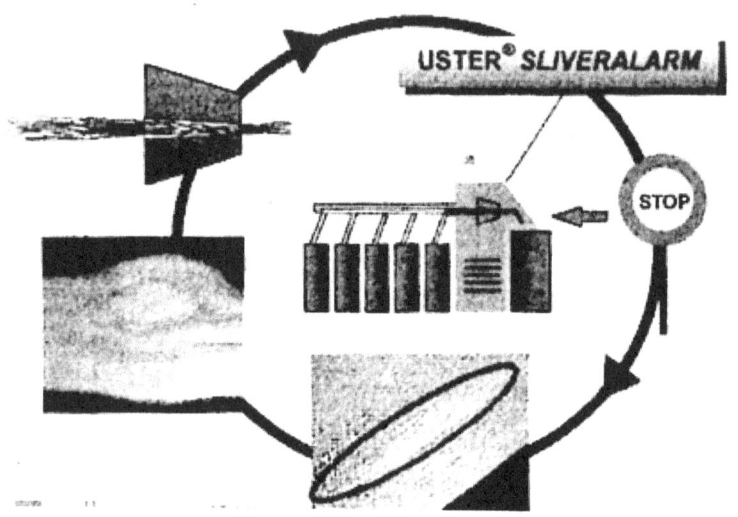

Figure 9.22 Uster sliver alarm

The FP (fibre pressure) sensor technology is able to detect very short defects of 2 cm, like thick places which are not possible to be detected by T & G (Tongue and Groove) measuring methods. T &G units are limited to detect defects longer than 15 cm.

Benefits of USTER sliver data

- Warning lamp in the machine which informs the operator if the thick places exceeds the nominal value.
- The production machine is stopped automatically when larger overruns occur. From the analysis of the report, the operator can find the cause and can take appropriate measures to eliminate it.
- Count variation (A%) can be checked online which is costly quality defects in downstream processes. Typical causes are incorrect machine settings, auto levellers not functioning properly and incorrect sliver feeds.
- Sliver evenness (CV%) which is frequently caused by mass variations due to faulty drafts, incorrect draw frame settings.
- Periodic faults are detected through spectrogram analysis. These mass deviations are not random like CV% but have a typical periodicity in a range from centimetres to meters.

- The values shown online correlate well with off line testing in the laboratory.
- Detailed information on a single machine shows production values, efficiency, loss of efficiency due to various types of machine stops, spectrogram, 24 hours count diagram and CV% results.
- USTER sliver alarm is an optional and optimal starter solution. It works as an independent unit on the draw frame and monitors the same quality parameters maintained by USTER sliver data.
- USTER sliver alarm and USTER sliver data are the modular building blocks for a quality management system.
- Improvements in the spinning process with less end breaks and less quality defects.
- A constant and consistent sliver quality is an important pre-requisite not only for ring spinning but also for rotor and air jet spinning machines.
- Number of clearer cuts in winding is reduced due to this preventive system of detecting thick places in the draw frame.

9.20 Integrated monitoring system

The task if it is rational in operation over time, then, in addition to the automation of activities of attendance and transport, it is necessary to include the monitoring equipment also in the overall analysis. Until recently, such systems were thought of being a difficult one, but due to the integration of various activities the application of is wide open. One such system developed by Zellweger Mill Data is an excellent example. The complete system comprises the following important structures.

- Decentralised process data systems for the individual departments.
- KIT data for carding machines.
- Sliver data for spinning preparatory process.
- Ring data for ring frames.
- Roving data for speed frames developed by Premier Evolvics.
- Cone data for auto coners.
- Rotor data for rotor spinning machines.
- Lab data.
- The central, comprehensive MILL DATA for coordinating all the decentralised process data systems.

9.21 USTER sliver guard

Uster sliver guard is an on-line quality monitoring system installed in spinning preparatory machines. Before going in detail about this on-line monitoring equipment, it is mandatory to discuss about the yarn faults.

Case study

A typical quality problem in a leading spinning mill. The yarn is wound on a tapered board with obvious count deviations. Such defects lead to costly quality problems in knitwear and woven fabrics. From the analysis of the board, the defect is thick places in the sliver which is the potential cause for the quality problem.

Zellweger Uster's latest on-line monitoring equipment – Uster Sliver Guard provides many salient features and it is available either as a monitoring system or with auto levelling also depending upon the mill's requirements. As a monitoring system, USTER SLIVER GUARD can be fitted besides draw frames, cards and combers also. The function of this monitoring system is continuous monitoring of all major sliver quality parameters which can be easily seen in the system's terminal on the machine.

Some of the quality parameters displayed on the machine are:

- Sliver count (A%)
- Evenness (CV%, CVL%)
- Short thick places which are greater than 1.5 cm length
- Periodic faults in Spectrogram form.

Yarn faults can be broadly classified as following:

- Unevenness (frequently occurring yarn faults like Imperfections and periodic faults).
- Seldom occurring yarn faults which are objectionable like 10 classes of yarn faults.
- Count variation in the yarn.
- Strength and variation in strength.
- Elongation and variation in elongation.
- Contaminations in the yarn due to the inherent defects in raw material, introduced in in-process in the spinning and spinning preparatory processes.

  Controlling of count and count variation imperfections like thin, thick and neps, periodic faults in the yarn starts from the second process stage of the spinning process. (i.e. Carding). The auto levellers fitted on the carding machine helps to control the long-term variation in the card sliver delivered. Most of the carding machines of long-term auto leveller type as short-term variations get corrected in the draw frame passages. Some card manufacturers give an option of short-term auto leveller in carding, but it is quite expensive and complex. Hence, carding machines are fitted with long-term auto leveller which monitors the variation in the sliver count material and if any deviation is detected, the signal are fed to the control unit and the feed motor speed is altered accordingly.

Until few years back, there was no auto levelling system in draw frames. Due to the stringent requirements laid down by the customers across the globe, monitoring/controlling of hank variation in draw frames became a must. Moreover, since it is the last stage of sliver production in the process, it is the last chance for the quality correction. There are no sliver doublings which can minimise the possible defects like thick and thin places in the sliver after this draw frame process. In addition to that, the maximum delivery speed is higher in draw frames, say 500 to 650 m/min, the defective material delivered will also be on the higher side.

Further, the periodicities introduced in the sliver material by comber or defective mechanical parts and unevenness due to poor fibre control, etc., can be monitored/ corrected by final Draw frame more effectively. The quality of the sliver determines to a great extent the quality of the yarn produced. An irregularity in the sliver can cause end breaks in spinning or increased clearer cuts in the winding. In weaving and knitting, we face difficulties in warp preparation which lead to efficiency losses. The final product even can have visible faults caused by defects in the spinning preparation process. This explains the growing interest among spinning mills to opt for on-line sliver quality monitoring.

10
Monitoring in combers

10.1 Introduction

Combing process is an expensive process to manufacture high quality yarns meant for specific end uses. The objective of the comber is to remove the short fibres from the cotton material, improve the uniformity of the sliver produced for further processes. However, attention must be paid to the comber for their sliver quality, stopping the machines instantly in the event of any choking of fibres in the delivery, table funnel, etc. In order to determine the quality of the comber sliver, on-line monitoring systems have been developed by USTER UQM. UQM is an optional accessory to the combers. Uster Quality Monitor (UQM) is a system for monitoring the quality of the delivered combed sliver. If UQM is integrated in any combers, it is operated via user interface.

10.2 Function of UQM

UQM uses a special measuring device which permanently measures the cross-section of the sliver delivered. UQM compares the delivered sliver value with set quality limits and it stops the machine when the set quality limit is exceeded.

Uster quality monitor (UQM)

UQM (Figure 10.1) is a system for monitoring the quality of combed sliver in combing machines. It is an optional accessory integrated in to the comber which is operated via the comber interface.

Principle

UQM utilises a special measuring device which permanently measures the cross-section of the sliver being produced. It compares the measured data with the set quality limits. If the quality is not within the prescribed limits the machine is stopped.

Function of measuring device

The measuring device in the comber is shown in figure. The sliver delivered is compressed between the cover plate (3) and the guide plate (4). When the sliver is compressed, it deflects the hard metal sensing element (5) to a greater or lesser extent depending upon the variation in the sliver cross-section. This bends the leaf spring (6) on the leaf spring (6), there is a strain gauge (7) which emits a different voltage depending on the volume of the sliver.

The FP – measuring device is supported by an automatic cleaning unit. In this cleaning unit, a gentle flow of compressed air helps prevent deposits of fibres and dust from forming in the measuring device. A short sharp blast at high pressure removes deposits dirt that may have formed eventually in the measuring device.

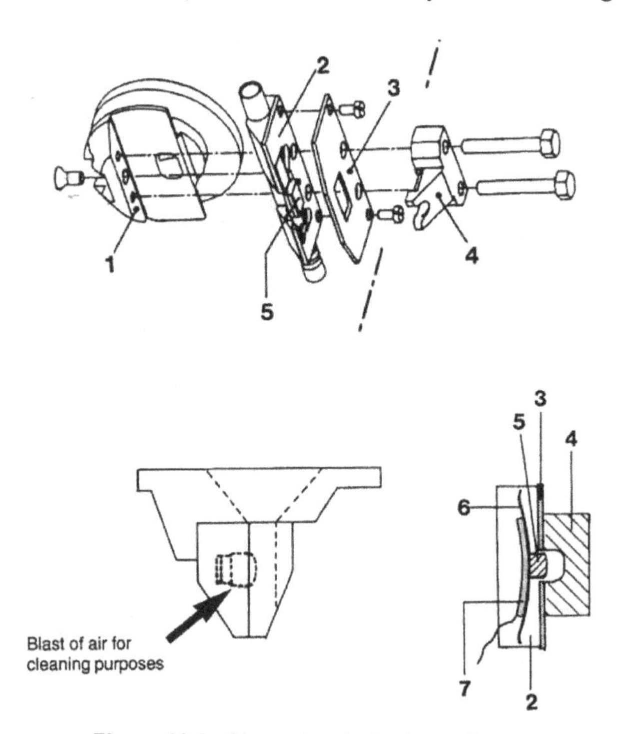

Figure 10.1 Measuring device in combers

1 = Trumpet funnel, 2 = Measuring unit, 3 = Cover plate, 4 = Sliver guide, 5 = hard metal sensing element, 6 = leaf spring, 7 = Strain gauge.

10.3 Delivery speed measurement by proximity switch

The delivery speed of the comber is monitored by means of pulses received on the cog wheel of the calendar shaft (1) as shown in the Figure 10.2 (A) and (B). The distance (A) shown in the Figure 10.2 (A) and (B) affects the signal emitted by the proximity switch. Hence the proximity switch should be set at 1.8±0.2 mm.

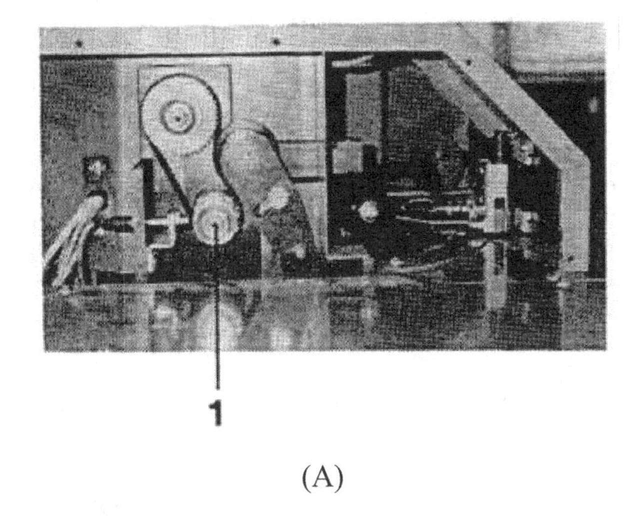

(A)

(B)

Figure 10.2 Delivery speed monitoring in combers

10.4 Light barrier and light scanner

In modern comber machines, automatic changeover of can is necessary so as to avoid the time taken to replace the full can with empty can by the operator. Moreover, the sliver length per can is also fixed, say, 4000 m, it is easy to go for batch creeling in next process called draw frame. This avoids the so called "piecing fault" in the sliver due to bad piecing which will reflect in classimat faults. Hence, light barriers and light scanners are integrated in modern combers as shown in the Figure 10.3.

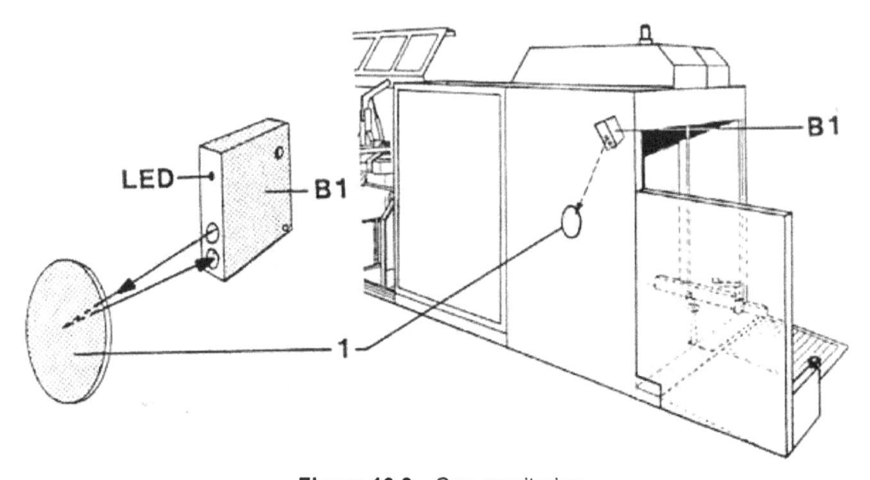

Figure 10.3 Can monitoring

In the delivery side of the comber, a photo detector (B1) is provided which senses whether an empty can for change over is ready or not. If there is an empty can in its place, the light beam is directly reflected on to the reflector (1). At the same time, LED lights if the light beam is reflected.

10.5 Light barrier for sliver congestion

Light barriers are provided at the delivery side of the can to detect if there is any sliver choking or congestion. As soon as there is no sliver choking, the light reflected by the light barriers B2 and B3 reaches the photo detector. The reflector for the photo detector is attached to the cover. If the light beam is obstructed by choking of slivers, the photo detector will not receive any signals and the machine stops at once. In this way, sliver choking and damages to the machinery component is avoided (Figure 10.4).

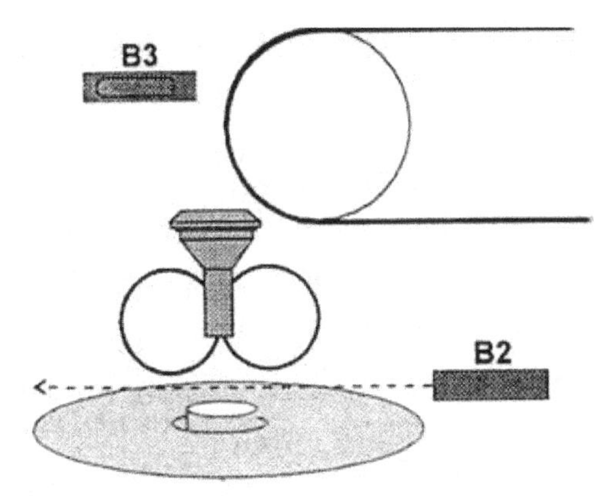

Figure 10.4 Light barriers to detect sliver choking in the delivery side of the combers

10.6 Light barriers at the suction duct

In the suction duct of the comber, there are two light barriers. One B5 is the light receiver and B6 is the light transmitter. The light beam shines along the duct. If there is no obstruction in the duct, the light emitted by the light barrier (B6) is received by the receiver (B5). If there is an obstruction, the light beam is interrupted and, after a time lag, of few seconds, the machine is shut down (Figure 10.5).

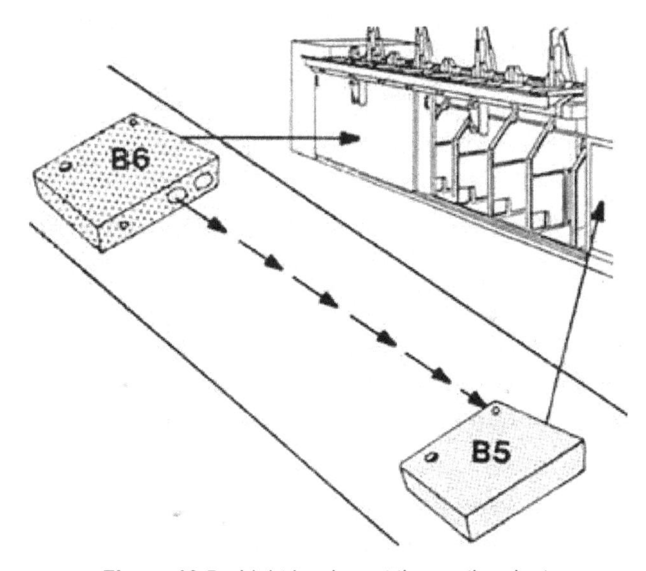

Figure 10.5 Light barriers at the suction duct

10.7 Light scanner for monitoring the single lap

In the comber, each head is provided with a light scanner on the lap side. As there are eight heads in a comber machine, there are eight light scanners provided one for each head. The purpose of this light scanner is the light beam reflected by the lap. When the material is present, the light reflected by the lap is received by the LED (Figure 10.6).

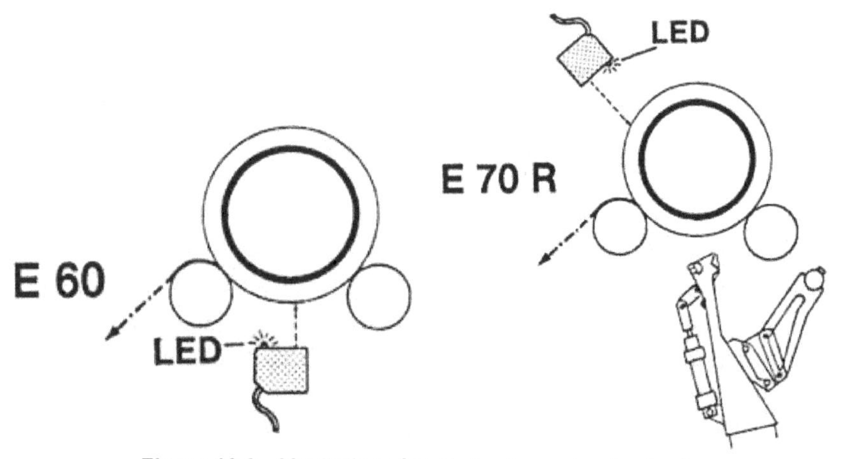

Figure 10.6 Monitoring of single lap in comber lap feed

When the lap get exhausted as the machine runs for producing combed slivers, the black tube or spool in which the material is wound appears. As soon as the black tube or spool appears, the light beam is no longer reflected and the machine recognises that the lap has run out. This stops the machine and prevents any "singles" in sliver material delivered.

10.8 Light barriers for individual head monitoring

The function of this table funnel monitor is to monitor the combed fleece condition i.e., fleece too thick or too thin. If the fleece is too thick, the table is pulled down and if it is too thin, the spring (5) will push the table up. In both occasions, the machine gets stopped. The machine should only be able to produce sliver if the table is in central position as shown in Figures 10.7 A, B and C.

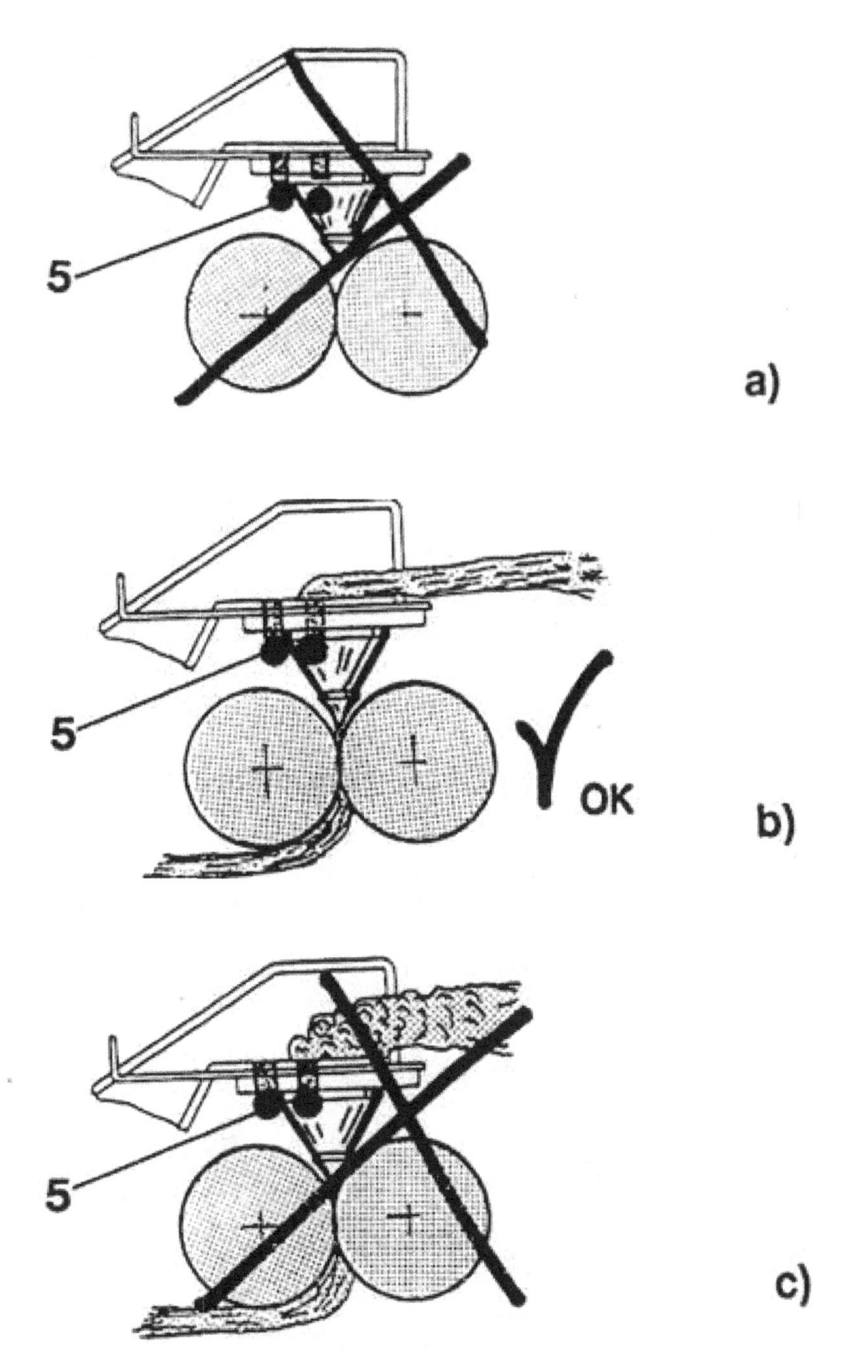

Figure 10.7 (A, B and C) Individual comber head monitoring

10.9 Table funnel monitor with one light barrier

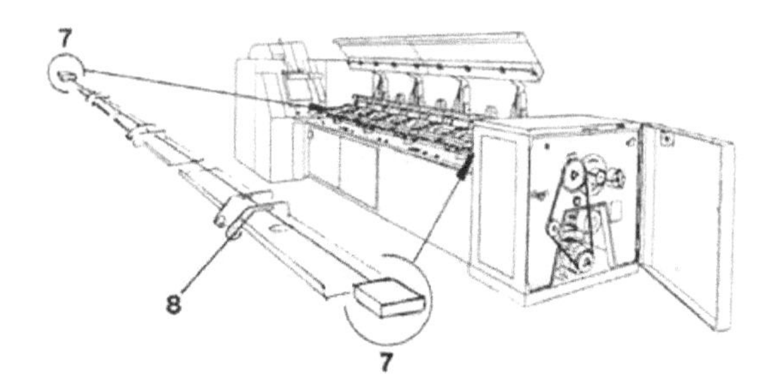

Figure 10.8

The table funnel monitor shown in the Figure 10.8 functions similar to the individual head monitor for lap. The combed slivers delivered by the detaching rollers are drawn-off and pass through table funnel to the drafting system. All the eight slivers from the individual heads collected must pass through and the draw-off tables must be in central position. This is monitored by the light emitter and receiver in the machine. If there is any obstruction in the funnel or if the table is not in central position, the machine gets stopped.

10.10 Monitoring of thin places in the comber sliver

The function of the sliver monitor is to stop the machine whenever the delivered sliver is too thin or not according to the set value. In the Figure 10.9 A, B and C, the sliver monitor (3) must recognise if the sliver is too thin between the two calendar disk rolls (4). The trip cam (2) provided and hence the cut-out point can be varied by adjusting the screw (1). The screw (1) must be adjusted according to the sliver count. At the start of the comber production, the sliver count values.

Figure 10.9 (A) Thin place monitoring in comber delivered sliver

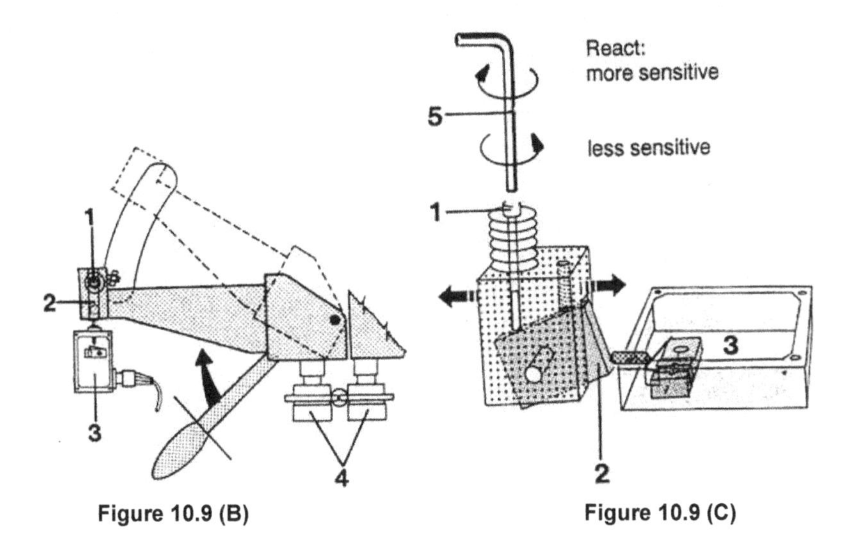

Figure 10.9 (B) Figure 10.9 (C)

must be regulated. If there is no thin place between the calendar roll disk (4), the signal lamp H1 will be in "off" position. As soon as the sliver monitor recognizes a thin place, the signal lamp H1 is "on" and the machine is stopped. This can be tested by tearing some portion of the sliver at any combing position and observe the same at the calendar roll disk (4). If the sliver is thin, the signal lamp H1 flickers and the machine are stopped.

Similarly, lap-up monitoring in drafting system, pressure monitoring in noils suctioning and pressure monitoring in detaching rolls have been provided which stops the machine if there is any mal functioning.

11

Monitoring in speed frames

11.1 Introduction

The quality of the yarn is influenced by many attributes like raw material, process parameters, machinery maintenance, material handling and so on. The quality of the yarn is significantly dependent on the quality of roving produced on the speed frame. Hence on-line monitoring system is very essential for the roving machines in addition to the number of sensors to monitor and stop the machine in the event of a roving break. Many developments have taken place in the machinery to improve the working performance, increase the productivity and also reduce the wastage of the raw material.

11.2 Purpose of online monitoring

The primary concern for any spinner is to produce yarn of an acceptable quality at all times consistently and maintain the same at a minimum cost. It means that the quality of yarn is significantly dependent upon the quality of roving. Hence online monitoring system for roving frame is very essential for a spinning mill. The Premier Roving Eye 5000 provides necessary solution for the spinners which satisfies the requirements by continuous information on production and quality through various forms and also ensures good yarn quality which meets the latest specifications laid down by the customers.

Figure 11.1 shows the yarn quality influencing factors. The yarn quality is influenced by various parameters like yarn count, CV% of count, unevenness, cut length CV%, periodic faults, yarn strength, weak places in the yarn, CV% of strength, etc., all these parameters are primarily influenced by three main factors such as

Figure 11.1 Influencing factors of roving material on yarn quality

- Quality of the input raw material.
- Machinery conditions of the ring frame in terms of their mechanical condition and process parameters.

The environmental conditions of the ring frame section

11.2.1 Online monitoring devices in speed frame

Online monitoring devices in speed frames helps to continuously monitor the production data, quality data. Besides, photo sensors in the machine is helpful in stopping the machine whenever the sliver breaks in creel section, roving breaks in the delivery thus avoiding the fibres lapping over the rolls and also reduces the operator patrolling time due to the presence of indication on individual spindles.

11.2.2 Photo sensor in the creel section

In the creel section of the speed frames, the photo sensor is provided which consists of a light emitter at one end and light receiver at the other end of the machine. In general, the sliver is withdrawn from the cans and enters the drafting system of the speed frame. When there is no sliver break, the light emitted by the photo sensor reaches the other end of the receiver and the machine runs continuously. If there is any sliver break, the slivers obstruct the beam of light from the emitter to the receiver of the photo sensor and hence the machine is stopped indicating the sliver breaks with a glow of "RED" light on the machine. This helps the operator to easily identify the sliver breaks in the creel.

11.2.3 Photo sensor at the front roll delivery

The sliver from the cans are withdrawn and enters in to the drafting system of the speed frame. The drafting system may be 3 over 3 or 4 over 4 pressure or spring drafting system. After the sliver is drafted and the size is reduced according to the linear density of the material required, the rovings are wound on the plastic bobbin in uniform manner. During the delivery of the roving from the front roller of the drafting system to the flyer top, the photo sensor senses the presence of material.

The photo sensors consist of an emitter and receiver and it is provided for each and every spindle. The principle of operation of this photo sensor is, whenever the roving material is delivered from the drafting system to the flyer top the material interrupts the beam of light from the photo sensor to the receiver and the machine runs continuously. If the roving is not delivered from the drafting system due to roller lapping or breaks at the flyer top, the light beam from the photo sensor is not interrupted and the light from the emitter reaches the receiver and hence the photo sensor blinks with "RED" on the respective spindle position. Furthermore, the information is sent to the central control unit which displays on the screen the spindle number which cause stoppage of machine. With this monitoring system on individual spindles, it is easily possible for the operator to identify the spindle number and mend the same as quickly as possible without wasting his time in patrolling and identifying the spindle number.

11.3 Individual roving monitoring

Rieter has introduced in its latest version speed frames the individual monitoring of roving in the machines. This sensor senses the individual spindles by photo sensor principle which stops only when the roving ends are down and will not stop due to any hindrances upon the machines like fibre fly. Monitoring of individual spindles is shown in the Figure 11.2.

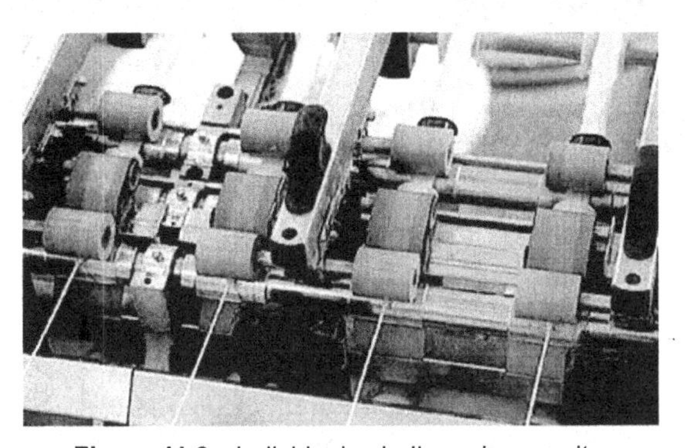

Figure 11.2 Individual spindle roving monitor

It is an optional attachment and especially suitable for man-made fibres, blends and dark coloured fibres where the visibility is difficult to ascertain the roving breaks. With this individual monitoring of the spindles, there are no incorrect stoppages, logging of the data on roving ends down, lapping is prevented and considerable increase in roving frame efficiency.

11.4 Roving tension control

Rieter has technological development in the drafting zone in the speed frame machine. In this pneumatically loaded drafting arrangement, constant pressure on the spindles is monitored by a sensor and ensures consistent roving uniformity along the entire length of the machine. Pressure is centrally adjustable and infinitely variable. The load relief on the spindles is automatic and centralised which prevents the permanent deformation of the top rollers. Roving tension monitoring is shown in the Figure 11.3.

Figure 11.3 Non-contact monitoring of roving tension

This non-contact roving tension monitor consists of two sensors and a microprocessor which monitors the roving tension throughout the bobbin build which ensures high yarn uniformity and avoid false drafting which in yarn stage introduced as long length faults especially long thin places in the yarn.

11.5 Roving stop motion

In the roving stop motion, the light barriers monitor the delivery of the material from the drafting arrangement. In this method, the light beam is directed straight past the flyer tops. In case, if there is any roving break, the broken roving end whirls around the flyer top and interrupts the light beam and cause the machine to stop (Figure 11.4).

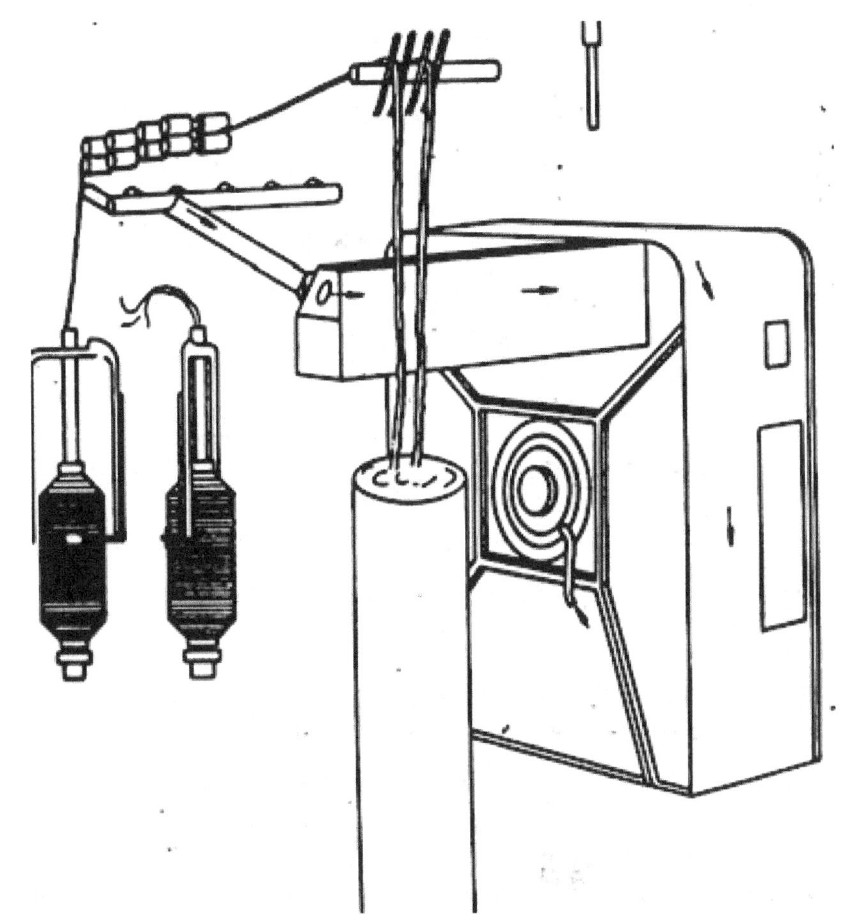

Figure 11.4 Roving stop motion

An alternative method developed by "Luwa" is the pneuma stop. In this system (Figure 11.5), the capacitive detection method is utilised in the monitoring unit. The device is accompanied with a pneumatic suction system in the delivery end of the drafting system. This suction system is essential to avoid series of roving breaks along a bobbin row after the first break in a row. If there is any roving break, the suction system draws the broken sliver material in to a large collector duct at the end of the machine. Fibres entering the collection duct pass through a capacitive detection head shown in the Figure 11.5. In this capacitive detection unit, an electrical field generated between the two comb electrodes. If any fibre material passes between these two comb electrodes, it causes a change in capacitance which generates a signal to stop the machine.

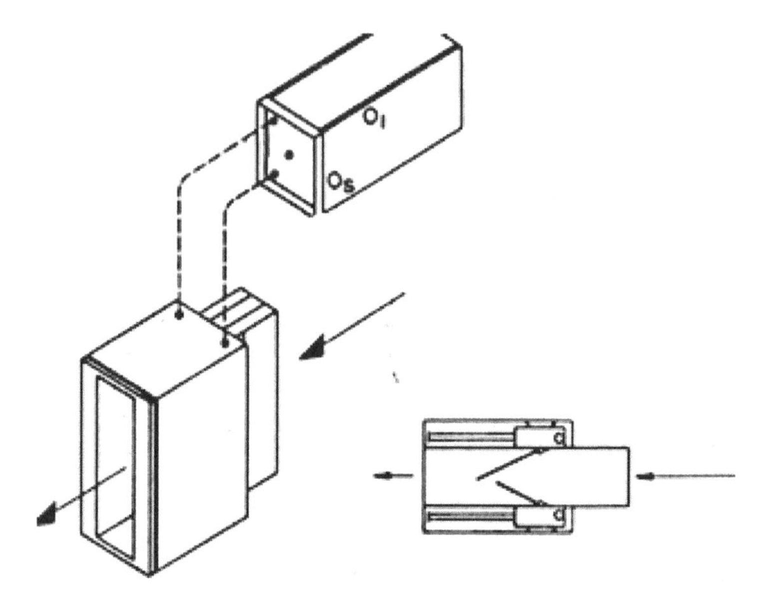

Figure 11.5 Pneuma stop suction system

11.6 Pneumatic loading in the drafting system in speed frame

The loading support is stamped from steel sheet and is mounted on a continuous hexagonal section tube behind the rollers. The tube contains the compressed air hose connected to a central compressor unit. Three top roller holders mounted on two bearing slides are accommodated in the loading support itself. The two bearing slides form a double lever system. Depending on where a pin is inserted in one of the three holes as the pivot at "**m**", the total pressure coming from the compressed air hose and acting on the entire pressure arm via a cam is applied more strongly to the back roller or the two front rollers. Pressure can also be distributed differently between the two front rollers via a second pin/hole system in the bearing slide of these two rollers at "**n**" (Figure 11.6). The total pressure on the top rollers is changed by simply adjusting the pressure in the compressed air hose via a reducing valve at the end of the machine, and distribution to the individual rollers via the system of levers already referred to. The main advantages of pneumatic loading are:

- simple and very rapid, centralised changes in pressure.
- simple and rapid pressure reduction to a minimum in the event of machine stoppages, so that the roller covers are not deformed during prolonged interruptions to operations.

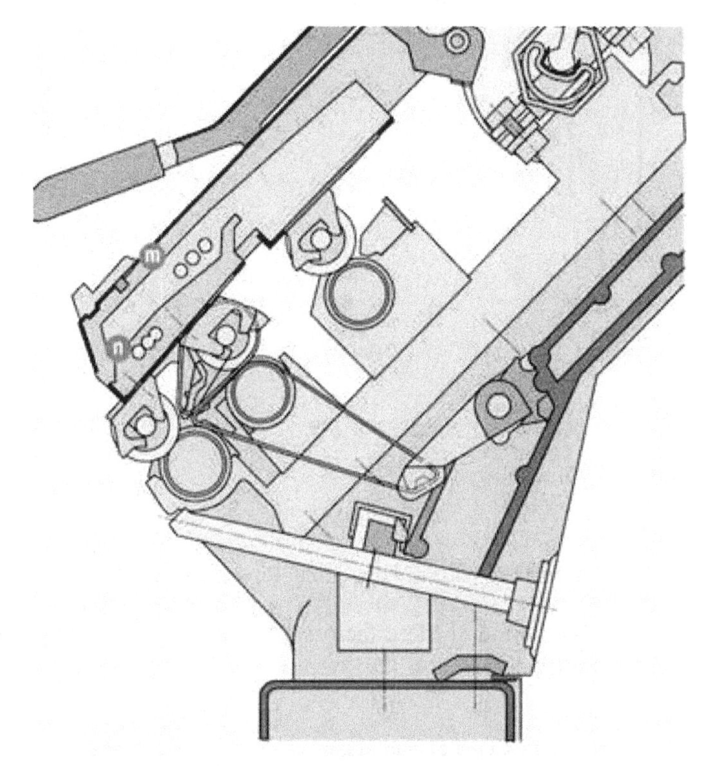

Figure 11.6 Pneumatic loading in speed frame

11.7 Quality monitoring in speed frames – significance

Quality control at the roving stage is very important since it is the final stage before the insertion of permanent real twist to the yarn and hence the chances of correction in the subsequent stages are restricted. Research workers have studied that the contribution of roving irregularity is 15%–20% of the yarn irregularity. This is indicated by an expression connecting roving U% and yarn U% derived by SITRA states that

$$V^2 = 0.88\,(V_r^2 + V_B^2)$$

Where V is the yarn irregularity
V_r is the irregularity introduced in the ring frame and
V_B is the irregularity of the roving material.

Furthermore, the quality of the roving is determined by various factors like sliver quality either carded sliver or combed sliver, machinery conditions and process parameters as shown in the Figure 11.7.

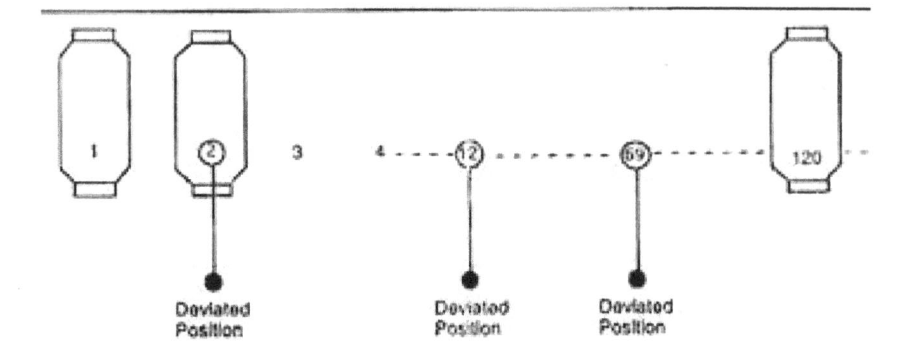

Figure 11.7 Quality control with Roving Eye 5000

In discrepancy in some of these factors result in the following problems as shown in the Table 11.1.

Table 11.1 Effect of inadequacy in the roving machines

Problem definition	Influence in yarn	Final consequences
Sliver stretch and sliver splitting	Long thin place	More cuts in winders. Unacceptable fabric appearance
Ratching (roving stretch)	Within and between bobbin count variation	Poor fabric appearance
Periodic faults	Unevenness and poor yarn strength	Poor performance of yarn in knitting and weaving
Ineffective stop motion	Higher count cv due to spinner's doubles	Poor fabric appearance
Piecing faults	Classimat faults	Poor fabric appearance

11.8 Quality monitoring in speed frames – present status

At present, the quality parameters of the roving is monitored by taking few representative samples with the spindle numbers according to the test plan in the laboratory. The quality parameters normally checked in roving frames, the frequency of observation and sample size are shown in Table 11.2.

Table 11.2 Roving quality parameters

S. No.	Quality parameters	Frequency of testing	Sample size
1	Roving Hank	Once in 2 weeks	Eight wrappings/sample
2	Evenness	Once in a week	Four readings/sample from all frames
3	Hank variation	Once in a week	Evaluated from samples collected from one frame per day (sample collection one bobbin per frame) two wrappings per bobbin
4	End breakage rate	Once in a month and whenever new lot change occurs	All frames
5	Ratching (Tension difference)	Once in a month	One sample of 15–30 yards. (Initial and final build of the bobbin.)

Although modern test equipments are available in the laboratory to check the roving quality, it suffers from the following disadvantages.

- The first disadvantage is the problem of taking enough samples which is not adequate to represent the population.
- A second more disadvantage is the delay factor. It means with the latest speed frames running at speeds more than 1300 rpm, there are many chances for the faulty material get manufactured before a fault is detected, analysed for their reasons and corrected.
- Material handling of the samples also plays a vital role during transportation to the laboratory which will eventually lead to inaccuracy in the test results.
- Above and all, the practical difficulties associated with the off-line testing is the influence of Rogue spindles. Rogue bobbins with abnormal quality can result in a marked deterioration in the quality of the yarn.

The difficulties associated with the off-line monitoring can be encountered with the help of an online quality monitoring system in the speed frames. With the continuous on line monitoring system, the existing manual method of collecting the samples and testing in the laboratory can be minimised to occasional checks on the quality parameters which serve as verification of the functioning of the online monitoring system.

11.9 Premier Roving Eye 5000 – On line monitoring system

Premier Roving Eye 5000 is an online monitoring system for speed frames is conceptualised to continuously monitor the roving for its various quality characteristics. In addition, the system also provides information on production and stoppages. It identifies the spindles which deviate from the set limits and restrict the use of off standard bobbins to be used in ring frames which will avoid customer complaints.
The Premier Roving Eye system has the following configuration.

- The sensor on each spindle.
- The collector circuit module for every eight spindles.
- One machine processing station for every 20 collector circuit modules.
- One personal computer.

Figure 11.8 Schematic arrangements of Roving Eye 5000.

The salient features of the system are:

- Technology

  Roving Eye 5000 is the totally indigenous technology for the online monitoring system in speed frames in the world.
- Quality characteristics

  Premier Roving Eye 5000 provides a wide range of quality parameters like
  - Cut length CV% for 0.5 m, 1 m, 2 m, 5 m, 15 m and 60 m. Hank 5 m, 15 m and roving stretch.
  - All these parameters are monitored for each and every spindle on continuous basis for every doff.

Reports are presented in numerical and graphical formats and can be viewed either on the machine itself or on a central computer.

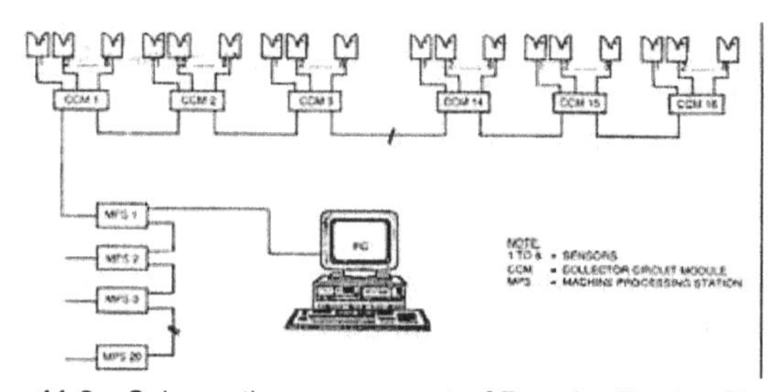

Figure 11.8 Schematic arrangement of Premier Roving Eye 5000

- Warning system

  There is a possibility of entering unit values with reference to the quality characteristics, the system can be made to give an alarm which would serve as a warning to the user so that corrective action can be taken immediately for off standard bobbins.

- Analysis of Rogue spindles

  All the spindles are monitored for their quality characteristics according to the set alarm limits set by the user. Rogue spindles are identified if the spindle exceeds the set alarm limits in all quality characteristics which exceed a fixed number of times.

- Spectrogram analysis

  The system analyses each and every spindle for the presence of periodic faults and identifies with spectral peaks in graphical form.

- Variance length curve analysis

  Variance length curve is an important quality tool which gives an indication of the long term non-periodic variation. The preparatory departments of roving that contribute the variations are automatically identified and displayed.

- Production reports

  At the end of every doff, PREMIER EYE 5000 provides information on hanks produced, production in kgs, stoppages with code identification, end breakage rate, etc.

- Long-term storage of data

  Long-term storage of data helps the user to assess the data history whenever required. All the machines are connected to a computer for long-term storage of quality and production data. This would enable the spinner to assess the consistency in yarn quality over a long period.

- Lay out analysis

  The information monitored by Roving Eye 5000 is also presented in the form of a layout which enables easy analysis of machine wise and spindle wise performance with respect to the roving quality.

11.10 Quality assurance with Premier Roving Eye 5000

Premier Roving Eye 5000 provides all the information on quality characteristics provided by the latest generation off-line evenness testers. Hence extensive quality control facilities are available with this system.

1. Determination of the final yarn quality required and fixes the same.
2. Determination of the corresponding roving quality which will meet the yarn quality.
3. Assigning the limit values for various quality parameters of the system.
4. Continuous monitoring of the roving quality with the set limit values.
5. Segregate the spindle positions with the deviated values at the end of the doff.
6. Process only the standard quality bobbins in the ring frames.
7. Identification of the causes for the deviated bobbins and taking corrective action.

12

Monitoring systems in ring frames

12.1 Introduction

In the present international textile market, survival and progress of any spinning mill is fully dependent upon producing quality yarn at competitive prices and also maintaining the consistency in performance. It is not easy to achieve this consistency since there is no guarantee on raw material characteristics especially natural fibres like cotton. Secondly, frictional characteristics of moving machine parts cannot be totally avoided. Another factor is that the present day spinning and other processing machines are of high productive machines. Any deviation in the machinery parts will result in sub-standard material. Hence, time lapse in occurrence, detection and correction of the faults must be as early as possible. It is quite impossible to maintain the minimum time gap with manual monitoring of the process. One suggested method is online continuous monitoring round the clock with the help of instruments. Ring data system is one such instrument made available for continuous monitoring of ring frame performance.

12.2 Salient features of modern online instrument

Instrumentation is a technology of measurement. To understand this, knowledge of any parameter largely depends on its measurement. Thus measurement provides a base for real and sustainable improvement. Measurement is done in two ways. One is static and dynamic. Measurement in static is easy and any simpler instruments can do but with slow response. However, measurement in dynamic state conditions is quite difficult and more sophisticated, complex instruments with quick response are needed for such measurement.

In modern instruments, measured values or the collected data displayed have to be analysed for logical conclusions and for controlling the same in process. Analysis of the data in quick time is possible nowadays due to the development in microprocessor technology. Control can be exercised in two ways – online auto control and off-line manual control. In the first case, online control exercised through instrument like auto levellers in draw frames and card. In the second case of off-line control, the instrument provides signals like sound, light signals and printed reports, suitable controlling action is done manually by adjusting the process parameters like speeds, settings, etc. Thus, measurement, display, analysis and control are the four main functions of any modern instrument. In textile processing machineries, instruments for online, round the clock monitoring of the process are found essential. However, these instruments are required to be reliable, maintainable, sensitive and moreover should be user friendly. It means light weight instruments which consume less energy are preferred. Considering all these factors, modern instruments are based on other engineering disciplines like electronics, mechanical, pneumatics, hydraulics, etc.

With the developments in the machine design and metallurgy, the importance of electronics and monitoring systems has also gained advantage in every process stage. It means there is process monitoring in blow room stage like contamination detector, auto levelling systems in carding and draw frames, monitoring systems in speed frames and so on. Similarly online monitoring systems also have been developed in ring frames for production monitoring. One such kind of this is "Ring data". The purpose of this development is to monitor the production, end-breaks, efficiency, etc. Conventionally, and still now, the end-breaks in ring frames are assessed, manually by conducting the number of ends down per hour or full doff by quality assurance investigator. The number of ends down like spindle breaks, multiple breaks, traveller fly, etc., will be noted on the data sheet and the ends down percentage are evaluated per hour. This process requires skill of the investigator and will not give consistent report since it may be manipulated. Hence in order to have a continuous monitoring for 24 hours running, ring data is helping the textile spinners to exercise better control over the process, quality and wastes can be envisaged.

12.3 Elements of ring data system

Ring data system can be installed in three different versions – namely pilot, mobile and overall. Although the basic functions carried out in each version are the same, the span of control varies from version to version. Among these, "overall" is the most developed version.

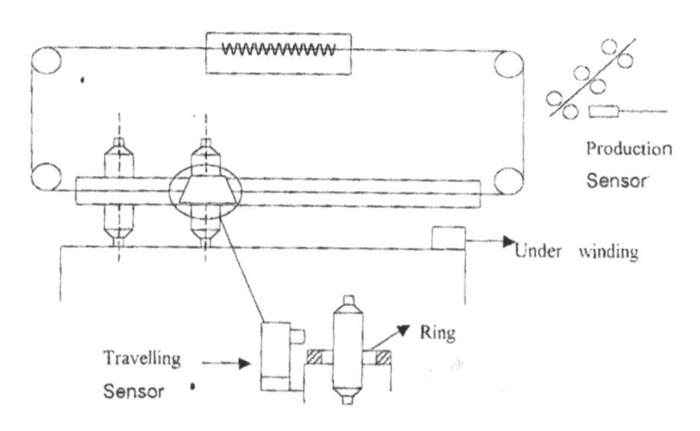

Figure 12.1 Sensors in ring data system

In each ring data system, there are three types of sensors (Figure 12.1) employed in each ring frame.

a) Sensors for determining drafting system front roller delivery speed.
b) Doffing sensor for registering the duration of the doffing time and the number of doffs taken out.
c) Travelling sensor for sensing the traveller speed in terms of revolutions per minute (rpm). Traveller speed is very much close to spindle speed of the ring frame and slightly less due to the concept "Traveller lag".

12.4 Functions of travelling sensor in ring data system

A travelling sensor runs continuously back and forth at the height of the ring rail on each side of the machine. On a single machine (pilot unit) or on all machines in the mill. Such travelling sensor is required one on each side of the ring frame. It continuously moves on ring rail from gearing end to pneumafil end and back. During this journey the sensor senses whether the traveller is running on ring. This generates a magnetic field that is affected by rapidly rotating traveller. If a yarn break occurs, the traveller ceases to rotate and the sensor displays the resulting impulse as an end down, also recording the spindle number. As a result of its rapid motion back and forth it registers the spindle several times until the end down is repaired. The spindle sown time is also recorded. Such incidences are sensed and the information is conveyed to microprocessor controlled machine station.

On every ring frame, one microprocessor controlled machine station is employed. The function of this station is to accept the data signals through different sensors and controls the movement of the travelling sensor. Another sensor fitted on the front roll records the delivery speed and machine stoppages and another sensor records the number of doffs and their duration. All the information collected is ultimately transmitted to a computer with monitor and printer, which performs the necessary analysis and stores the data for preset periods.

The following parameters are calculated by the CPU with the information received from travelling sensor.

- Ends down: Ends down at any instance which is received through systems PC.
- End breakage rate: End breakages per 1,000 spindle hours for a particular ring frame is classified in to three types.

    - Start-up breaks (STU): Breaks that are recorded at the start of the machine after doffing is completed within a pre-selected time.
    - Normal breaks (EB): Breaks that are recorded during the spinning operation from the start-up of the production till the end of the doff.
    - Other breaks (PRT): Breaks other than STU and EB, normally called as idle spindles due to various reasons like back stuff shortage, maintenance defects like broken bottom apron, lappet missing, etc.
- End mending time (EM): End mending time indicates the mean duration of normal end breaks in minutes. This parameter is an indicator of operator's efficiency.
- Momentary stops position (MSP)
    - This parameter indicates momentary stoppage of spindles at any instance.
- Idle spindles: Non-productive spindle positions of a ring frame are called as idle.
- Spindles: It means these spindles which do not produce anything for a particular period of time. In general, spindles which remain non-productive for more than 30 minutes are considered as idle spindles.
- Rogue spindles: Rogue spindles are the spindles which cause unusually higher end breaks than the nominal and alerts the user for remedial action.
- Worst spinning positions: This report shows a maximum of 10 spindles with spindle number/machine number which cause higher number of end breaks.
- Actual spindle speed: The rotational speed of each and every traveller is measured by the travelling sensor and the mean revolution per minute (RPM) is stored in the system.
- TPM/TPI: Twist per meter/ Twist per inch measured of the produced yarn is stored in the system.
- Slip spindles percentage: The individual traveller speed of each spindle is constantly compared with the "Mean RPM of the machine". A spindle will be recorded as slip spindle, if its traveller

speed is less than the set limit, say 3%–5% for more than the number of times set.

12.5 Central unit of the ring data system

All the ring frames which are equipped with ring data are connected to a central unit which is the heart of the ring data system. It is connected to machine stations from different ring frames. Microprocessor in this computer system processes and analyses the data received from different machine stations. From this, logical conclusions can be drawn from such analysis which can be used for controlling the ring frame performance. Ring data system is also equipped with printing and video terminals. Printed reports can be made available as per need through the printing terminal. Video terminal is used where actual information is rapidly required. Through proper functioning of ring data system, the following information is available for each ring frame under observation:

1. Machine number
2. Date and time
3. Period monitored
4. Production period
5. Spindle speeds in rpm
6. Yarn twist
7. Production output in kg
8. Efficiency of the machine
9. Down times
10. Doffing times
11. Front roller speed in rpm
12. Idle spindles (%)
13. End breakage rate
14. Slipped spindles. i.e. spindles with lower twist/ speed
15. Rogue spindles. i.e. spindles giving repeated end breaks
16. Operator cycle time
17. Number of cops doffed
18. Mean duration of ends down
19. Preset maximum number of ends down
20. Hank production per unit time.

With the ring data system, shift wise, count wise and machine wise analysis can be made by taking printouts.

12.6 Analysis of ring data reports

Reports generated by ring data system are more beneficial for the spinners to have an idea and thorough analysis about the machine performance. It is not only for having control over productivity and also on end breaks and waste control which is quite an important factor for yarn realisation. The reports of the ring data will be dealt in detail in the following sections:

a) Report on repeated end breaks

   End breaks in a ring frame are a truly random occurrence. But in actual case, it is not the case so. The reasons for end breaks are many like man, machine, materials and methods. It is because of poor quality of feed material, poor maintenance of mechanical parts, poor work methods due to unskilled operators. Repeated end breaks occur at some specific spindles. Such spindles which cause potential increase in end breaks per 100 hours are called as "Rogue spindles". In general, the average end breakage rate rise up to 15%–20% because of such rogue spindles. Moreover, manual identification of such rogue spindles is time consuming and quite impracticable. However, the identification of such rogue spindles by the ring data system is very easy.

Print outs can be taken machine wise, count wise with the number of spindles in each machine which cause repeated end breaks in a shift or more. Based on the reports, it is possible to direct the maintenance crew to attend such rogue spindles for necessary corrective action. After the corrective actions are taken, such spindles can again be observed for such reoccurrences. It not properly attended in time; such rogue spindles will certainly increase the average end breakage rate of the machine.

Useful corrective actions are of two types:

i) Checking the quality of the feed material and the yarn delivered.

ii) Necessary corrective actions can be taken in machinery parts like proper spindle and lappet gauging, traveller cleaner settings, condition of traveller, etc.

Proper attention if made after observing the ring data report will help the spinners to keep the end breakage rate under control.

A sample report of ring data system is shown in Table 12.1.

Table 12.1 Sample report on ring data system

Ring frame M/c no.	Side	Spindle no.	End breaks	Spindle no.	End breaks	Spindle no.	End breaks	Spindle no.	End breaks
1	L	123	5	247	5	504	6	432	7
1	R	165	4	265	5	432	7	501	6
2	L	120	8	287	6	401	6	500	4
2	R	65	8	143	3	342	5	427	6
3	L	32	4	154	6	283	5	391	5
3	R	54	5	172	6	275	6	356	5

b) Report on idle spindles

Idle spindles are non-productive production spindles in speed frames or in ring frames. Such spindles are a loss to the management since the production of this spindle is not taken in to account. The reasons for idle spindles are due to the maintenance or due to operator's negligence. Assuming that if a spindle production is 180 g per 8 hours and if the spindle is idle for 3 shift hours of 24 hours, then the loss in production is 180 × 3 = 540 g. In general, in spinning mills the idle spindle percentage should be nil or 0.1%.

According to the ring data system, idle spindles are those spindles which remain non-productive for a certain period and exceeds pre-selected time limit. Mills can select time limit to suit their practices and spinning performance. In many good textile spinning mills, the time limit is 30 minutes. The efficiency of the ring frame operator directly affects the end breakage rate and percentage of idle spindles. If the operator is not efficient, the time taken for mending an end break will be more and higher will be the percentage of idle spindles. Furthermore, it will also increase the pneumafil waste of the machine which will have an impact on yarn realisation. Higher incidence of roller lapping and creel breaks will also increase the idle spindles.

A sample report on idle spindle is shown in Table 12.2 frame wise.

Table 12.2 Report on idle spindles

M/c No.	Side	Spindle no.	Spindle no.	Spindle no.	Spindle no.	Spindle no.	Spindle no.
5	L	19	45	67	156	272	345
5	R	54	87	136	345	376	446
7	L	64	43	185	234	321	500
7	R	23	36	57	68	139	456

In the above idle spindle report, the idle spindles are mentioned frame wise. Thus it becomes easier for the spinning manager to identify the corresponding operator who gives lower piecing rate, more idle spindles and more pneumafil waste in the spinning department. It also helps to tackle the problem and reduce the production of irregular shape bobbins like bottle bobbins, lean cops and half cops, etc.

Another advantage of the ring data system is that it helps in minimising the idle spinning positions by means of indication bulb. According to this feature, the ring frame side is divided in to different sections and one signal lamp is provided for each section. If there is an end break, the corresponding bulb gives indication. This helps the operator in patrolling in a better way by unnecessary walking and searching for an end break. This indirectly helps in more efficient piecing, lower percentage of idle spindles and lower pneumafil wastes.

c) Report on slipping spindles

Ring frame spindles are usually driven through tin roller pulley and spindle tape arrangement. Although it is expected that all the spindles are expected to rotate at the same spindle speed set or calculated, some spindles will rotate at lower spindle speed than the standard speed selected. This is because of some short comings in the driving mechanism. As a result of this, the yarn produced in such spindles will be having lower twist per inch or meter than the standard, such spindles will give lower single yarn strength and elongation and also affect the dye pick-up. In general, such spindles may be fewer in number, in knitted goods it is sufficient for the customers to complain about the quality. Manually identifying such spindles is quite laborious and impossible. In such situation ring data system helps in identifying "slipping spindles".

The principle is the travelling sensor senses the speed of the each spindle. If any spindle whose speed is lesser by 5% than the standard speed, it is identified as "slipping spindle".

A sample report on slipping spindles is shown in Table 12.3.

Table 12.3 Report on slipping spindles in ring frame

M/c no.	Side	Spindle no.	% drop in speed
12	L	45	7
12	R	86	6
15	L	190	5

Slipping spindles in some ring frames are shown in Table 12.3. Slipping spindles with maximum drop in speed is shown as percentage drop in speed of the spindle. With the help of this report, it becomes easier to identify the slow running speed spindles and bring back them to normal running speed. Slipping spindles are due to slack spindle driving tapes, improper positioning of the tape on the spindle wharve, accumulation of fluff around the wharve and obstruction to the free running of spindle. This report is generally taken once in a shift.

d) Report on machine stoppages

One of the important parameter in a spinning mill is machine utilisation percentage. Machines remain stopped for various reasons. Some reasons are avoidable and others are unavoidable. Ring data system provides cause-wise report on machine stoppages. In this report, number of incidences and total duration of stoppages are indicated in this report. Thus this report will be very useful for analysis and can be put an indirect pressure on maintenance crew and production crew for delay in the doffing system and delay in completion of maintenance work like wheel changes, etc. Analytical studies can also be conducted to find out the possible reasons for machine down time and for improvement in machine utilisation.

e) Report of production and efficiency

Ring data system also provides production and efficiency reports like conventional ones. Count wise, machine wise data can be made available through this report for subsequent analysis. Information like nominal yarn count, spindle speeds, twist level, hanks produced, cause wise details of machine stoppages, number of doffs taken out, average end breakage rate is also included in this report.

f) Report on malfunctioning

Ring data system has its own inherent difficulties which cannot be rule out. The system can malfunction due to the following reasons.

- Any interruptions in travelling sensor movement.
- Improper levelling and alignment of ring rail.
- Sticking of traveller to the sensor.
- Accumulation of dirt and fly in the path of the travelling sensor.
- Very high or low humidity in the department.
- Frequent changes in voltage and frequency of supplied electrical power.

Such malfunctioning can be on any individual ring frames or in group of ring frames. Hence, it is advisable to study about the malfunctioning report of ring data system before studying other reports for necessary corrective action. This helps to avoid drawing wrong conclusions on production and maintenance personnel.

g) Benefits of ring data system

- Ring data system helps in controlling end breaks, yarn quality, pneumafil wastes, machine utilisation, working efficiency of the ring frame operators, doffers and maintenance gangs, etc.

- "Rogue spindles" which give repeated end breaks are analysed and associated faults can be rectified.
- Reduction in the waste generation and continuous monitoring of the machines in ring frame section helps the management to improve the working performance.

12.7 Stationary single spindle monitoring system

With the developments of electronics, nowadays machinery manufacturers have also incorporated built-in stationary single spindle monitoring sensors which is placed on each spindles.

Production monitoring sensor

A production monitoring sensor is attached to the delivery roll (front roll) of the ring frames. The sensor monitors the front roll speed and the same is processed through CPU. From the CPU, the following parameters are calculated:

- Front roll speed: Actual front roll speed of the spindle in rpm is indicated.
- Delivery speeds: Yarn delivery speed in meters per minute calculated from front roll speed.
- Total ring frame production: Total ring frame production at any instance of time or for a prescribed shift period of 8 hours in kgs.
- GMS per spindle per shift: Shift wise GPSS for count-wise, group-wise and ring frame wise.
- Total down time (STM): Total down time in minutes since the beginning of a shift.
- Efficiency: AEF% Actual efficiency, PEF% Production efficiency, NEF% Now (present) efficiency of the machine can be obtained through the system.
- Number of doffs in a shift: Total number of doffs for a particular ring frame can be obtained through the system. This report is helpful for ascertaining the work load of the ring frame doffers in case of manual doffing for a particular count group.
- Doffing time. (DFM): Total doffing time in minutes since the beginning of the shift. This report helps to monitor the efficiency of the ring frame doffing team.

12.8 Sensors and proximity switches in ring spinning machines

Under winding position of the ring rails
The function of the under winding is the ring rails go in to the under winding position (bottom position), where the under winding thread is wound on the bobbin (Figure 12.2).

Setting of back winding and under winding coils

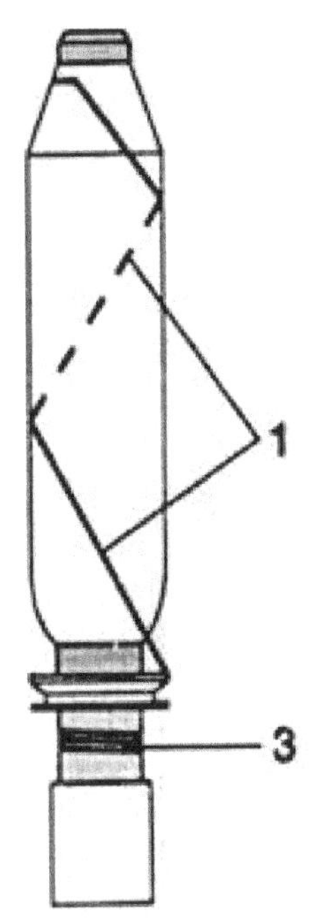

Figure 12.2 Under winding position of spindle in ring frames

Back windings in ring frames are set by the proximity switch B37. The number of turns for back windings depends upon the type of fibre material wound on the ring cop. The guide lines are:

Under windings

Cotton – 2 to 3 turns, Blends – 3 to 4 turns, 100% synthetic fibres – 4 to 5 turns.

Back windings

For cotton – 1.5 to 2.5 reserve windings
For blends and synthetic fibres – 1.25 to 1.75 reserve windings

If the back windings are too steep or too flat, adjust the selector switch S15 and proximity switch B37 higher. If the under winding coils are too short or too long, it is better to adjust the proximity switch B37 and allow the spindles to turn 5–10 tomes so that under windings will be correct.

12.9 Proximity switches for braking spindle drive motor

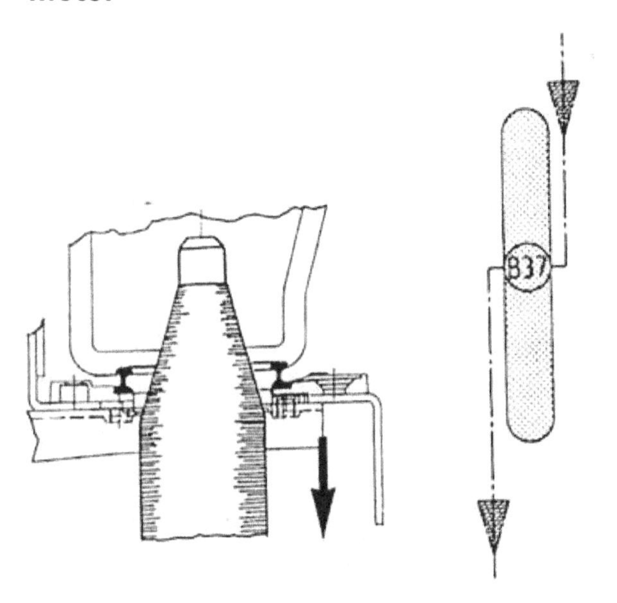

Figure 12.3 Proximity switches for spindle braking when the ring rails move downwards.

The spindle drive motor is sensed by the proximity switch B19 which transmits the pulses from the pulse disk on the motor shaft to the computer. If the computer receives no pulse from the pulse disk, within a given period of time, the ring spinning machine will be shut down. The proximity switch B19 is set with a clearance of 1–2 mm from the impulse disk (Figure 12.3).

12.10 Proximity switches for drafting system motor speed

The proximity switch B32 transmits the pulses from the pulse disk of the drafting system drive to the computer. If the computer does not receive any pulse from the pulse disk, for example, toothed belt broken, the ring frame machine is shut down. Further, if the number of pulses transmitted deviates by ±1.5% from the set value, the ring spinning machine is shut down (Figure 12.4).

The proximity switch B32 is set with a clearance of 1–2 mm from the impulse pulley.

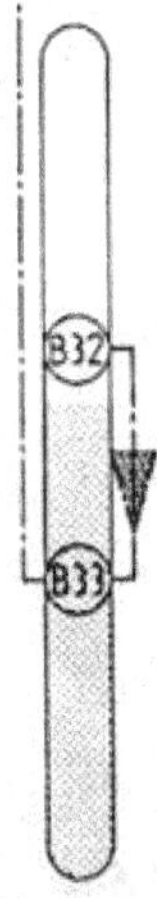

Figure 12.4 Proximity switch for drafting system motor speed

12.11 Proximity switch for spindle speed

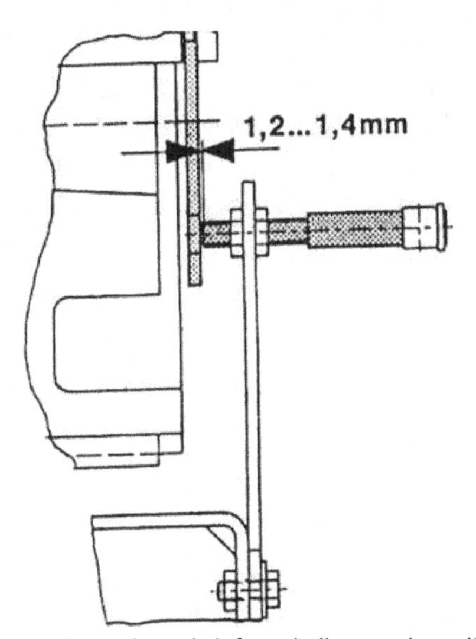

Figure 12.5 Proximity switch for spindle speed monitoring

The proximity switch B23 (not shown in the Figure 12.5) transmits the pulses emitted from the pulse disk shaft to the spindle speed display. The distance between proximity switch B23 and the pulse disk is 1.2–1.4 mm.

12.12 Proximity switch for delivery drafting roller

Figure 12.6 Proximity switch for delivery drafting roller

The proximity switches B24 in the Figure 12.6 transmits the pulses to the computer which displays the delivery speed of the bottom delivery roller. This is important during the under winding process. If the number of under windings is too short, it will cause more start-up breaks during the start of the ring frame.

12.13 Solenoid valves for supplying air pressure to the drafting system

The function of the solenoid valve shown in the Figure 12.7 is to supply the operating pressure required at the drafting system. The compressed air is reduced by means of reducing valve before supplying to the drafting system. If the machine is idle, the operating pressure is reduced after 60 minutes in order to avoid wear and tear to the top synthetic rollers. This is indicated by means of flash lamps which show the operating pressure is reached. If the operating pressure is not reached within 10 seconds, the ring spinning machine will not start and a malfunction code will be displayed on the screen.

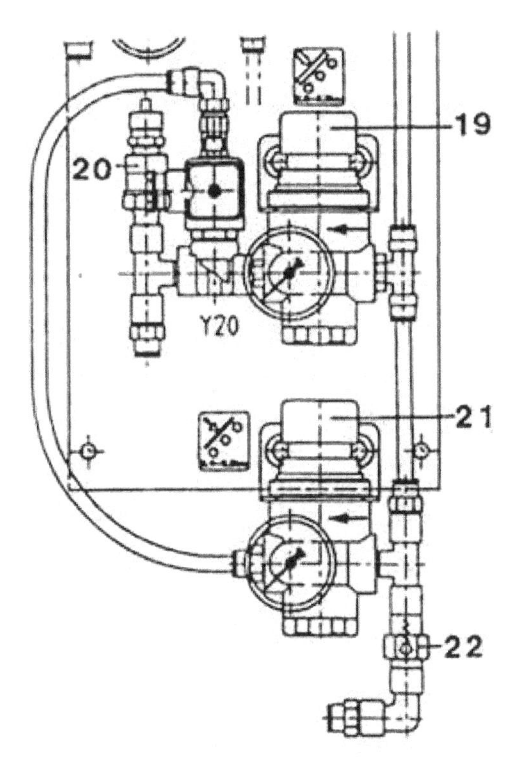

Figure 12.7 Solenoid valve for air pressure supply to the drafting system

13

Online monitoring system in automatic cone winders

13.1 Introduction

With the growth of the export market in recent years, spinners around the globe have to be competitive with respect to quality in order to meet the requirements of the customers. Furthermore, the export of yarns and finished fabrics has grown many folds in recent times. In addition to that, buyers in European markets lay down strict quality requirements of yarns for which the spinners have to be more vigilant to avoid customer complaints. On the other hand, the local yarn buyers/weaving mills also emphasise on strict quality standards. This has paved way for the need to develop the yarn clearers in winding machines in spinning units.

Since the winding process is the last stage in the spinning mill, this section is responsible to identify and remove defects before the product goes to the market. In spinning mills, most of the faults are classified in to raw material faults, faults created by the defective machine parts, improper maintenance, liberation of fluff IJ the department, cleanliness of the machines and other factors contribute for the faults which are objectionable to the customers.

The faults which are objectionable to the naked eye and also which downgrades the appearance of the fabric are called "Objectionable Faults". In the earlier generation yarn clearers, it was designed to eliminate only short thick faults and spinner's doubles in the yarn. Subsequently, the elimination of long thick places and long thin places in the yarn became essential and later generation yarn clearers were able to remove such faults.

13.2 Basic requirements of yarn clearing installations

The basic requirement which a yarn clearing installation must fulfil is the reliable extraction of all yarn faults which would be considered disturbing in subsequent processes like weaving and knitting. On the other hand, the practical application of a yarn clearing installation leads to a number of problems which will be dealt in the following sections.

Clearers on winding machines

The change from manual operations to automatic processes in the textile industry has often boosted new developments and innovations in textile electronics since automatic machines require various sensors to control individual functions.

In the early sixties, the first automatic winding machines were introduced in the market. An automatic winding machine had to replace disturbing thick places or any defects by a knot. This had to be done with a mechanical knotter on the machine. In this situation, the "sensor" which had to detect thick places was the "Uster Slub Catcher". This consists of a small mechanical apparatus with an adjustable slot. The slot width was set in such a way that it acts on a yarn break at the arrival of a disturbing thick place. However, it was not successful for automatic winding machines.

Figure 13.1 Electronic yarn clearers in winding machines

The electronic yarn clearer (Figure 13.1) has been playing an important role in the market for the past three decades. It represents the last monitoring system in the spinning process where remaining disturbing faults can be eliminated and replaced by a "splice" instead of a knot which is objectionable in knitting and in weaving processes. It also represents a filter in which off-quality bobbins can be automatically separated from the yarn batch of good quality. The Uster Quantum and Loepfe are the sixth generation yarn clearers. The clearer for winding machine is an advanced sensor which is able to eliminate disturbing thick places, thin places, foreign fibres and outlier bobbins which do not fulfil the conditions like evenness, imperfections, hairiness and count deviations, etc.

13.3 Classifying systems

The correct settings of the yarn clearers of the first generation was quite complex. It was unknown what kind of yarn faults had to be eliminated and how many cuts per 100 km had to be executed. Therefore, an analysing system was developed which served as a reference for a previous analysis of the number and size of the disturbing yarn faults. In 1968, Uster technologies have introduced this classifying system. The USTER CLASSIMAT with which disturbing thick places could be analysed in detail. With this system, the disturbing thick places in the yarn could be determined which had to be cut in the winding machines. With the classified faults and the "clearing curve", the spinner could also determine the number of cuts per 100 km to eliminate the disturbing faults, and hence the efficiency of the winding machine.

Figure 13.2 shows how the disturbing yarn faults could be categorised in 23 thick place classes.

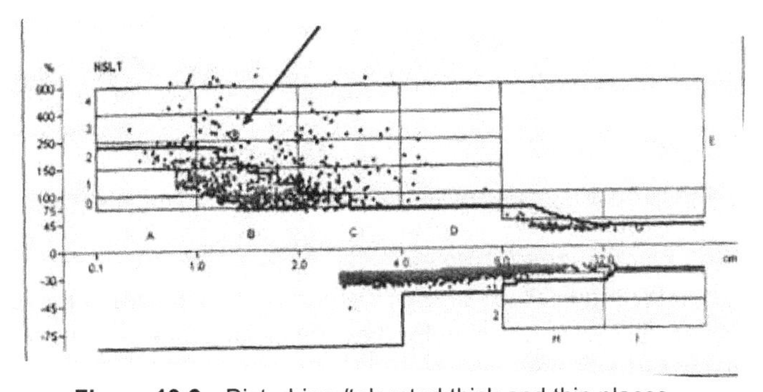

Figure 13.2 Disturbing /tolerated thick and thin places

Horizontal scale: Fault length in cm; Vertical scale: Mass increase in %.

These grades were used to determine what kind of thick places had to be eliminated by the clearer and which faults could remain in the yarn to keep the efficiency of the machine at a high level.

13.4 Setting of the clearing characteristics

It is particularly important to set the yarn clearer accurately so that the yarn faults whose presence in the yarn cannot be tolerated. Moreover, it should be purposeful and economical. With the clearing installation, the clearing of the various yarn faults must be coordinated. In this aspect, it is important that once a pre-determined clearing characteristic can be repeated, even with a much later clearing.

Based on these requirements, a newly developed unit of measurement, the so called USTER CLASSIMAT grades are now available which provide for an objective determination and classification of all yarn faults. This method of determination is based on the physical and measurable characteristics of yarn faults i.e., fault cross-section and fault length (Figure 13.3).

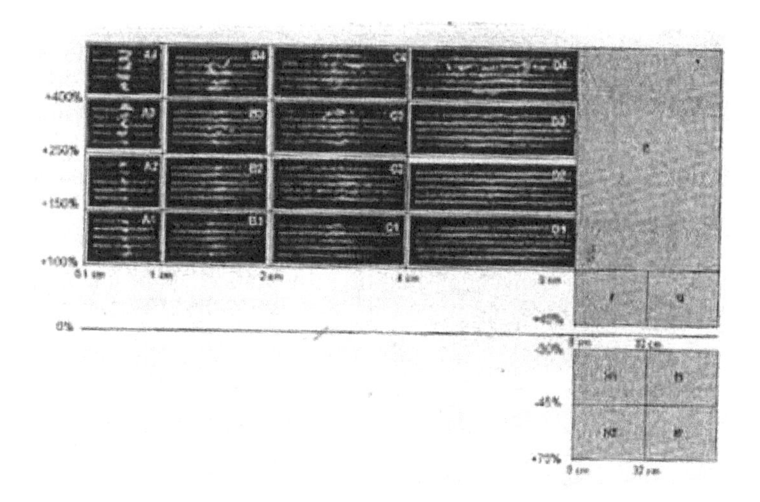

Figure 13.3 Classes of the Uster Quantum 2 clearer in winding

Based on the grades, the range of disturbing thick places in the yarn is divided in to 16 fault classes whereby they are arranged in to 4 cross-sectional sizes and 4 lengths.

With the grades system, both the yarn producer and yarn processor can assess the faults in qualitative and economic point of view. The grades of the CLASSIMAT system are shown in Figure 13.3. Accordingly, a clearing limit is set on the grades, which divides the yarn faults which are to be extracted from those faults which can be tolerated in the yarn. From this clearing limit, the setting values at the yarn clearing installation can be directly undertaken.

13.5 Determination of optimum yarn clearing conditions

An important pre-requisite for economic yarn clearing is the availability of values with respect to the frequency of particularly disturbing yarn faults. A too fine clearing of yarn is not required as it only affects the appearance but every cut in the auto coners is replaced by a knot or spice. A rational and accurate determination of the number of faults to be extracted is available by means of the newly designed CLASSIMAT Yarn Fault Classifying Installation. With this installation, faults are detected and counted with respect to 23 classes according to fault size and length. The results are normally based on a reference length of 100,000 m of yarn. If for example, a very high number of splices which would result in uneconomic winding then by means of a respective correction of the clearing limit, the number of faults to be extracted can be reduced quite considerably without noticeably affecting the quality of the yarn.

The following parameters can be monitored through an electronic yarn clearing installation at winding machines.

- Classification of seldom occurring yarn faults.
- Periodic defects in the yarn.
- Foreign matter classification. (if foreign fibre sensors are installed)
- Classification of yarn joints and its quality.
- Count variation detection.
- Winding defects.
- On-line yarn imperfections on the running yarn.
- Alarms due to bad quality or off standard ring frame packages.

13.6 Auto coner informator

Present day auto coners information system provides details on on-line running production as follows:

- Single drum or machine production at any instance.
- Machine and production efficiency.
- Cause wise production down times as follows:
    - Red light: Drums which are stopped due to operational or mechanical failure.
    - Yellow light: Drum stopped for package doffing after it has attained pre-selected length of yarn on cone.
    - Recycling or Repeaters cycles. Repeater cycling for splicing.
    - Splice failure: Shows the number of splice failures.
    - Bobbin change cycles: Shows the number of bobbin changes.
    - Rejected bobbins: Shows the number of rejected bobbins.
    - Clearer cuts per 100 km of yarn.
    - Clearer cuts per kg of yarn.
    - Clearer cuts per bobbin.
    - Tension breaks.
    - Yarn breaks – Clearer cuts + Tension breaks + stop cuts.
    - Total yarn joints.

The information mentioned above may be available in specified units or in percentage.

13.6.1 Advantages of online monitoring system

- Quick and instant report at any moment for production and quality performance.
- Monitoring of whole production quantity thus assuring 100% quality control.
- Sample size is too large, hence more reliable and accurate measurement.
- Quality stops ensure that no defective material is produced.
- Storage of data on large scale for retrieving the information at any time possible.
- Dependence on human and human errors is avoided.

13.6.2 Setting of the yarn clearing installation

The yarn clearing installation should be able to be programmed according to normal textile units. Twist in the yarn, type of material, colour and humidity can affect the performance of the clearers to varying degrees. If any one of these factors is too influential, then the function of the clearers is uncertain and the setting also becomes too complicated. In addition to the setting of winding speed, the control panel of electronic yarn clearer has the following four adjustments:

- Yarn count (Ne) is set as a basis for the sensitivity control and for the detection of spinner's doubles.
- Sensitivity (SE%) is the percentage level of yarn thickness beyond which the yarn will be cut for the removal of yarn faults.
- Reference length (RL) refers to the length over which the weight of the yarn will be determined.
- Material number (M) is set as per manufacturer's recommendation for different fibres and blends depending upon the moisture content.

13.7 Classifications of yarn clearers

In general, there are two classifications in yarn clearers. They are

- Capacitance type clearer
- Optical type clearer

Capacitance type clearers are manufactured by Zellweger Uster Technologies and optical type clearers are manufactured by Loepfe Bros. Swiss.

13.7.1 Capacitance type clearer

In the capacitive principle of measurement shown in the Figure 13.4, a magnitude is measured which is proportional to the mass of the fibre mixture located between the electrodes. All those places in a yarn which impart to the condenser a change in capacitance which exceeds a given permissible value will then be cut out. This involves a roundabout route. This means thick place = increased mass = increased capacitance value. Thin place = decreased mass = decreased capacitance value. The capacitive clearers on its nature, does not react to extraneous light, on the other hand, differences in the moisture content of the material cause variations in the result due to the variations in the relative humidity of the air. Loose fibres or moving fibres cause unjustified cuttings in both the types of clearers.

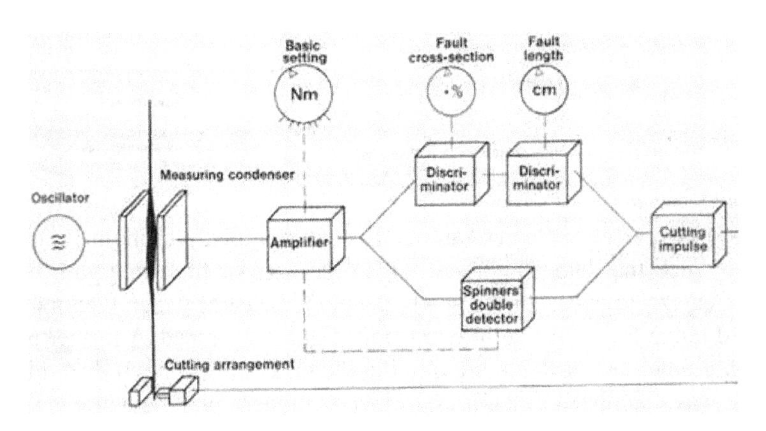

Figure 13.4 Capacitance type clearer

13.7.2 Photo electric clearer

Measuring principle of photo electric type clearer

In the photo electric type yarn clearer (Figure 13.5), the photo electric cell provides a signal which corresponds to the diameter of the yarn. With this, the signal changes according to the variation in the diameter of the yarn. It means thick place = increased yarn diameter = increased signal value and vice versa for thin places. This clearer is independent of changes in humidity.

The drawback of the photoelectric type clearer is

- A basis for the clearer sensitivity is not available which makes the setting difficult.
- A value is used which is derived from the yarn diameter and the properties of the material like twist, colour, brightness, etc.

Hence, the setting of the thickness and length of the faults to be extracted is carried out by means of optically determined sizes.

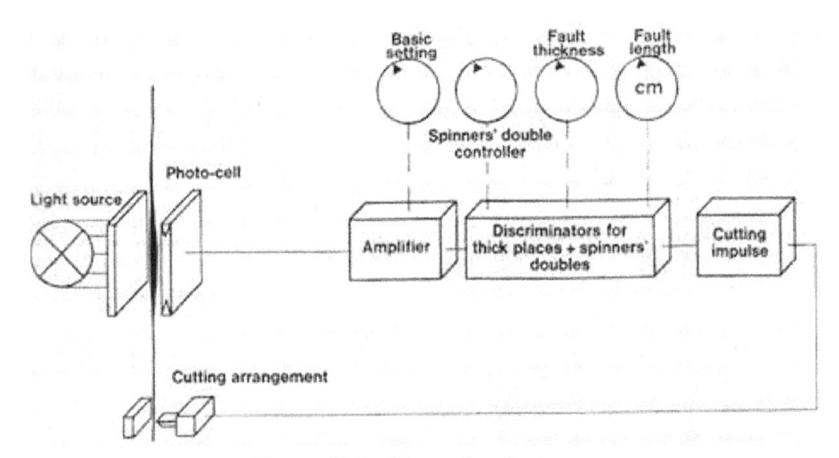

Figure 13.5 Photo electric clearer

The photo electric clearer can be affected by quickly changing extraneous light such that large faults are left in the yarn or fault free yarn is cut by the clearer. However, in the newly developed clearer which works in the infra-red range the influence of extraneous light is excluded. A further problem is presented by dust and dirt entering the measuring zone. If dust, fibres, paraffin wax, etc., become built up in layers, modern photo electric clearers control the strength of the light value within a certain range only fully automatically. Outside this range, it is quite serious which lead to incorrect yarn fault evaluation can result. Machine vibrations can bring about disturbances of the light filament which eventually lead to unjustified cuttings.

13.7.3 Disturbing factors in both methods of measurement

Besides the most important measured sizes of yarn cross-section (Capacitance principle) and yarn diameter (Optical Principle), various disturbing parameters affect both with capacitance and photo electric clearers. These in turn influence the basis and the thereby the evaluation of the yarn faults.

The aim of this comparison tests is to provide an insight to how strong the disturbing parameters like humidity, twist, colour, (matt, transparent) have an effect on the measuring principles.

As a basis of comparison of test results, yarn count is taken as the suitable measure. The signals of the capacitance clearer refer to the weight per unit length and therefore it is related directly to the yarn count.

(a) Twist

Yarn s produced for various end uses is reflected by its important properties like strength, elongation, bulkiness, etc. This is affected by changing the number of turns per unit length. For example, a soft twisted tricot yarn and hard twisted warp yarn is presented to the photo electric clearer, there is change in the measured signal. It means that degree of twit has certainly has direct influence on the diameter of a yarn. A softly twisted yarns will have a larger diameter than a strongly twisted warp yarn of same yarn count. In other words, twist always runs in to finer places in the yarn and in thick places there will be less number of turns per unit length.

(b) Colour/brightness

Natural colours as seen by the human eye show a variable photo-metric condition due to reflection and absorption. The same differences also occur in the raw materials between cotton (natural fibres) and variably treated synthetic fibres i.e., bright or dull transparency.

Consequently, these factors have an effect on the photo electric clearers which show a weakening of the light beams by the yarn. This has an influence on both visible and infra-red ranges because each textile material exhibits different degrees of light absorption which is a function of wave length. If in an extreme case, if white and black yarn is presented to the photo electric clearer, the measured signal can change in the relationship of 1:2.8. Furthermore, if bright and dull viscose yarns are presented to the clearer signal differences were obtained with 1:1.5.

Similarly, in the case of capacitive clearer also, small signal differences were noticed due to the chemical influence on the di-electric constant during dyeing.

(c) Humidity

Humidity is defined as the mass of water vapour present in dry air. This is an important property in the case of cellulosic fibres like cotton, viscose, and the protein fibres like casein, wool. It is their capacity to accept the moisture. The moisture content of the material changes according to the humidity of the surrounding atmosphere. For example, cotton has a moisture content of 6.6% with 50% R.H. 8.2% with 65% R.H. 10.2% with 80% R.H. In the capacitive clearers, the capacitors react to moisture content changes in the yarn. If the clearing installation with a setting according to 65% R.H. is presented with a relatively dry yarn (e.g., corresponding to 40% R.H.) or a very damp yarn (e.g., corresponding to 90% R.H.), the signals measured by the capacitive clearers change by maximum ±17% with respect to effective yarn count of Nm 40.

In the same sense, the effect of humidity has little influence on the measured signal. However, if one considers, the swelling of cotton yarn due to high humidity, dimensional differences can be brought in the yarn in proportion to the moisture content of the material. Such dimensional differences refer to length, width, cross-sectional area and volume.

(d) Influence of operating conditions on the performance of capacitive and photo electric yarn clearers. There is a definite influence on the working of both the clearer installations on operating conditions. In other words, disturbances like light, dust, vibrations, etc., lead to incorrect yarn fault determinations or cause unjustified clearer cuts in auto winders. In either case, it is dependent upon the working principle.

(e) Stability

Stability is the result of the electronic components and circuits used and the conception of the complete installation. With both photo electric and capacitive clearing installations, equivalent components are used in the power supply and sensitivity setting instruments as well as in the analysing circuits so that stability over many years can be expected. The light source used in the infra-red clearer uses a light diode (transistor) from which a much longer time is expected. The capacitive type of clearers consists of elements (oscillator, condenser) in measuring unit have stable properties for long operating conditions. Hence calibration is not necessary.

13.8 Economical and technical problem on yarn clearing

The third generation Loepfe yarn clearer FR-3 (earlier model) is suitable for yarns within the range Nm2 to Nm200 and is characterised by its adaptability as regards selection of faults, a feature which has a decisive effect on economic performance of yarn clearing. This result in lowest possible incidence of splice, even when clearing demands are extremely high. The principle of operation used is based on the following:

- The length and thickness of a yarn fault are detected independently.

- A fault is eliminated only if its length and thickness are in a certain ratio, the value of which can be selected.

This provides sharp discrimination which can be established as desired by the adjustment of the clearer between admissible faults and faults which has to be removed from the yarn. This is illustrated diagrammatically in Figure 13.6.

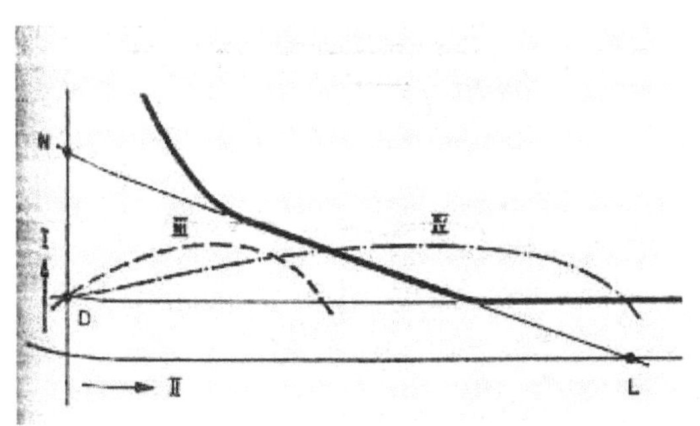

Figure 13.6 Adjustment of Loepfe yarn clearer for elimination of inadmissible yarn faults
I = thickness, II = length, III = tolerated faults, IV = disturbing yarn faults

From the Figure 13.6, every point on the thick continuous boundary line denotes the permitted thickness for a fault of a given length. It should be mentioned here that, unlike other methods hitherto used, the length adjusted on the clearer is in fact always equal to the length which the eye perceives as the length of the fault.

13.9 Method of cutting the fault in the yarn clearer

Figure 13.7 illustrates the type of fault discrimination.

In one scanned length, a fault is cut out only when the weight of the whole length of yarn of length (L) exceeds the set limit. In this instance the set limit is +200%. If the fault is longer than (L), it is cut out whenever its weight exceeds the prescribed limit. It is shorter than (L), it is cut out only when the *weight exceeds a corresponding value.*

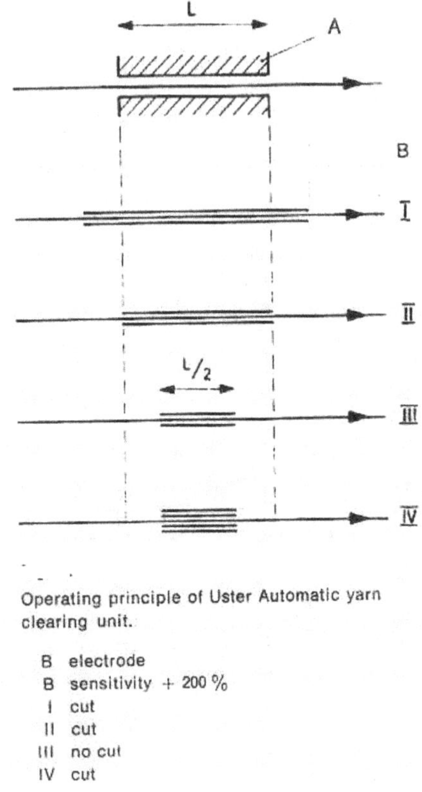

Figure 13.7 Operating principle of Uster automatic yarn clearing unit

The drawback of this capacitance principle is

- Lightly twisted or roughened places in the yarn are not detected because they do not result in increased mass.
- Furthermore, the change in capacitance produced by the yarn will be influenced to some extent by the dielectric constant of the material, its moisture content and finish. Certain models are therefore fitted with suitable compensating devices.

13.10 Explanation of the data output of the classifying system

The scatter plot

The scatter plot is a very important feature of USTER CLASSIMAT QUANTUM which helps the user in analysing the exact place of each event in the classification matrix and also indicates the yarn faults of both the standard classes and the extended classes as points in the classification matrix. The exact length and cross-section increase or decrease of the individual yarn faults can easily be determined with the horizontal and vertical scales.

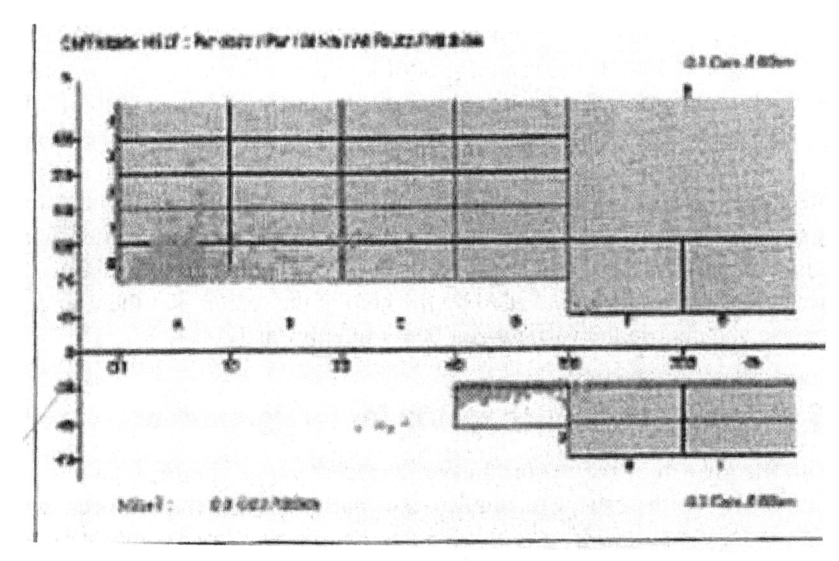

Figure 13.8(A) 100% carded cotton yarn of Ne 30 (low number of faults)

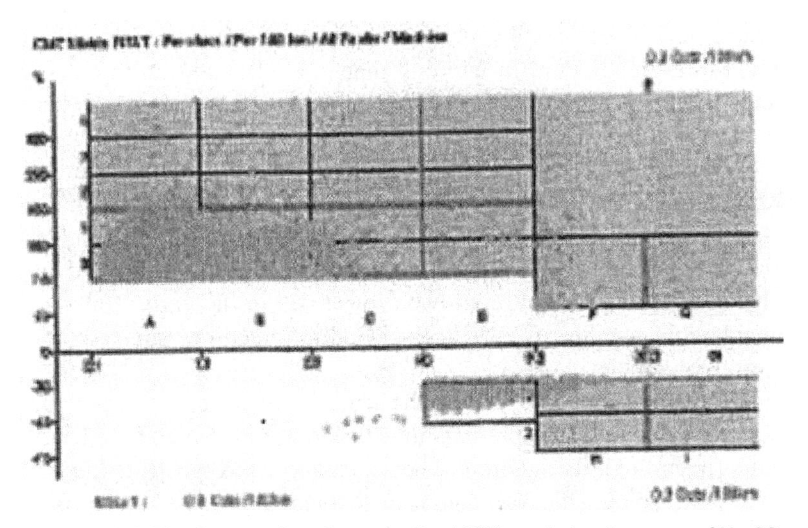

Figure 13.8(B) Scatter plot of yarn faults, 100% carded cotton yarn of Ne 30 (high number of faults)

The usage of this scatter plot is shown by an example. Two yarns of two mills of count (Ne 30, 100% carded cotton). A 100 km of these two yarns were measured with USTER CLASSIMAT QUANTUM and the neps, short thick places, long thick and long thin places (NSLT) were counted and classified. In Figure 13.8 (A) the scatter plot of a yarn produced in a mill with a good quality management having less number of yarn faults. These faults are shown according to the CLASSIMAT classes (NSLT).

In Figure 13.8 (B) the scatter plot of a poor quality tarn is shown having high number of yarn faults. This can be illustrated by an example. In the B1 class of the first yarn, it had only nine short thick places, whereas in contrast, in Figure 4, the yarn with 488 short thick places in the same class. The faults in the B1 class can be the result of wrong raw material, fibre ruptures in preparatory processes and in spinning, badly maintained wire points in carding, etc. From these two scatter plots, the difference between these two yarns is enormous, and by the help of the scatter plot the user can analyse and choose the yarn having better quality easily.

13.11 Classification matrix for foreign fibres

With the growth of the export market in recent years, spinners have to be competitive with respect to quality especially to meet the requirements of the global market. In recent years, contamination in cotton has posed many problems to the spinners. Contaminations with foreign fibres like jute, plastics, coloured fibres, polypropylene (white/ colour) have made the spinners to think about the removal of the same. Foreign fibre sensors have been developed by yarn clearer manufactures which solve this problem to a great extent in winding machines. In addition to the standard classification of yarn faults, the system also offers the user to foreign fibres and vegetable matters in the yarn and classifies these faults in 27 foreign fibre classes. With vegetable filters, it is possible to differentiate between organic and synthetic foreign fibres. Based on the fact that vegetables mostly do not have a disturbing effect on the appearance of the fabrics. It is because they can be bleached or can absorb the same dyestuff and hence these particles are allowed to remain in the yarn. It also saves a considerable number of cuts in the winding machines and reduces the formation of splices.

Figure 13.9 (A) and (B) shows the structure of the classification matrix for foreign fibres, which represents the appearance in (%) and the length of the fault (in cm) of a foreign fibre. The appearance corresponds to the visibility of a fault. No classification data is available in A1 class, because there is no significant fault in this class.

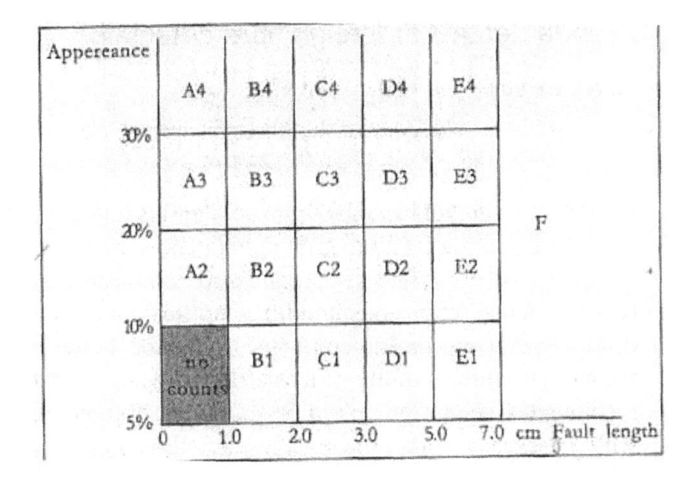

Figure 13.9(A) Structure of the classification matrix for foreign fibres (coarse setting)

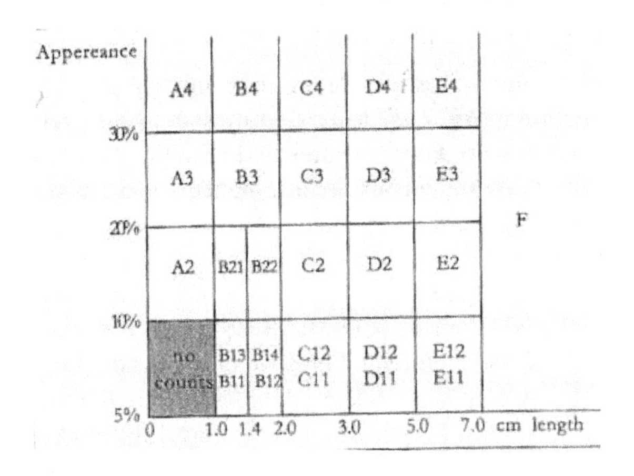

Figure 13.9(B) Structure of the classification matrix for foreign fibres (fine setting)

Another system of detecting the foreign fibres in the yarn in auto coner is the opto electronic yarn clearer system synonymous for dependable detection and classification of foreign fibres in the yarn. Foreign fibres are detected by the SIRO principle developed by Loepfe bros, Switzerland. In this system, the classification of foreign fibres is based on evaluation of contrast differences. The LOEPFE foreign fibre standard presents a classification field with various examples of foreign fibres for each class. The foreign fibres are assigned to a classification field so that the class related differences (brightness, length and appearance) can be easily recognised within a yarn.

13.11.1 Fundamentals of foreign fibre detection

The foreign fibres are classified in the following way:

- By the *brightness difference* of the fibres from the base
  - Colour of the cotton

    Cotton fibres do not have the same colour. It is influenced by several factors during the fibre growth, influenced by various factors like rain, frost, attack by insects and also during storage of cotton (extreme temperature and humidity conditions) in the cotton goes down. Depending on the colour of the cotton the clearer takes the basic brightness of the yarn as criterion. The contrast between a coloured foreign fibre and a bright cotton fibre is greater than with a darker one. This means a foreign fibre is classified differently when processing cottons of different origins.
  - Yarn count

    A dark cotton colour spun in to a coarse yarn contrasts less in brightness with a foreign fibre.
  - By the different fibre length

    The foreign fibres are not all tied similarly in the yarn structure. They are more likely to be tied in to a coarser yarn than in to a fine spun yarn. Foreign fibres tied in to the yarn so that they constitute "interrupted fibres" and hence can be classified by their actual length only with recourse to classifying algorithms.

Undetected foreign fibres

- All the foreign fibres do not differ from the basic brightness of the cotton. There are some foreign fibres which show no difference in brightness. This is particularly clear when processing the yarn in to textile surface structures. This problem may be made clear in a bleached single jersey knitted fabric inspected visually before and after finishing.
- Grey cloth inspection.
- Finished goods inspection.

In the grey state of the fabric, it is very difficult to recognise the objectionable foreign fibres. After the bleaching process in wet processing of the knitted fabric; it is possible to recognise the objectionable foreign fibres. However, only very fine, light coloured foreign fibres spun in to the yarn are revealed in the bleached fabric. In auto coners, foreign fibres of this delicate nature cannot be detected by the clearer, because their contrast with the raw material is non-existent or insufficient.

Foreign matter creates three problems.

- More ends down in spinning and disturbs the spinning process.
- Result in second quality in knits or woven goods in quality claims.
- Reputation of the yarn or fabric producer is damaged.

13.11.2 Measuring principle of yarn Master 900

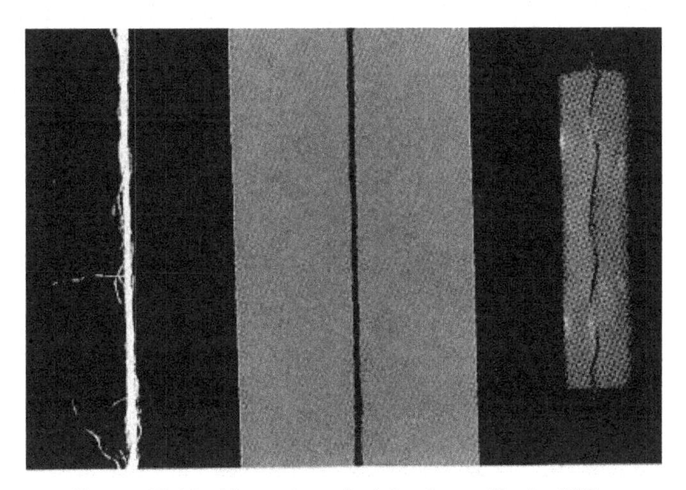

Figure 13.10 Measuring principle of yarn Master 900

Principle

The yarn is irradiated from all sides with infra red light and the obstruction due to the yarn diameter is evaluated by a photo detector. To detect foreign fibres the yarn diameter signal is made to disappear by a suitable light influence, enabling any differences in contrast to be recognised. In order to detect diameter or cross-section related yarn faults during winding, the yarn is imaged with the highest possible accuracy in the measuring field of a sensor.
The figure shows the foreign fibre imaging (from left to right) (Figure 13.10).

- Human eye
- Infra red sensor and
- Yarn Master sensor

13.11.3 Classification of foreign fibres in Loepfe Yarn Master

The LOEPFE foreign fibre standard is based on the coordinate table of the class clearing for foreign fibres (Figure 13.11).

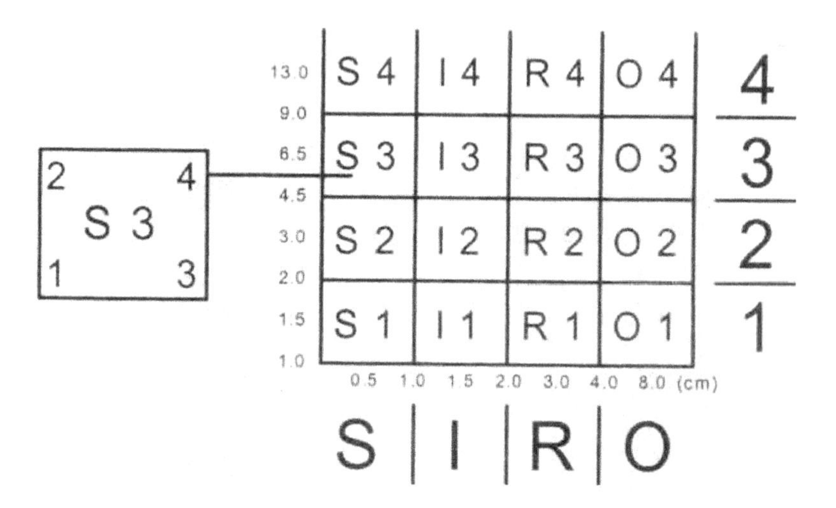

Figure 13.11 Classification matrix of foreign fibres

In class (S3) additional fine classification is shown separately.

Here the foreign fibres are classified according to the following example:

The horizontal (x-axis) shows the length classification in (cm) and the vertical axis (y-axis) shows the classification based on diameter.

- Sub-division of the length classes in the horizontal S-I-R-O.
- Sub-division of the darkness grades in the vertical (1-4).
- Additional sub-division of each class in to 4 sub classes (i.e. fine classification).

In the Figure 13.11 no classification data is given in class S1. This is because of the high sensitivity of the class. Seed trash from the cotton plant as well as neps is more conspicuous in class S1. Hence it will not give any representative conclusion regarding the amount of foreign fibres in the yarn.

The foreign fibre matrix is divided in to 21 (Figure 13.9 (A) or 27 (Figure 13.9 (B)) foreign fibre classes depending on a coarse or fine setting.

13.12 Scatter plot for foreign fibres

With the scatter plot, the distribution of foreign fibres in the yarn can be seen at a glance. In addition to that, vegetables are separated from actual foreign fibres with the help of vegetable filter and displayed separately.

In the Figure 13.12 (A) and (B), we can see the foreign fibre scatter plot of the two yarns.

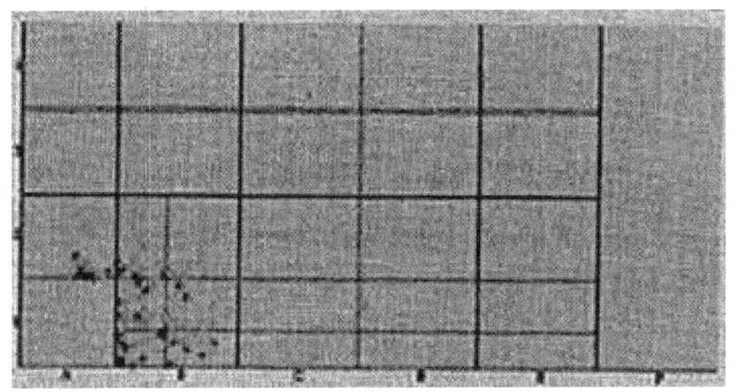

Figure 13.12(A) Low number of foreign fibres

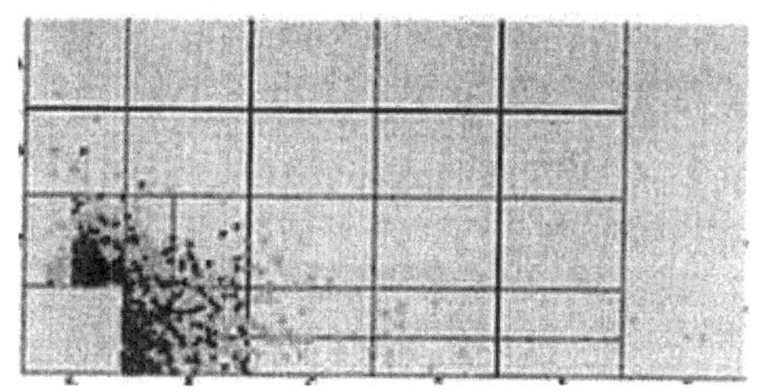

Figure 13.12(B) Higher count of foreign fibres

The scatter plot (Figure 13.12 (A)) shows the yarns with a low number of foreign fibres and vegetable matters. In Figure 13.12 (B) the scatter plot shows a yarn with higher number of foreign fibres and vegetables. With this scatter plot, the user can analyse and choose the yarn having a lower number of foreign fibres. The difference between these two yarns in terms of foreign fibres presence is enormous and will have considerable consequence on the appearance of the fabrics.

13.13 Sensors for monitoring all the bobbins in winding machines

With the introduction of automatic winding machines in the spinning mills, yarn clearers were developed which monitor every bobbin. At the starting of the introduction of the electronic yarn clearers in winding machines, it was designed for the elimination of disturbing thick and thin places in the yarn. However, in the past 40 years, the yarn clearer has become a multi-functional sensor with capabilities to minimise the remaining quality problems on the cone.

(a) First task of a yarn clearer/elimination of disturbing thick and thin places

Clearing of yarn faults is always a compromise between the machine efficiency and quality. On one hand, a spinner wants to eliminate as many as thick and thin places to avoid quality complaints from the fabric buyers. On the other hand, every replacement of a disturbing yarn fault by a splice reduces the efficiency of the machine and increases the number of splices per bobbin. The yarn quality can be demonstrated in the Figures 13.13 (A) and (B).

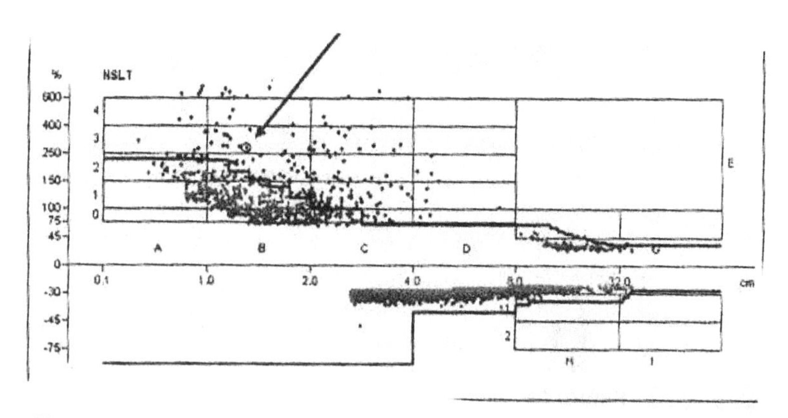

Figure 13.13(A) Disturbing and tolerated thick and thin places in the yarn

The dark lines represent the yarn body and all the yarn faults above the "clearing curve" have to be eliminated and replaced by a splice. If the quality specialist does not know the relationship between the faults which have to be cleared and the faults which need not be cleared, then the number of splices will be more and the efficiency of the machine will get reduced. Figure 13.13 (B) shows the selection of faults simulated on the screen.

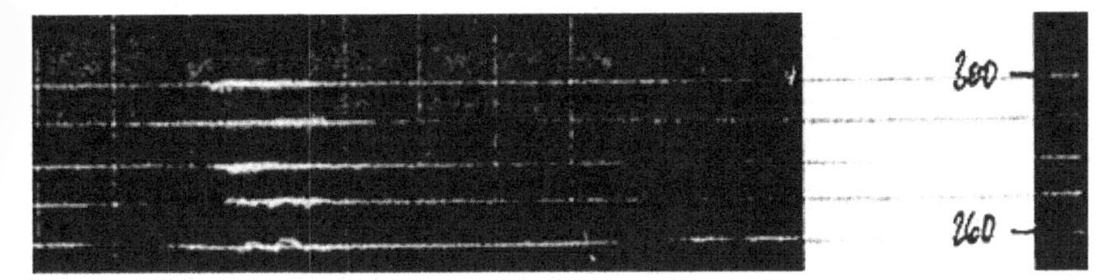

Figure 13.13(B) Size of the disturbing thick places

(b) Second task of yarn clearers

The presence of contamination in the cotton is inevitable and cause a big threat for the spinners. Although the yarn quality is good in terms of unevenness and irregularities, if the yarn after knitting contains many contaminations like foreign fibres, the chances of rejection will be high or accepted with some compensation. Therefore, many efforts have been taken in many mills to eliminate the foreign fibres.

Various trials have shown that the manual or automatic elimination of foreign material can reduce the number of large foreign particles, nevertheless cannot reduce the amount of foreign fibres. Therefore, the best method to eliminate the foreign fibres is to keep away the large particles from the carding machine, but to eliminate the frequent foreign fibres by a clearer.

The frequency and the size of the foreign particles and the domain of foreign matter recognition and elimination systems is shown in the Figure 13.13 (C).

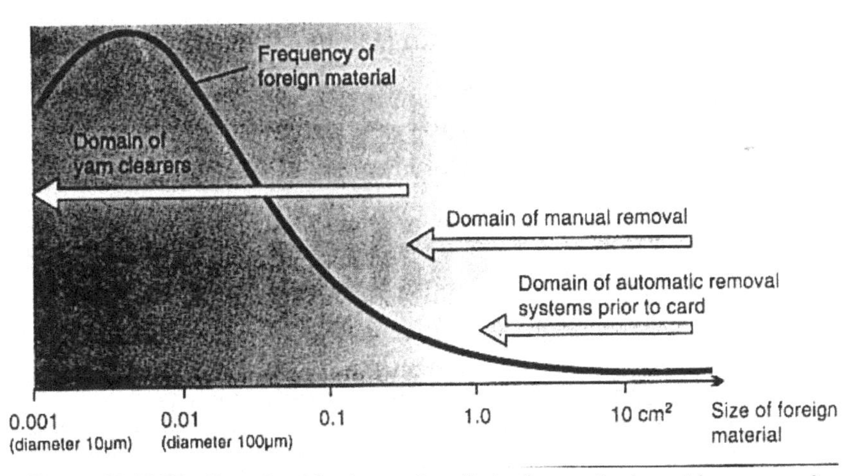

Figure 13.13(C) Domain of foreign matter elimination methods and frequency of foreign material

Since, the presence of small foreign matter particles such as single plastic fibres, human or animal hair, fragments of bird feathers, etc., have mostly occurs at high frequency, the cut rate of the electronic clearers is hardly affected when the automatic recognition and elimination systems prior to the card are switched-off. In many cotton growing areas, the packing material used is polypropylene. As a result, many bales are contaminated with the polypropylene fibres which get mixed with the cotton fibres during the opening of cotton bales of mixing. Polypropylene fibres, however, do not absorb dyestuff and remain "white" after dyeing. Such fibres cannot be sensed by optical sensors in the raw cotton yarns because there is no colour difference between cotton and polypropylene. Hence, a particular type of sensor is designed to eliminate this problem.

(c) Third task of a yarn clearer/ separation of off-quality bobbins

In a spinning mill, bobbins are produced from the ring frames and the yarn faults are removed in the winding machine according to the pre-set yarn clearing curve. If the bobbins produced at the ring frame are beyond the pre-set standard with respect to yarn evenness, imperfections, hairiness, etc., such quality problems cannot be corrected on the winding machine because the fault may be available throughout the bobbin. Hence, these outliers or off-quality bobbins have to be separated and ejected on the winding machine. Figure 13.13 (D) shows the monitoring of hairiness on the winding machines.

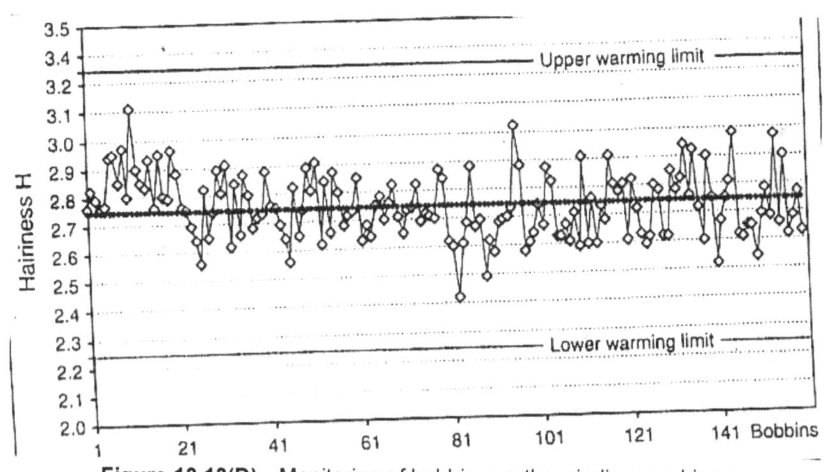

Figure 13.13(D) Monitoring of bobbins on the winding machines

(d) Fourth task of the yarn clearers (Feed back control system to correct defective machine parts)

The number of production positions in the spinning mills is very high. Therefore, a well recognised spinning mill should have a repair crew to rectify the yarn faults which may arise due to many reasons including defective machinery parts. The repair crew, however, requires information from the laboratory analysis to carry out their works in a systematic way. It means the bobbins from the spinning machines are identified with a number to identify the production positions where the ejected bobbins come from.

The ejected bobbins from the winding machines are brought back to the testing lab where the quality problems are evaluated. The findings are listed in a separate sheet and the information is given to the repair crew. The objective is to bring back the "outliers" to the normal distribution as shown in the Figure 13.13 (E).

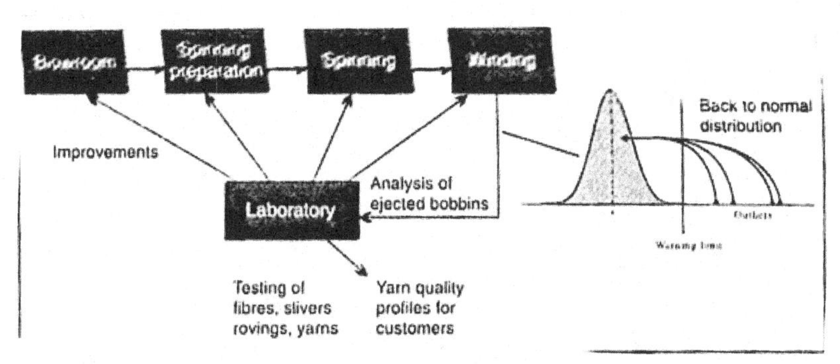

Figure 13.13(E) Outliers in the bobbins tested in the laboratory

From the lab test report on the ejected bobbins which are called as "outliers", the repair crew has to undertake the repair works in the corresponding spinning positions in the machine, tested again to confirm to the standards. The systematic quality management is shown in the Figure 13.13 (F).

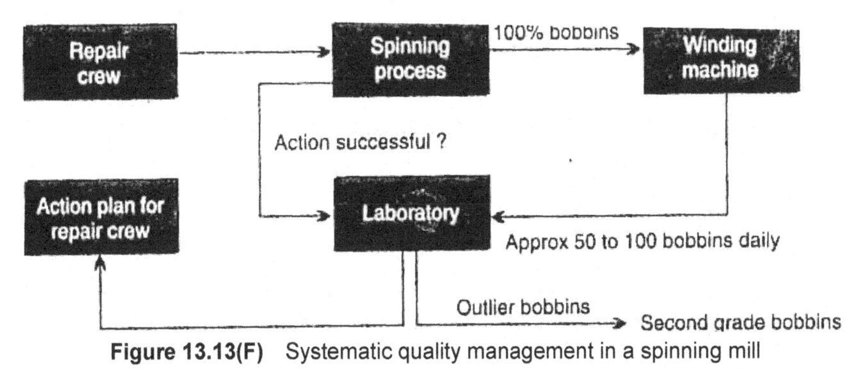

Figure 13.13(F) Systematic quality management in a spinning mill

Bobbins which are identified as having tolerated quality characteristics will go back to the first quality yarn batch and the outlier bobbins will be used as seconds.

13.14 Bench marking of Classimat classification

Uster Technologies introduced Uster Statistics in the year 1978. Uster Technologies have carried out many tests with yarns coming all over the world and the results were published based on the level in which the mill stands. Uster Statistics have published statistics for foreign fibres also in the year 2007 and were introduced in the market (Figure 13.14).

In Figure 13.14 the symbol on the right hand side explains the significance of statistical values and indicates the range from 5% to 95%. For example, a foreign fibre content on the 95% level means that 95% of the spinning mills worldwide are below this figure. A value on 5% level, however, indicates that only 5% of all the spinning mills worldwide could achieve this quality level.

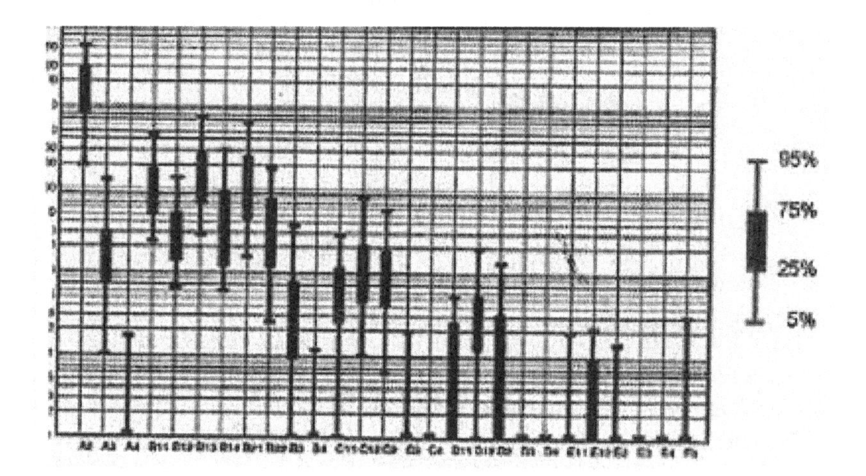

Figure 13.14 Uster statistics on foreign fibres (2007)

13.14.1 Advantages of yarn clearer installations in auto winders

The following are the benefits of the advanced version of yarn clearers in auto winder section.

- Q data gives information on quality parameters like Unevenness, Imperfections and Classification of yarn faults.
- Point distribution of faults.
- Pearl chain clearing channel to detect and remove places with frequent yarn faults.
- Classification of foreign fibre faults like FD and FL channel. FD channel clears darker faults in lighter yarns and vice-versa for FL channels.

- MF channel to clear yarn having multiple foreign fibre faults.

13.14.2 Computer aided yarn clearing (CAY)

Zellweger Uster has also developed an intelligent software USTER CAY that can automatically analyses the yarn fault distribution and optimise the clearing curve according to the spinner's requirements.

- CAY precisely measures each individual yarn faults and display the frequency and distribution of the defects as a basis for effective optimisation.
- Quantum clearer helps to fine tune the clearer limits to achieve the highest clearing efficiency with the least number of cuts.
- Quantum detects the slightest foreign fibres beyond the former sensitivity limits to meet the highest quality requirements.
- The setting of the basic clearing curve and foreign fibre curve are user-friendly and with the availability of help points, the user can optimise the clearing curve according to his requirements. Hence, the user can achieve the required clearing with the least possible cuts which results in maximum production.

13.14.3 Online monitoring of the automatic cone winders

With the introduction of the advance electronic yarn clearing installations in auto cone winding positions it is possible to detect:

- Rogue winding positions by means of alarm systems.
- Possibility of detecting "count mix-up" in cone winding department that is being wound in to a batch.
- Simple method of setting the yarn count in all winding positions and of holding the yarn count constant.

13.15 USTER CONE DATA production data information system

The management of textile industry requires information about the know-how, flexibility and control. It requires the ability to co-ordinate the continuously changing effects of the material, the machines, the ambient conditions and personnel and this at any time with little prior notice. In the age of computer technology, it becomes necessary that the information should be obtained by an automatic data collection installation. For this purpose, the management of the winding department requires information with respect to the winding process. In addition to the integrated process of yarn clearing, information with regard to yarn faults is of primary importance. One such information system is the USTER CONE data system.

The USTER CONE data system was first introduced in the year ITMA 1979. It offers a real and economic alternative according to the following basic conception.

- Specially-arranged de-centralised data system for the winding department.
- Consequent standardisation and hardware and software.
- Easy to operate and easy to analyze the reports.
- Modular principle and extension possibilities in stages.

System hardware

The signal transducer for all required signals is directly undertaken by the UAM control unit. This has the advantage that no extra cabling is required. The general arrangement of the data collection installations with rows of winding positions is shown in Figure 13.15.

In the Figure 13.15, the signals obtained at the control unit by the section to section cabling are collected together in a connection box. The connection boxes are arranged one per machine. The connection boxes lead the signals from individual machines to the central unit of the data collection installation by means of a single cable.

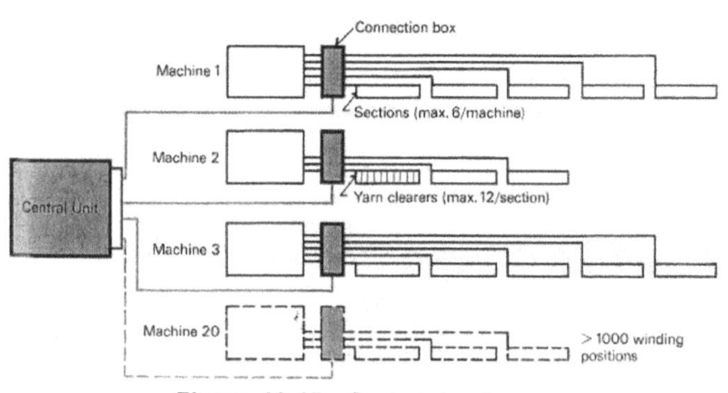

Figure 13.15 System hardware

Reports from USTER CONE DATA

The interpretation of USTER CONE DATA reports is extremely simple to analyse as only two information groupings contain all the
more important parameters of winding:

- The system report with specific machine data.
- The production report with yarn specific data.

The system report contains information with respect to all the connected machine sections of a winding installation. The production report is collected together the production data with respect to all winding positions. Furthermore, for purpose – oriented evaluation, three types of report are available. These serve as a concentration of the data according to the following criteria:

- Machine
- Article
- Group.

These reports contain both system and production reports. The printing of the reports can be taken for a specific value or according to the choice of all values in increasing succession. Besides, there is a possibility available of only printing out the mean value characteristic.

13.16 Knotters and Splicers in auto winders

Winding is the process of winding the yarn from the ring frame cops which weigh from 45 to 175 g in to a large conical shaped or cylindrical packages in winding machines. In this process, the yarn faults are removed by electronic yarn clearers depending upon the settings in the yarn clearing installation. Whenever a yarn fault is detected by the yarn clearer, it is cut and the broken ends of the yarn are joined by knitters or splicers to continue the winding operation.

End breakages are the main reasons for the loss of productivity especially in spinning and in weaving processes. It is undesirable here as it produces more waste. However, it was desirable in winding process to some extent as the end breakage occurs due to objectionable yarn faults. In winding process, it removes the weak places in the yarn and improves the efficiency of the subsequent processes. The most important of these processes is the weaving where "value addition" is made. It means a higher end breakage rate at winding indicates the poor quality of the yarn and the preparatory spinning processes.

The first automatic winders came in to the industry in 1930s. Abott and Barber – Coloman are the machines introduced in 1935 with a stationary spindle and a travelling knotter. The improved CC model was introduced in 1959. The advent of automatic winders and automatic knotters in the fifties brought the winding process in to limelight. The development of the knotter however posed certain inherent draw backs.

- The time taken by the knotter which moves from one spindle to another spindle is more for mending the end break which reduces the efficiency of the winding machine.
- Since only one knotter carriage is used for one winding machine, if there are more end breaks in the machine, many spindles remain idle and wait until the knotter comes and mend the breaks and then restarting the winding process.
- The knots produced in the winding machine is also an objectionable fault in the yarn as it hinders the free movement of the yarns in the healds of the weaving and also cause damage to the knitting needles in the knitting machine if the knot size is more than the required.
- Various types of knots like Fisherman's knot, Weaver's knot were considerably thick in the diameter which introduces more friction, breakages and affects the fabric appearance.

This led to the development of splicers in the auto winding machines. In the modern auto winders, each spindle is provided with individual splicers instead of having one or two knotter carriages in the previous version. This improved the efficiency of the machine and the splicing quality is good. In addition, the splices are also checked by the electronic yarn clearer before winding the yarn on the package. The various methods of splicing are:

- Wrapping
- Glueing
- Welding
- Electrostatic splicing
- Mechanical splicing
- Pneumatic splicing

Among the various methods mentioned above, the first four methods were not in practical applications due to some disadvantages. Mechanical splicers do splicing mechanically with the disk systems and it is utilised in Savio Autoconer machines.

Mechanical splicer – Twinsplicer

The Twinsplicer is a yarn splicer that operates by a purely mechanical principle. The untwisting – re-twisting action on the yarn is carried out between two self-compensating interfaced disks. Three adjustments must be set on the Twinsplicer unit, untwisting – re-twisting – drafting. The presence of wax on the yarn, does not affect the characteristics of the splice. The unit is equipped with a dust protection cover; the cover is opened during splicing only.

In brief, the splice is carried out in the following way. After the two threads have been introduced between the two discs, the threads are untwisted and drafted, the threads are brought together to create the centre and the excess yarn on the ends is removed, the yarn ends and the spliced yarn are then joined together and the yarn is re-twisted, the discs are then opened and the yarn gets spliced.

Pneumatic splicers

Pneumatic splicers are extensively used in all the modern winding machines owing to their advantages in splicing. The pneumatic splicing system used in auto winders is shown in the Figure 13.16.

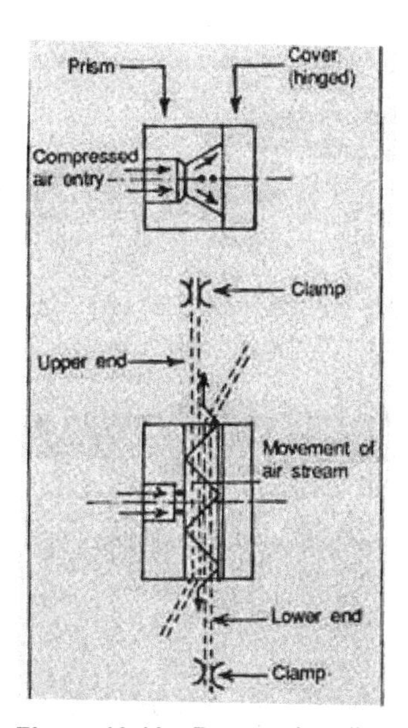

Figure 13.16 Pneumatic splicer

A splicing mechanism is shown in the Figure 13.17. Two ends of the yarn are kept parallel and face opposite directions. To condition the ends, the yarns are gripped and fibres are sucked from the exposed ends to taper them. Splicing is carried out after the two conditioned yarn ends are laid inside the splicing chamber as shown in the figure. So they are parallel, facing opposite directions and appropriately spaced without the tips of the conditioned ends protruding. One way to do this is to withdraw loops of yarn. The splicing chamber of an air splicer is sometimes made in two parts that open to allow easy insertion of the yarn ends and then permit closure for the splicing phase of the operation. The two ends are spliced together by a rapidly rotating body of turbulent air inside the splicing chamber. The turbulence is induced by air that enters the cylindrical chamber tangentially. The air blast first intermingles the fibres and then causes the newly made joint to rotate to produce false twist. The yarn is then removed from the splicer and winding is recommenced. The diameter of the splice is 20%–30% thicker than the parent yarn diameter and the strength of the spliced joint is around 80%–90% of the parent yarn strength.

(a) Two broken yarn ends are fed to the splicing chamber

Figure 13.17(A) Steps in air splicing

Figure 13.17(B) yarn ends are opened by opening nozzle in the splicing chamber

Figure 13.17(C) Splicing of two yarns by twisting nozzle in the splicing chamber

The splice quality is assessed by

$$\text{Retained splice strength (\%)} = \frac{\text{Strength in spliced yarn}}{\text{Strength in original yarn}} \cdot 100$$

$$\text{Splice breaking ratio} = \frac{\text{No of breaks in splice zone } (\pm 1 \text{ cm})}{\text{Total number of tests}}$$

Aquasplicer

Special feature of Aquasplicer is that air and water are used to splice a single and plied cotton yarns and blends by applying the appropriate mingling chamber. Mesdan is offering such types of splicers for almost all count range and all types of yarns.

13.17 Tensor constant density winding process by Savio

The tensor tension sensors are incorporated on Savio automatic winder for control of yarn tension This device continuously detects actual winding tension and is positioned immediately before the drum. The tensor takes the signal from head computer and varies the pressure on the yarn as required through the yarn tensioner. The operating range of the yarn tensioner can be set on the computer. When the yarn tensioner reaches the minimum value of this range it can, if so required, activate the preset speed reduction curve. The tensor has no moving parts. It also acts as an anti-wrap device.

Autotense Yarn Tension Control by Schlafhorst

Autotense is yarn tension control system incorporated on Schlafhorst automatic winder. The yarn tension sensor continuously measures the yarn tension. The measured values are transmitted in a closed control loop to the tensioner, the pressure of which is increased or reduced as required. Thus yarn tension is maintained at constant level throughout the complete winding process. This regulation system not only prevents an increase in yarn tension towards the end of the bobbin but also compensates for the lower yarn tension during acceleration by increasing the applied pressure thus helps in uniformity of the package build from one package to the next.

Schlafhorst Propack® system

This is a cradle anti-patterning system incorporated by Schlafhorst in Autoconer 338 winding machine (Figure 13.18). As already mentioned patterning occurs when the number of coils laid on the package per double traverse is a whole number.

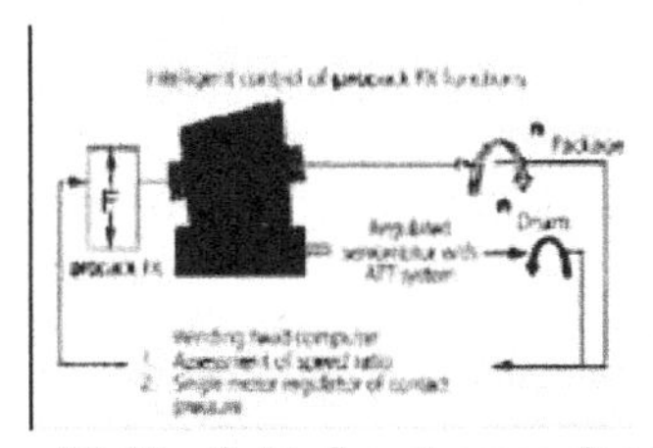

Figure 13.18 Schlafhorst propack system

In other words, pattern zone is reached when the ratio between drum speed and package speed reaches a critical value. Winding unit computer of the propack® system constantly determines the ratios between the drum and the package rotational speeds (Figure 13.18). As soon as the critical speed ratio which produces patterning is about to reach, the propack® system reduces the pressure on the cradle by a pre-determined amount. Thus, the package runs at slower speed below the critical patterning speed till the package diameter is adjusted to a value above the pattern zone. Then the propack® cradle anti-patterning system is turned off.

13.17.1 Semi-conductor device (TRIAC)

Muratec Mach Coner Automatic winder uses Semi-conductor device (TRIAC) for pattern breaking. The device is illustrated in Figure 13.19. Anti-patterning is achieved by means of DIAL setting for ON and OFF durations of the drum motor. Thus the speed change cycle is adjusted. As the drum and therefore, the package is constantly being accelerated and decelerated, the ratios of their speeds never reach critical value to produce patterning.

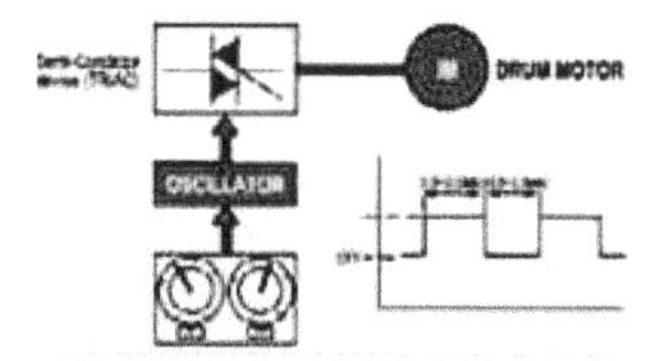

Figure 13.19 Semi-conductor device (TRIAC)

13.18 Wax disc sensors in winding

In general for the yarns meant for knitting end use, the yarns are waxed in auto winders to reduce the hairiness and coefficient of friction during the knitting process. If the coefficient of friction of the yarns is more, it causes more needle breaks in knitting thus affecting the productivity and quality of the fabrics. During the use of wax, in many occasions, wax run-out will not be noticed by the operator and sometimes the yarns without waxing will be wound on the package. This presents difficulties in the knitting even though the other quality parameters are within the standards. Hence, a new system developed by SAVIO in the auto winders is the monitoring of the wax run out and the information is given to the operator by a sensor.

Wax consumption is monitored by a sensor, and the system signals the operator when it's time to load a new disc. The disc is mounted 90° to the traditional direction and the traverse motion of yarn allows it to rub against the wax disc. The amount of wax required on the yarn can be controlled by the disc's rotational speed, direction of rotation and angle of contact. The sensor is a magnet that detects the movement of a spring-loaded element named the wax adopter. As the wax is consumed, the adopter, which is pressing on the wax disc, moves backwards when the thickness of the wax disc reaches a small, critical thickness and a signal is shown to the operator to replace the wax disc. The winder is equipped with several motors to independently drive the package, wax disc, upper splicer arm and lower splicer arm.

14

Sensors in rotor spinning

14.1 Introduction

Rotor spinning is the process of manufacturing yarns from wastes like cylinder fly, sweepings, flat wastes, sliver and pneumafil wastes to produce coarser yarn counts within Ne30. It is also utilised to produce yarns from 100% cotton, blends with synthetic fibres like acrylic, polyester and with regenerated fibres like viscose, modal, etc. However, it is limited to only coarser counts due to practical and technological limitations. The advancements in the spinning technology have taken place from blow room to winding. Most of the machineries are having electronic controls and sensors in the respective process for better control over the process and improved quality of the product with less waste. Similarly, the rotor spinning technology have also undergone developments in the areas of rotor design, feed tube, package build-up and higher speeds.

14.2 Sensors for monitoring the running yarn in rotor spinning

Figure 14.1 (A) and (B) shows the sensor for monitoring of the running yarn in rotor spinning position. Each spinning position is equipped with an optical yarn switch (3) which is located above the delivery shaft. This monitors the running yarn. The yarn switch (3) works with the principle of a light barrier. A light barrier (11) and a light receiver (12) are positioned in such a manner that the running yarn creates a signal of (24VDC).

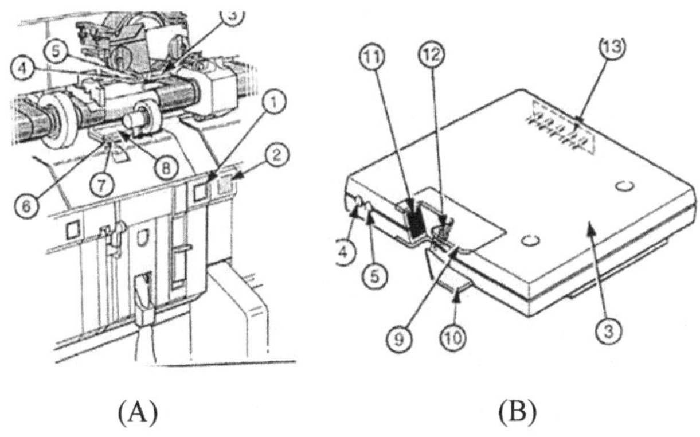

Figure 14.1(A) and (B) Monitoring of the running yarn in the rotor

As soon as the yarn is missing (VDC), the signal is sent to the feed roller so that the feed clutch is stopped. There is a plug through which running yarn signal is down loaded, the operating power of 24VDC is supplied and the signal for LED'S (4) and (5) is supplied. The ceramic guides, (top – 9 and 10) are designed in such a way that the running yarn and oscillating yarn is lead very close to the light emitter (11) and light receiver (12) but without touching them. The electronic of the yarn switch (3) is designed so that the depositing parts such as fibre fly or dust particles cannot create any false running yarn signal. Furthermore, an interference in outside light source is also not possible.

14.3 Auto-piecing process in rotor spinning

In modern rotor spinning machine, auto piecing of yarn is carried out with robots. Robots normally move on rails provided in the machine and sense each winding position whether there is any yarn break or not. If there is any yarn break sensed by the robot, the robot stops in that winding position and the following operations are carried out (Figure 14.2).

- The robot looks on to the spinning position.
- The robot pushes a suction duct (5) against the opening of the deflection channel (3).
- The robot then activates a magnetic valve of the section panel and creates an air pressure inside the membrane valve (6).
- The elastic membrane closes the suction pipe (2).
- Now the feed is started and the fibre supply is vacuumed off by the robot through the deflection channel (3).

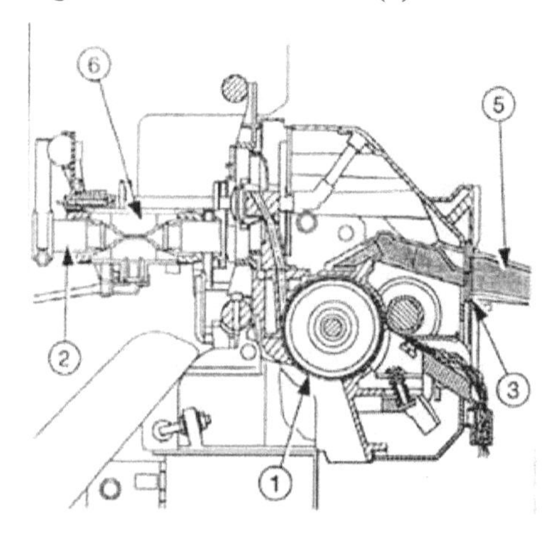

Figure 14.2 Auto piecing by Robot

- The yarn end is "dropped-off" by the robot in to the delivery tube and rotor.
- Shortly, the robot stops the vacuum to the deflection duct and opens the membrane valve.
- The fibre flow is directed in to the rotor and connects to the "dropped" yarn end.
- The yarn is drawn off the rotor.
- The spinning process starts and the package starts winding the yarn.

14.4 Online monitoring of yarn faults in rotor spinning machine

Electronic yarn clearing installation on OE rotor spinning machines.

In 1967, the first OE rotor spinning machine was introduced in the textile industry. In 1982, the OE rotor spinning machine was automated, i.e., the yarn breaks caused by remaining disturbing faults could automatically be mended by a piecer carriage. The automation of this machine has helped the textile professionals to utilise a sensor on this machine to detect disturbing yarn faults and to eliminate such yarn faults. The replacement of the yarn faults by a piecer was done by a robot. With the invention of the robot, the OE rotor spinning machine became the first fully automatic spinning machine.

With the introduction of electronic yarn clearing installations in rotor spinning like in winding, it is possible to control and monitor the yarn faults which occur in every rotor spinning positions. There are possibilities to control the "rogue" spinning positions and also the immediate assessment of quality and production conditions. The UPG 5- installed in many rotor spinning machines operates via the operator interface (BOB) of the Rieter rotor spinning machine. A communication interface is used for the connection of UPG 5-R1 to the operator interface (BOB).

The USTER QUANTUM for OE rotor spinning machines is the fourth generation of clearers for this machine. This clearer is capable of detecting thick places, thin places, foreign fibres, and also it can determine the quality level of the yarn package with respect to its evenness, imperfections, hairiness, count deviations, Moiré defects, etc.

14.4.1 Measuring head

The measuring head of the UPG (USTER POLYGUARD) works on the principle of capacitive measuring principle. It sends an electrical signal to the evaluation unit which is proportional to the yarn cross-section.

The measuring head versions vary for different yarn counts which are shown in the following Table 14.1.

Table 14.1 Measuring head range

MK-C15	Nm 12 120	Nec 7 70	New 10.6 106	Tex 85 8.5
MK-C20	Nm 6 60	Nec 3 35	New 5.3 53	Tex 170 17

14.4.2 Evaluation unit

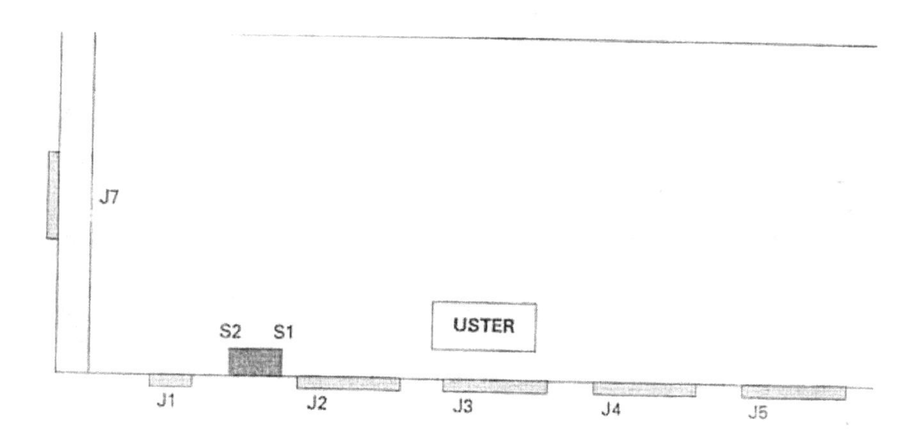

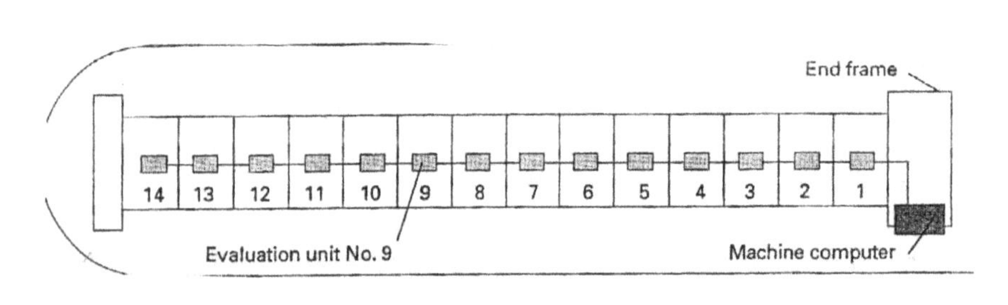

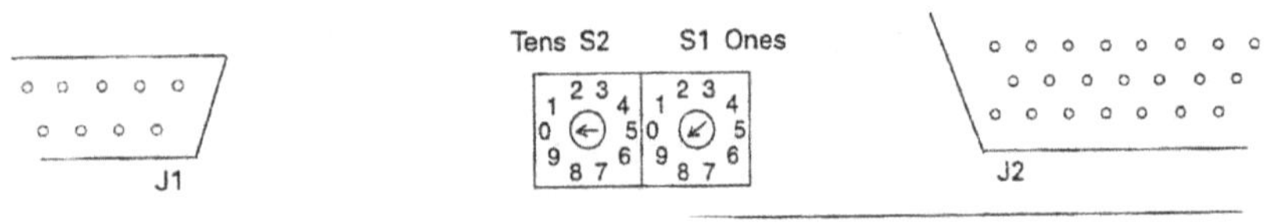

Figure 14.3 Evaluation unit

One evaluation unit can handle the signal evaluation for the 20 UPG measuring heads of one machine section. In order to permit a systematic data exchange between the operator interface (BOB) and the evaluation units, each evaluation unit is given an address. The addresses of the evaluation units are set on the addressing switches, corresponding to the numbering of the sections of the machines as shown in Figure 14.3. The addressing switches must engage correctly to the set number.

14.4.3 Virtual yarn clearing

- Virtual yarn clearing is set for the purpose of detecting optimum settings of the active yarn fault channels like N,S,L,T.
- The benefit of this virtual yarn clearing is the number of disturbing yarn faults can be reliably determined.
- Virtual clearing is useful to evaluate the yarn quality.
- The virtual clearing limit curve can only be set the same or more sensitive than the active yarn clearing limit curve.
- Yarn clearing and virtual clearing operate independently of each other.
- A less sensitivity of the virtual clearing limit curve is of no value as the rotor will be stopped automatically, if a disturbing yarn fault is detected.
- The shape of the virtual clearing curve is dependent on the settings of the following fault channels:
    - VN – virtual for N – channel
    - VS – virtual for S – channel
    - VL – virtual for L – channel
    - VT – virtual for T – channel

14.4.4 Yarn fault classification

In the present context subjective evaluation of yarn fault is discarded and e-customers require objective evaluation of yarns. HE CLASSIMAT-UPG is an objective evaluation tool which counts HE seldom-occurring yarn faults in staple fibre yarns. The classification matrix is shown in the matrix form in Figure 14.4.

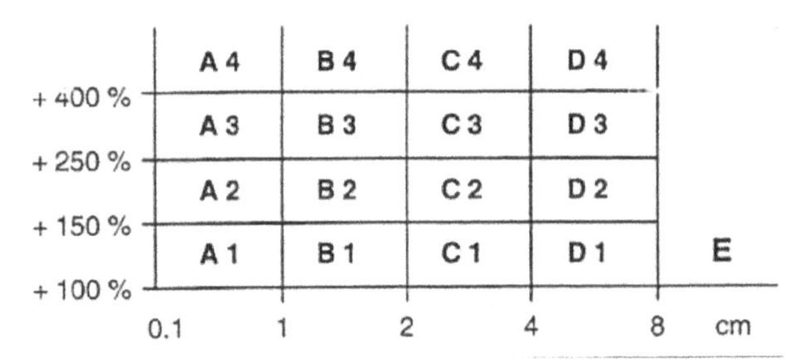

Figure 14.4 Yarn fault classification matrix (I generation)

- The yarn fault classification is limited to the counting of short thick places and long thick places with a cross-sectional increase of +100% or more.
- The CLASSIMAT II system classification according to 17 classes. In the third generation yarn clearers, 23 faults including long thin faults (T) can be evaluated.
- The UPG system classifies the yarn faults simultaneously at all spinning positions of the machine.
- The classifications of the yarn faults are available for the every individual spinning position and for the entire machine.

14.4.5 Setting of sensitivity levels

For the determination of yarn clearing limit in the yarns, the second generation yarn clearers like Uster Polymatic in winding and Uster Polyguard in Rotor spinning machine, Translators are used to know about the clearing limits and also to adjust the settings according to the faults shown in the Translator. In the third and fifth generation yarn clearers, this can be seen on line in the winding machine and can be adjusted based upon the yarn body. In modern winding machines, such on-line classification and adjustments of clearing curve with scatter plot is very helpful for the spinners to identify the yarn faults which are disturbing and settings can be adjusted within a second. Hence, it does not require off-line testing of yarn faults using CLASSIMAT system in the laboratory. However, the basis of the settings can be understood with the help of Translator shown in the Figure 14.5.

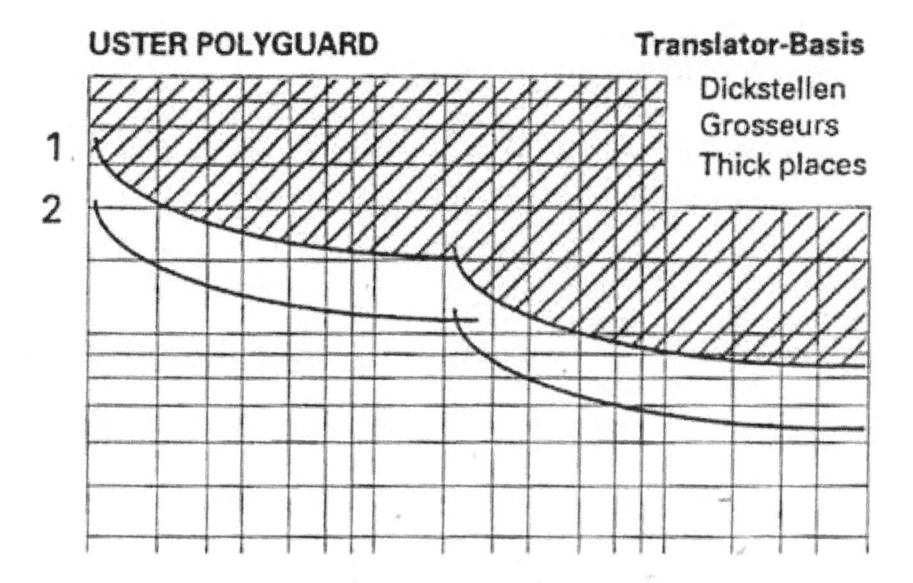

Figure 14.5(A) Setting of the sensitivity levels in the machine for yarn fault clearing

Yarn fault channels for virtual yarn clearing VN,VS,VL, VT.

Virtual yarn clearing is for the purpose of determining optimum settings of the active yarn fault channels N,S,L and T. This system was introduced when the electronic yarn clearers have come to the textile industry especially for spinning mills to ascertain the faults which need to be cut and faults which are not disturbing can remain in the yarn. However, nowadays with the development of the software and hardware, scatter plot diagrams help the spinners to identify the yarn faults near to the yarn body and the settings can be adjusted on line on the auto coner machine itself within seconds. However, it is better to have knowledge on the virtual clearing.

With the help of the auxiliary virtual clearing limit curve, the number of disturbing yarn faults can be reliably determined. Furthermore it is possible to evaluate the yarn quality. The virtual clearing curve can only be set the same or more sensitive than the active yarn limit curve.

In the Figure 14.5 (A), the light shaded area refers to fault sizes and lengths which will be detected by the fault channels S and L.

In the Figure 14.5 (B), all the yarn faults with sizes and lengths above the virtual clearing curve (2) in the dark shaded area will only be counted but not cleared.

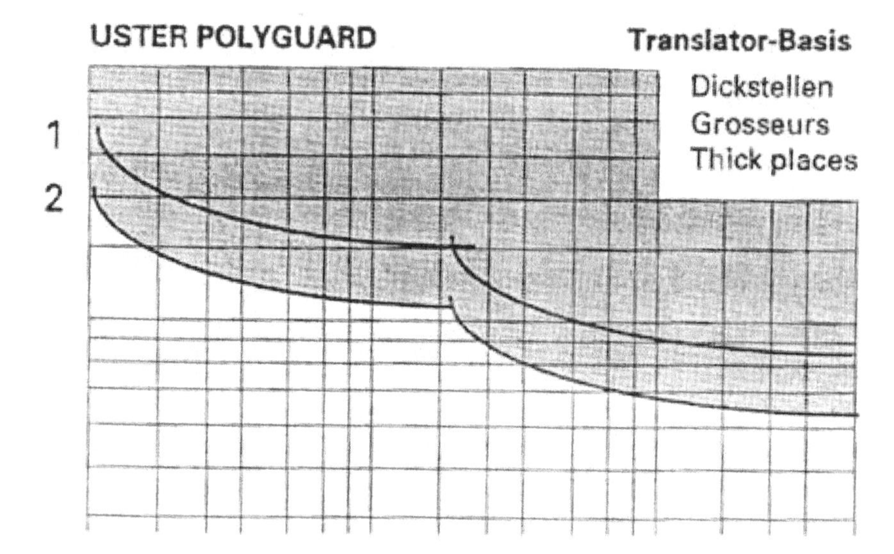

Figure 14.5(B) Virtual clearing curve

Yarn clearing and virtual clearing operate independently of each other.

Yarn clearing limit

The yarn clearing limit is a dividing line between

- Yarn faults which remain in the yarn.
- Yarn faults which must be removed.

14.4.6 Translator

The Translator is a transparent template with the curve of the yarn clearing limit. In the present day machines, it is not required as the settings can be adjusted in the machine monitor itself thus saving the time. From the yarn body, the faults which need to be extracted can be entered depending upon the requirements (Figure 14.6).

- The vertical axis (%) indicates the deviation in cross-section of the yarn fault i.e. sensitivity.
- The horizontal axis (cm) indicates the length of the yarn faults i.e. reference length.

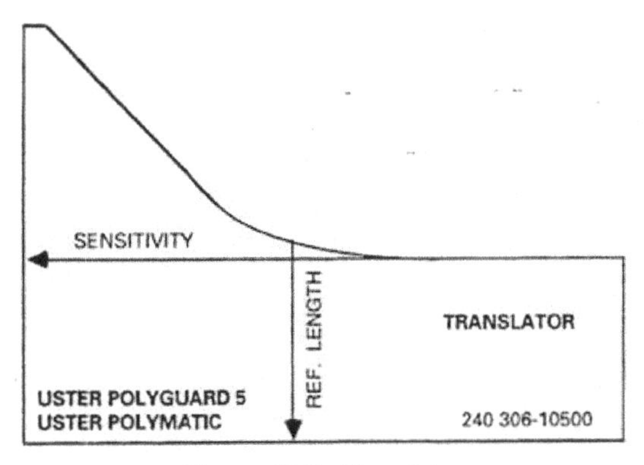

Figure 14.6 Translator

14.4.7 Setting values for short thick places N and S

Practical example for the determination of the yarn clearing limit.

- Select three or four yarn faults in each of the ranges N,S,L and T which would just qualify for the elimination.
- The short thick places in the S – range are represented by the Uster CLASSIMAT grades.
- The yarn faults in the L and T ranges can only be represented on the basis of their length and cross-sectional deviation.
- Enter these yarn faults on the Translator-Basis in Figure 14.7 (thick places).

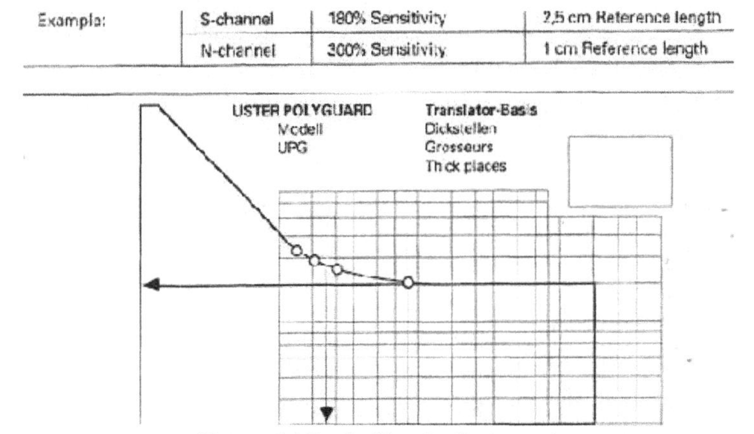

Example:	S-channel	180% Sensitivity	2,5 cm Reference length
	N-channel	300% Sensitivity	1 cm Reference length

Figure 14.7 Setting for S channel
The two arrows indicate the settings in the yarn fault channel.

14.4.8 Settings for long thick places (L)

Long thick places are the disturbing factors in the yarn, say, coarse and double threads.

Example:	L-channel	50% Sensitivity	100 cm Reference length

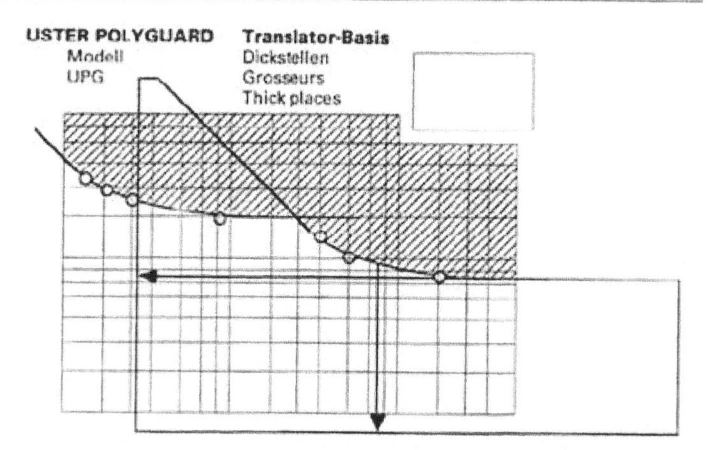

Figure 14.8 Setting for long thick places (L) in the yarn

- Place the Translator on the front of the Translator-Basis.
- Now slide the Translator parallel to the Basis, until the curve covers the L-faults entered directly on the right hand side of the sheet as shown in the Figure 14.8.
- The two arrows on the Translator indicate the settings in the L- channel.

CVm – UPG

The CVm – UPG is the measure of the yarn unevenness. Yarn evenness is influenced by

- Raw material.
- Spinning devices.
- Condition of machinery.
- Evaluation length is 100–1,000 m which is adjustable in steps of 100 m.

- With a CVm – UPG of 15%, in the case of a normal distribution, 68% of the yarn length would be within 15% of the mean yarn cross-section.
- The calculation of a CVm – UPG is carried out simultaneously at all the spinning positions of the UPG installation.

14.5 CVm – UPG alarm

In order to detect reliably the spinning positions with a deviating CVm and yarn with a too high mean CVm value, the system is provided with two alarms.

Spinning position alarm

The spinning position alarm monitors the deviation of CVm of single winding positions from the mean CVm of the UPG installation.

Limit of the alarm range = CV ±

Setting range = 0–±50% deviation from mean CVm value

This is illustrated by an example, spinning positions with a CVm deviating by ±20% from the current mean value of the system value will have to trigger the alarm as shown in the Figure 14.9.

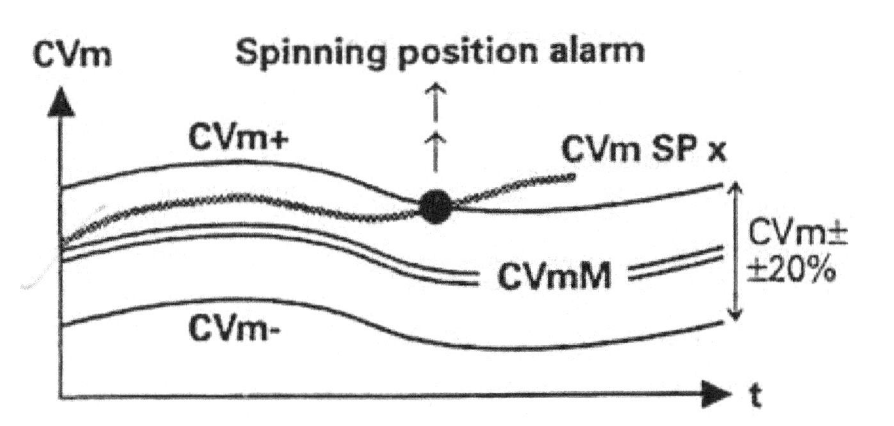

Figure 14.9 CVm – alarm limit

14.6 Mean value alarm

The mean value alarm monitors the mean CVm value of the installation. Setting range for Upper alarm limit and lower alarm limit ranges from 0 to 50%.This is illustrated with the example and Figure 14.10.

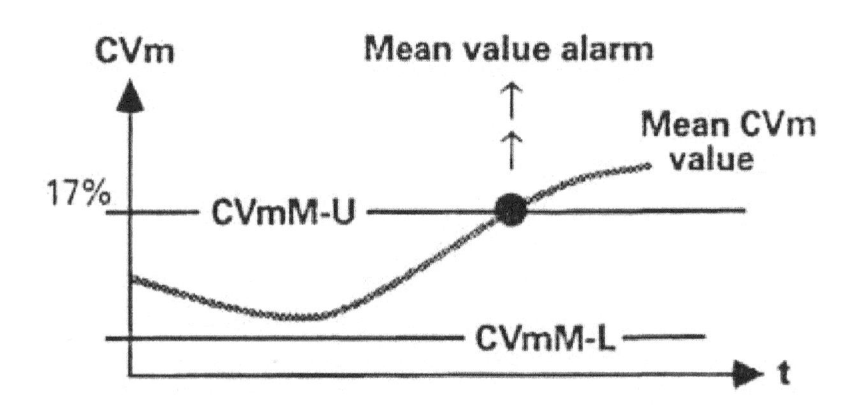

Figure 14.10 Mean value alarm

In the above example, alarm is to be triggered if the current mean value CVm value of the UPG installation exceeds 17%. Alarms are indicated for the particular positions at the operating section of the spinning machine.

14.7 Display of yarn faults detected

Detection of yarn faults in any particular rotor spinning position is generally indicated by a brief lighting-up of the red lamp on the respective measuring head. From the indication on the measuring head, yarn faults can be identified.

Yarn fault alarm types

The following are the yarn fault alarm groups:

- N.S – Neps and short thick places.
- L.T – Long thick and long thin places.
- MO – Moiré.
- C – Count of yarn fault.

Yarn clearing limit

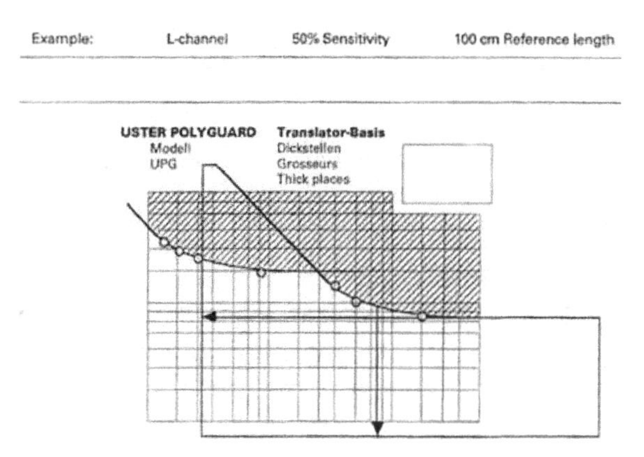

Figure 14.11 Yarn fault clearing limit

The yarn fault alarm can be set according to our requirements and if any spinning position exceeds the alarm limit, the respective spinning position is blocked for producing sub-standard quality of yarn. This alarm helps the maintenance crew to do specific maintenance in the respective spinning position (Figure 14.11).

14.8 Spectrogram analysis

Spectrogram is an analysis of the yarn unevenness with respect to the periodic yarn faults within the range of 2 and 1600 cm wavelength (i.e. distance between periodic faults). For the same length of yarn it is possible to detect simultaneously.

- CVm values
- Classify yarn faults
- Can prepare either spectrogram or count imperfections.

The spectrogram is a bar chart allows the detection of periodic yarn faults. A spectrogram is obtained by continuously monitoring the yarn cross-section with respect to the availability of wavelengths and measuring their amplitude (height). In the horizontal axis, the wavelength is represented in logarithmic scale and in vertical axis represents amplitudes of the yarn cross-section. In spectrogram, the shortest wavelength is 2 cm.

14.8.1 Spectrogram

A spectrogram is obtained by continuously monitoring the yarn cross-section with respect to their amplitudes and wavelengths. Spectrogram is a bar chart which detects the periodic yarn faults. In the spectrogram, horizontal axis represents wave length in logarithmic scale and vertical axis represents the amplitudes of the yarn cross-section. The highest amplitude corresponds to the maximum height. The shortest wavelength is 2 cm. The evaluation length is dependent on the maximum wavelength to be analysed as shown in the Table 14.2.

Table 14.2 Spectrogram wave length and evaluation length

Maximum wave length	Evaluation length
5 m	51 m
20 m	204 m
80 m	816 m
160 m	1632 m

The longest wave lengths represented at different scales have no statistically significant difference. Spectrograms are prepared at one spinning position after the other on a continuous cycle of a machine section. It means that if there are 10 sections, spectrograms will be prepared simultaneously for 10 spinning positions (1 spinning position per section). On the same piece of yarn, it is possible to simultaneously identify

- Either spectrograms or count imperfections
- Determine CVm values
- Classify yarn faults.

14.8.2 Spectrogram alarm

The spectrogram alarm limit (%) relates to the percentage deviation of one amplitude in comparison with the mean value of the four neighbouring amplitudes, two to the left and two to the right.

Optimisation of the alarm limits setting:

- Amplitude in the upper third
  = alarm limit × 0.5
- Amplitudes in the middle third
  = alarm limit × 1
- Amplitudes in the lower third
  = alarm limit × 2

With the spectrogram alarm, it is possible to stop, block the spinning position. This can be explained by an example (Figure 14.12).

Example: Spectrogram alarm = 60%

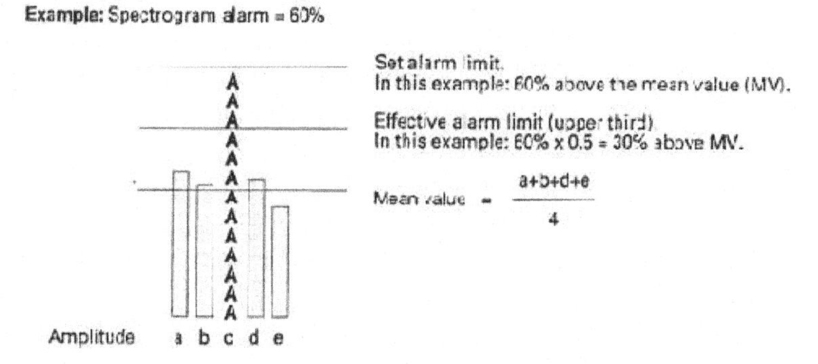

Figure 14.12 Spectrogram alarm limit

Set alarm limit. In the above example: 60% above mean value (MV).
Effective alarm limit (upper third)
In this example, 60 × 0.5 = 30% above MV.
Mean value = a+b+d+e/4

14.8.3 Attributes of Q – Pack in rotor spinning machine

The quality package UPG-Q-PACK extends the USTER POLYGUARD yarn monitoring installation by the on-line monitoring and determination of the following yarn characteristics:

- CVm – UPG (Coefficient of variation CVm of the yarn mass)

  The coefficient of variation is the measure for the yarn unevenness and it is the ration between the standard deviation and mean value of the yarn.
- CLASSIMAT –UPG yarn fault classification
  The CLASSIMAT-UPG counts and classifies seldom occurring yarn faults, The classification covers the range of the fault channels N,S and L. It classifies all the yarn faults and also the remaining yarn faults on the cone.
- IP-UPG (imperfections)

  Imperfections are frequently occurring yarn faults in staple-fibre spun yarns. The type and number of imperfections give indications on yarn characteristics, manufacturing processes and raw material.

- CVm-UPG alarm is the measure of yarn unevenness. It is reliably able to detect spinning positions with a deviating CVm and yarn with a too high mean CVm value.

14.8.4 Moiré channel (MO)

Moiré is a defect in rotor spun yarns which is defined as the succession of thick and thin places with a wavelength approximately equal to the rotor circumference. Moiré is in most cases occurs due to the accumulation of dirt or dust content in the groove of the rotor. In order to identify the Moiré defect in the rotor yarns, an entry of rotor circumference is recommended.

The Moiré channel (MO) in the rotor spinning machine continuously.

- Determines a typical yarn limit value from the yarn signals of all the spinning positions.
- Supervises the yarn signals of all the spinning positions with reference to the periodic faults with wavelengths corresponding to the entered circumference of the rotor.
- The Moiré channel actuates when the yarn signals of a periodic fault at a spinning position oversteps the limit value according to the entered sensitivity value.

For example, if the sensitivity level of MO is 100% and the if the yarn signal of the periodic fault exceeds this value (100%), the rotor will be stopped due to Moiré defect.

Setting range of the Moiré channel:
Rotor circumference: 20–70 mm
Sensitivity: 50%–500%

C: channel for yarn count
C-channel: The C-channel monitors every individual spinning position for yarn count deviations. In the first instance it calculates the mean value of the yarn from all the yarn signals. If a yarn signal exceeds the mean value over the set reference length and the set sensitivity, then the C-channel will initiate the stopping of the particular spinning position.

Setting range of the count channel
Sensitivity: ±5%–±30%
Reference length: 10–100 m

Function of the C-channel

- At the start up of the spinning operation, reference length is fixed to 30 m.
- If the sensitivity is set at less than ±10%, it will be increased to ±10%.
- After the start-up phase, the settings of sensitivity and reference length become effective.

15

Sensors and their application in spinning machines

15.1 Introduction

Electronic sensors in the spinning machines play a major role in the control systems of the process and the machineries. In the modern and advanced textile machineries developed by leading textile manufacturers, the importance is given not only in the machine design and concepts but also in the instrumentation and control systems in the machineries. It means besides the improvement in productivity, quality, a system is there to control the process flow from one machine to another machine.

15.2 Electronic sensors

Any instrumentation system consists of three major elements:

- Input device
- Processing device and
- An output device.

Electronic sensor is a transducer which converts any physical variable in to an electrical output. The input quantity for most instrumentation systems are non-electrical quantity. Electrical methods and techniques are used for measurement, manipulation or control.

15.2.1 Classification of transducers

Transducers are classified according to

- Their applications
- Method of energy conversion
- Nature of the output signal.

Classification of transducer according to their electrical principle involves

1. Transducer that requires external power.
2. Self-generating type (no external power).

Table 15.1 illustrates the electrical principle of the transducer.

Table 15.1 Transducer based on electrical principle

S. No.	Resistance	Application
1	Potentiometer device	Pressure displacement
2	Resistance strain gauge	Force, Torque, displacement
3	Hot wire anemometer	Air flow, gas flow and pressure
4	Resistance thermometer	Temperature, Radiant heat
5	Thermistor	Temperature
6	Resistance hygrometer	Measurement of humidity
	Capacitance	
1	Variable capacitance pressure	Displacement, Pressure
2	Capacitance micro phone dielectric gauge	Thickness, liquid level
	Inductance	
1	Magnetic circuit transducer	Pressure, displacement
2	Resistance pick-up	Pressure, displacement
	Voltage and current	L
1	Photo emissive cell	Light and Radiation
2	Photo multiplier tube	Photo sensitive relay
	Self-generating transducer	
1	Piezo electric pick-up	Sound, vibration, acceleration and pressure
2	Photo voltaic cell	Light meter, solar cell temperature
3	Thermocouple and Thermopile	Heat flow radiation

15.2.2 Selection of transducer

The selection of transducers for specific applications depends on the following parameters (Figure 15.1).

- Physical quantity to be measured.
- Transducer principle can be utilised to measure this quantity.
- Accuracy of the measurement.
- Reliability under given dusty working atmosphere.

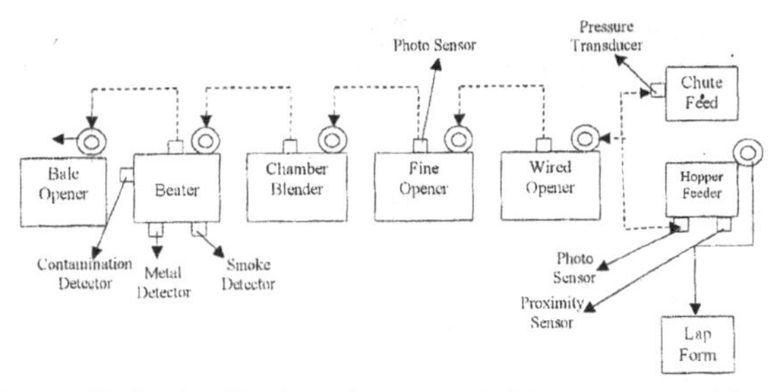

Figure 15.1 Application of sensors in blow room machines

Blow room line

The type of sensors and transducers used in blow room lines.

- In the material flow control in the blow room line, photocell control/ pressure transducers are used in Feed/Delivery control synchronisation between upstream and downstream machines.
- Cotton contamination detector installed in blow room line sequence for sensing and ejection of the contaminations like foreign fibres (jute, polypropylene, etc.) by ultra fast CCD cameras.
- Measurement of length of the lap wound in meters or yards on the lap spindle by means of a proximity sensor which counts the pulses and stops the machine as soon as the required length of lap is wound on the lap spindle.

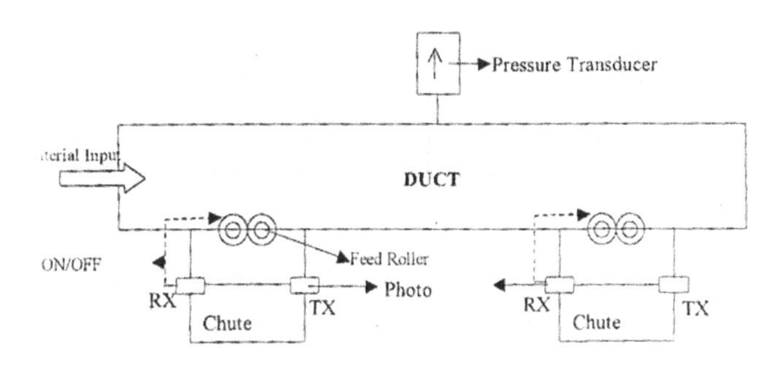

Figure 15.2 Pressure transducer used in Chute feed card for sensing the material filling in the chute

- In modern blow room line where the lap system is eliminated and replaced by conveying the opened cotton tufts directly to the carding machine via Chute Feed System. In this chute feed system, Pressure Transducer is used to control the feed material.
- Chute feed system uses photo sensor which controls the feed roller.

- Material pressure in the chute feed duct is sensed by pressure transducer which gives ON/OFF control (Figure 15.2).

15.3 Sensors in carding

- Figure 15.3 shows the modern card with auto levelling equipment and production rate control.
- Proximity sensors are used to determine the speeds of Lickerin, Cylinder, Doffer and Flats. Proximity sensors are also used in places of can changer, Lap thickness sensing for double lap.
- In the delivery side of the carding, where the card web is condensed in sliver form and deposited in the cans through calendar rolls. In the trumpet of the calendar rollers, Angular displacement transducer senses the sliver cross-section and sends signals to the feed roller for making necessary correction in the input feed material. This is achieved with the help of closed loop auto levelling system.
- Photo sensors are located between the entry of the coiler and the can to detect for sliver breakage. If there is no sliver breakage, the photo sensor light will not light up. In case where there is no sliver present or sliver breakage, the photo sensor light is "ON" which shows the problem.
- Servo amplifiers and servo drives are used for medium and long-term auto levelling.

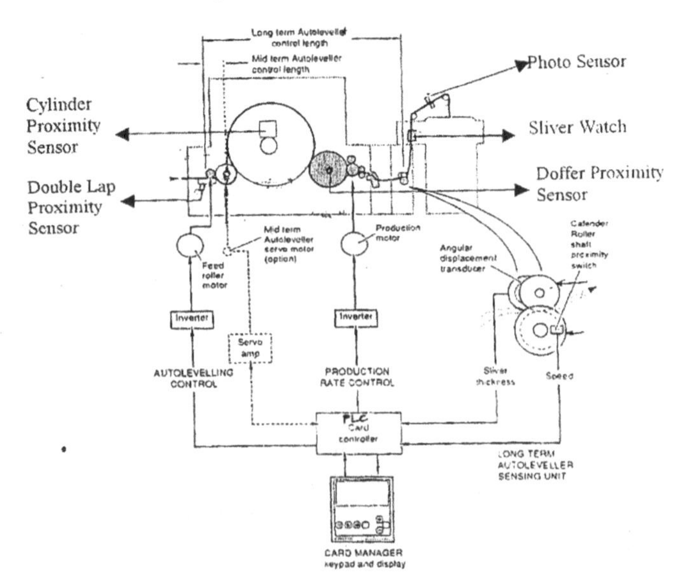

Figure 15.3 Sensors in carding machine

15.4 Sensors in draw frame

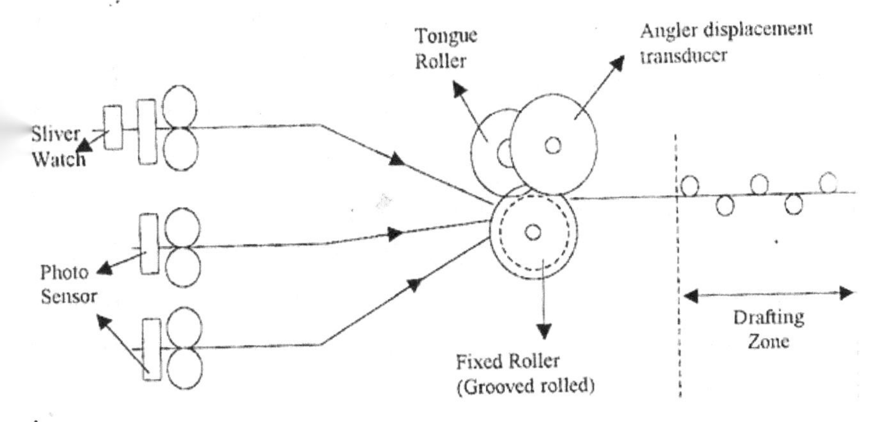

Figure 15.4 Sensing mechanism in draw frame

- Open loop or combined loop control is installed in draw frame which senses the delivered material and makes necessary correction if the linear density of the material delivered deviates from the set values (e.g. Tongue and Groove principle or non-contact type).
- Sliver watch is an optional attachment for detecting contaminations in the feed sliver, eliminate the same and prevents the spreading of contaminations spreading over several meters.
- Proximity sensor for sensing the delivery roller speed in meters/min. and also the length of sliver delivered in to the cans (Figure 15.4).

15.4.1 Sensors in combers

- Proximity sensors used for counting the length of slivers delivered in to the cans.
- Pressure transducer is used for measuring the suction drop.
- Pressure transducer is used for sliver monitoring.
- LED for sensing the feed lap run out to avoid singles in the comber sliver.

15.4.2 Sensors in speed frames

- Proximity sensor is used to measure the delivery speed of the front roller and also for length monitoring on the roving bobbin.

- Photo sensors are installed at the sreel section and also in the delivery roving side for detection of sliver and roving breaks.

15.5 Sensors in ring frames

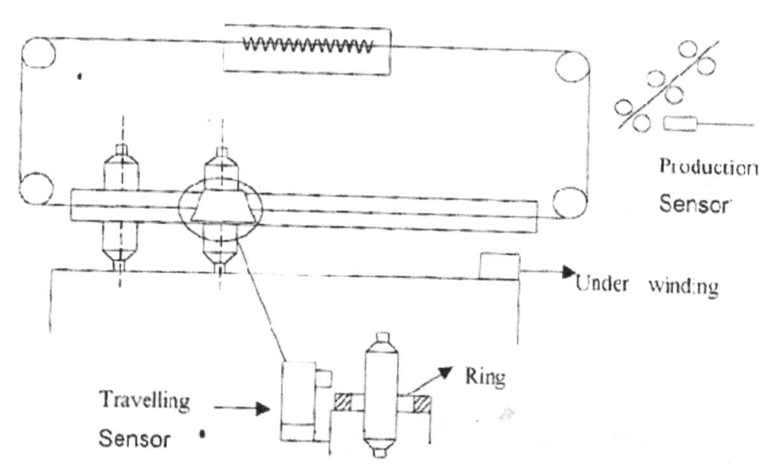

Figure 15.5 Production and travelling sensors in ring frame

- Proximity sensors are used for measuring spindle speed in rotation per minute (RPM), doffing and hank measurement.
- Opti spin offers detectors for individual ring spindle speed and real time detecting of yarn breakages.
- Ring data system consists of travelling sensors which senses the presence of yarn on the spindle. If there is no yarn on particular spindle, the sensor flashes light and the spindle is recorded on the main server. All the machines with ring data system are connected to a PC based central unit. From this, all information like production, efficiency, downtime of the spindles, idle spindles, ends down, slipping spindles can be found out for observation and analysis.
- Monitoring of twist direction (S or Z) (Figure 15.5).

15.6 Sensors in auto winders

Yarn tension sensors in winding

The principle of working of yarn tension sensor in modern winding machine is shown in the Figure 15.6.

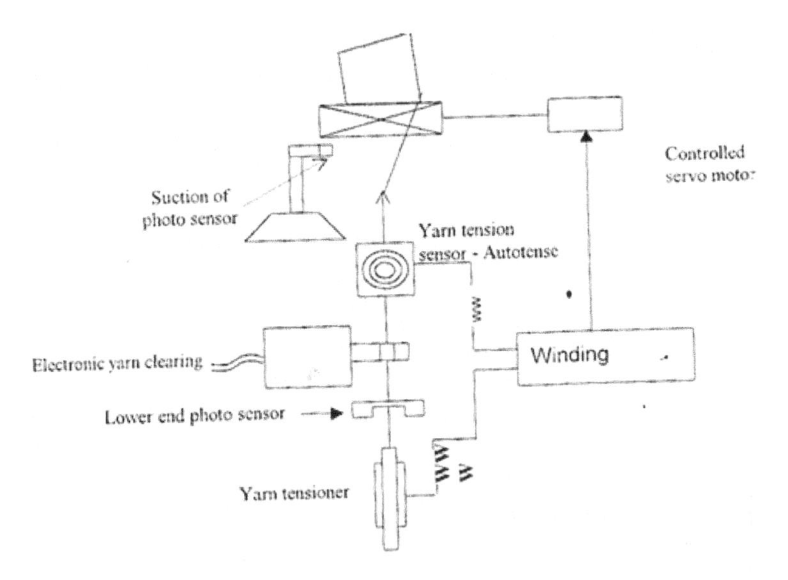

Figure 15.6 Sensors in auto winders

- Auto tense yarn tension control sensor not only prevents an increase in yarn tension towards the end of the bobbin but also compensates for the lower yarn tension by increasing applied pressure.
- Yarn tension sensor is situated in the yarn path after the clearer on each winding position which registers the actual yarn tension at the package and provides continuous direct measurement of the yarn tension.
- Sensors are also used in other places like pressure transducer (sensor) for sensing the suction pressure at the package and at the bobbin end.
- Similarly proximity sensors and photo sensors are used in package conveyor and in central neutral position.

15.7 Sensors in textile machinery

Inductive sensors: Inductive proximity switches are non-contact type electronic sensors ideally suited for sensing of metallic objects. These sensors are resistant to dust, humidity and oil. Inductive proximity sensor generates high frequency oscillations and any metallic object, which enters active area results, eddy current in the object and then stops oscillation that triggers output stage. Continuous development in the field has led to new application such as distance measurement with analogue types. Inductive sensors are available in the sensing range of 0.5–30 mm and different packages like Barrel Type, Flat Type, Box Type, Slot Type, and PCB Mounting.

- Capacitive sensors: Capacitive sensors are non-contact type electronic sensors ideally suited for sensing of metallic as well as non-metallic objects. With adjustable sensitivity almost any object can be detected with desired reliability. The primary function element of capacitive proximity sensors is high frequency oscillator with floating electrode in the transistor base circuit. When object appears in active area, high frequency oscillation begins and results DC signal, which triggers output stage. These sensors are available in barrel housing both NPN and PNP types.
- Photo sensors: Photo sensors are indispensable components of modern automated production processes. Optimum combination of precision optics with electronic and a rugged housing together result in providing versatile reflex and through beam sensors. Photo sensors are basically divided into two main categories. Photo sensors are available in different packages like Barrel Type, Slot Type, Box Type, Miniature type, etc.
- Through beam type sensor: Through beam type photo sensors are available with sensing range up to 5 m. Emitter and receiver are in separate housings facing each other with their optical axes matched. Sensor switches whenever light beam is interrupted. These are ideal for monitoring doors and gates, counting and monitoring objects over large distances.
- Reflex type diffused beam sensors: Diffused beam photo sensors are available with a sensing range up to 0.2 m. These are ideal for distinguish and sorting of objects according to their volumes and degree of reflection.
- Retro reflex type sensors: Retro reflex sensors are ideal for detection of height of stacked objects and controlling positioned objects on conveyers. They may be metallic or plastic with cylindrical, flat, slot or customised mounting certified for IP67, short circuit protection, over load, polarity reversal protection. These sensors are available with the range for 2 m.

Hall sensors/Magneto restrictive sensors: The field plate sensor is a non-contact proximity switch with two magneto resistor arranged in a magnetic bridge. A ferrous target influences the resistance values of the magneto resistors. The sensor is particularly effective for trouble free measuring of rotating objects. For example, using a gear with switching frequencies up to 20 KHz can be obtained.

15.7.1 Working principle

Inductive and capacitive

Their operating principle is based on a high frequency oscillator that creates a field in the close surroundings of the sensing surface. The presence of a metallic object (inductive) or any material (capacitive) in the operating area causes a change of the oscillation amplitude. The rise or fall of such oscillation is identified by a threshold circuit that changes the output state of the sensor. The operating distance of the sensor depends on the actuator's shape and size and is strictly linked to the nature of the material. A screw placed on the back of the capacitive sensor allows regulation of the operating distance. This sensitivity regulation is useful in applications, such as detection of full containers and non-detection of empty containers.

15.7.2 Photoelectric sensors

These sensors use light sensitive elements to detect objects and are made up of an emitter (light source) and a receiver. Three types of photoelectric sensors are available.

- Direct reflection – emitter and receiver are housed together and uses the light reflected directly off the object for detection.
- Reflection with reflector – emitter and receiver are housed together and requires a reflector. An object is detected when it interrupts the light beam between the sensor and reflector.
- Through beam – emitter and receiver are housed separately and detect an object when it interrupts the light beam between the emitter and receiver.

Photoelectric sensors are indispensable components of modern automated production processes. It combines precision optics with electronics and rugged housing to provide versatile reflex and through beam sensors. These sensors are designed for applications involving non-contact detection for measuring, counting and positioning (Figure 15.7).

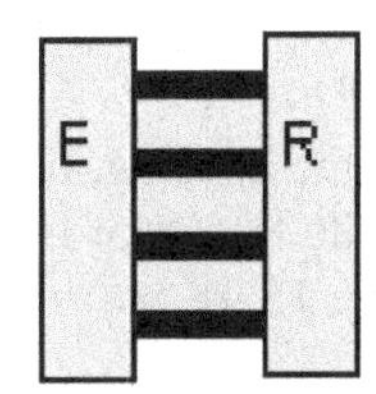

Figure 15.7 Photo electric sensors

Through beam types

These switches offer sensing range up to 5 m. Emitter and receiver are in separate housings facing each other with their optical axes matched. Sensor switches whenever light beam is interrupted.

Features

Switching point independent of surface nature. Narrow effective beam results in better repeatability largest sensing ranges.

Applications

Monitoring doors and gates. Counting and monitoring objects over large distances.

Reflex types

These are available in two types.

(a) Diffused beam reflex sensor (Figure 15.8)

Emitter and receiver are in same housing. Sensing range up to 0.2 m is provided. Emitter sends beam of pulsed infrared light. Only small portion of light diffused by object in all directions is sensed, which switches the sensor.

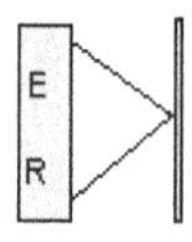

Figure 15.8 Diffused beam reflex sensor

Features

Sensing range depends largely on reflective properties of target. Suitable for distinguishing between black and white targets. Positioning and monitoring with only one active sensor.

Applications

Distinguishing and sorting of objects according to their volumes and degree of reflection.

(b) Retro reflective reflex sensor

With emitter and receiver in same housing, light reflected by triple prism reflector or reflecting tape opposite to sensor is sensed. Sensor switches when light beam is interrupted (Figure 15.9).

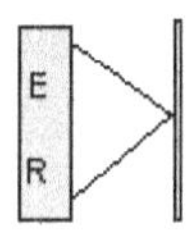

Figure 15.9 Retro reflective reflex sensor

Features

Large sensing range. Mat finished surfaces are sensed independent of surface properties.

Applications

Height detection of stacked objects. Control of randomly positioned objects on conveyor.

Magnetic type

Magnetic sensors are actuated by the presence of a permanent magnet. Their operating principle is based on the use of reed contacts, which consist of two low reluctance ferro-magnetic reeds enclosed in glass bulbs containing inert gas. The reciprocal attraction of both reeds in the presence of a magnetic field, due to magnetic induction, establishes an electrical contact. Typical applications include the detection, position, inspection and counting on automated machines and manufacturing systems. They are also used in the following machinery: packaging, production, printing, plastic molding, metal working, food processing, etc.

16

Electronic controls in weaving preparatory and weaving machines

16.1 Introduction

The aim of the spinning mills is to deliver a quality product to the weaving customer which should meet his requirements in fabric making weaving machines. The yarn should have acceptable unevenness, less number of thick, thin and neps, besides should have high strength and elongation with less number of so-called weak spots in the yarns. In the same manner, the aim of the weaving mills to produce high quality and value added products free from fabric defects and at the same time in an economic manner. In order to achieve these requirements, like in spinning mills where the in-line monitoring of production and quality monitoring have come. Similarly on-line monitoring is required for controlling the production and stoppages and thereby eliminating unnecessary loss in efficiency.

The emphasis on the weaving mills is therefore a continuous on-line monitoring of production and quality. The purpose of the on-line monitoring system is to collect various process information from all the departments such as weaving preparatory (warping, sizing, etc.), loom shed and fabric inspection.

At present data acquisition, automatic monitoring and display of trends can be performed with personal computers. This allows us to quickly identify and systematically eliminate weak points in the production process. There is need for quality product, which gives high standard, which is possible by use of automation in textiles. The automation has done a drastic change in textiles by the use of electronics and computers in it. Electronics and computers has given some the benefits in the textile industry such as high production rates, consistency in quality, and ease in monitoring, reduced maintenance, flexibility, reduction in man power, etc. Now-a-days modern weaving machines are equipped with the various control systems like electronic take up and let off, permanent insertion control (PIC), slim through-light sensor, electronic colour selector, ECOWEAVE /RTC, automatic picks finding, electronic yarn tension control, automatic pre-winder switch off, automatic start-mark prevention, E-shed, CAN-bus, etc.

16.2 Developments in sizing machine controls

Objective of sizing

The primary purpose of sizing is to produce warp yarns that will weave satisfactorily without suffering any consequential damage due to abrasion with the moving parts of the loom. The other objective, though not very common in modern practice, is to impart special properties to the fabric, such as weight, feel, softness, and handle. However, the aforementioned primary objective is of paramount technical significance. During the process of weaving, warp yarns are subjected to considerable tension together with an abrasive action. A warp yarn, during its passage from the weaver's beam to the fell of the cloth, is subjected to intensive abrasion against the whip roll, drop wires, heddle eyes, adjacent heddles, reed wires, and the picking element. The intensity of the abrasive action is especially high for heavy sett fabrics. The warp yarns may break during the process of weaving due to the complex mechanical actions consisting of cyclic extension, abrasion, and bending. To prevent warp yarns from excessive breakage under such weaving conditions, the threads are sized to impart better abrasion resistance and to improve yarn strength. The purpose of sizing is to increase the strength and abrasion resistance of the yarn by encapsulating the yarn with a smooth but tough size film. The coating of the size film around the yarn improves the abrasion resistance and protects the weak places in the yarns from the rigorous actions of the moving loom.

16.2.1 Controls and instrumentation

There are a variety of controls available on modern sizing machines. The essential functions of all these controls are to provide optimal quality of warp at a minimal cost. The controls usually act on the basis of information provided by the particular sensors placed on the machine. Figure 16.1 is a sketch of a typical two size box slasher with locations of various sensors and controls.

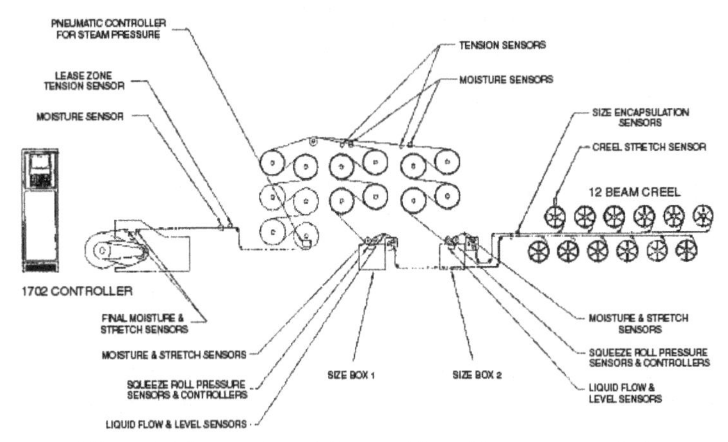

Figure 16.1 Two box sizing machine with sensors controls

Programming the machine. The easy to follow visualisation and recipe management, in which all the machine and textile parameters are stored, permit fast and simple programming.

Sizing monitoring. The sizing process is automatically monitored. All set points are specified with upper and lower tolerance limits. Deviations from the programmed value are displayed at once and instructions for their rectification are explained in the language of the operator.

Automatic tension control. In the direction of the yarn path, from the creel to the weaver's beam at the head stock, controls are placed to monitor tension and effectively regulate the speed of the sizing machine. The controls are placed in the creel to maintain uniform unwinding of the warp beams in the creel, in the size box for a smooth drive of the dry feed rolls, in the drying section to drive cylinders, and in the head stock to drive the weaver's beam.

Automatic size box level regulator, with a warning indicator for low size level or overflow.

Electronic stretch indicators and controllers with digital display for yarn elongation. Excessive yarn elongation (stretch) resulting from the applied tension is detrimental to the quality performance of the warp during weaving. The loss in elongation results in an increase in warp breaks on the loom. Surface speed sensors, mechanical or electronic, in direct contact with the warp sheet are placed from the creel to the front roll in each zone. The automatic controls adjust the size box roll speeds to maintain a constant stretch.

Electronic moisture detectors used to regulate the slashing speed automatically or steam pressure in the drying cylinders.

Steam pressure controllers in the cylinders which may be interfaced to drive the controller to reduce the steam pressure during the slow or creep speed operation.

Temperature controls for the drying cylinders which can be used for maintaining accurate temperature and effective condensate removal.

Squeeze roll pressure release system designed to decrease the squeezing pressure when operating the machine at creep speed or maintaining proportional pressure with respect to the operating speed of the sizing machine.

Size liquor filtration and circulation system designed to filter out yarn waste and fibres (wild yarns and short fibres) found in the sizing system.

Creel braking systems to decelerate the warper's beams effectively, thereby preventing over-run and maintaining the unwinding tension at a constant speed operation.

Microprocessor controls interfaced to a computer for effective management of the operating variables of the sizing machine.

Wet pick-up measurement and size add-on control. In this device, microwave energy which is absorbed by water is used to continuously measure the wet pick-up immediately after the yarn sheet leaves the size box. The on-line refract meter monitors the size solids in the size mix. The size add-on, which is the product of the wet pick-up and size solids in the size mixture, is automatically calculated by the microprocessor. The corrections in the size add on is made by automatically adjusting the squeeze roll pressure to keep the add-on practically constant throughout the sizing operation.

On-line size encapsulation measurement. Size encapsulation is the measure that defines the degree of reduction of the yarn hairiness due to sizing. One on-line yarn hairiness sensor is placed on the unsized and the other on the sized yarn sheet. The difference between the two, expressed as percentage, is the measure of size encapsulation.

16.2.2 Electronic control systems in sizing machines

a. Electronic stretch indicator

It consists of one input speed transducer for each size box and one output speed transducer. These transducers are either shaft driven or surface driven. It counts the revolutions and electronically computes the percentage of stretch. This value of stretch is displayed on a digital display. In order to obtain the desired value of stretch, the operator can manually adjust the differential variable speed transmission.

The electronic stretch controller also utilises the input and output transducers as the stretch indicator. However, the stretch controller system includes control motors to adjust the variable speed transmission driving the size boxes. The stretch controller enables the operator to preset a desired value of stretch expressed as percentage. The system thus adjusts the automatic adjustments of the size box drives to yield the required stretch.

b. Squeeze pressure control

The adjustable rate squeeze pressure control provides automatic and adjustable loading at run, slow and stop with minimum loading or release at slow and stop and proportionally increasing pressure as the slasher speed increases at maximum loading at the top, or gear-in, speed. Squeeze roll load increase with speed and can be pre-programmed to provide the desired proportional relationship of pressure with increasing speed.

c. Loom beam tension indicator and control system

This system measures and controls the web tension between the delivery roll and the loom beam in conjunction with the motor or transmission control for the beam drive. The system utilises load cells on the exit nip roll of the Head end to produce an electrical signal proportional to the load imposed by the tension of the yarns. An amplifier and a computing circuit convert the load cell signal in to a numerical value for display on an indicator and controlling the beam drive.

16.3 Computer controls in weaving preparatory and weaving machines

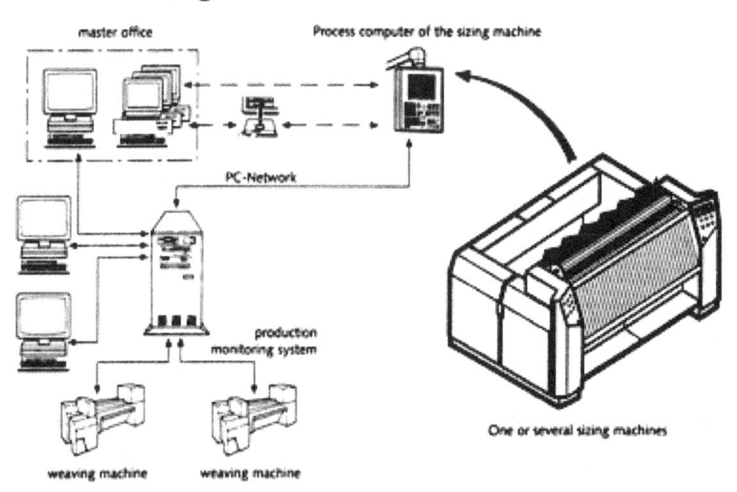

Figure 16.2 Computer control in weaving preparatory and weaving machines

Slashing

The ultimate goal of sizing is to eliminate or reduce warp breaks during weaving. Warp breaks are caused either by high tension or by low strength in the yarn. High tensions in the warp are caused by large shed openings, lack of proper tension compensation, high beat-up force and inadequate let-off. Knots, yarn entanglement and high friction also cause tension build-up. Slasher creel tension control is critical especially with Murata jet-spun (MJS) and open-end yarns. Maximum tension should not exceed 5% of breaking strength (15–20 grams (g) for ring-spun yarns and 12–15 g for open-end, MJS and MVS yarns). With coarse yarns, sometimes 30 g is allowable. The amount of size picked up is affected by the viscosity of the size mix as well as the yarn structure. The viscosity of the mix is controlled by the recipe, amount of solid content in the size liquor and the type of sizing product, mechanical mixing level, temperature and time of boiling. Flat filaments, textured and spun yarns pick-up size differently. The critical parameters to watch in the sizing process are size homogeneity, constant speed of the sizing machine, constant size concentrations and viscosity.

Flooding or dry zones should be prevented in the size box. Temperature of the size box is important for proper size pick-up. For 100% polyvinyl alcohol (PVA) sizing, a temperature of 160°–170°F is recommended. Today modern sizing machines dynamically adjust the degree of sizing. Expert software packages calculate sizing values on-line as a function of the warp weight. Size application measuring and control systems are used to measure and calculate sizing parameters automatically instead of time-consuming laboratory test procedures. Based on the calculated parameters, the squeezing pressure at creep and normal speed is controlled via computer. A byrometer measures the density of the mix and controls the supply rate of the ingredients.

The purpose is to keep the warp sizing degree constant. High-speed weaving machines require minimum hairiness in warp yarns. During slashing, yarn hairiness is affected mainly by the spacing between adjacent yarn ends in the size box and the slasher dryer configuration. In practice, the size box occupation may be used to determine yarn spacing. The Teflon® coating on all of the dryer cans should be in good shape to prevent dry can sticking (also called shedding) which may be a problem. Since open-end and MJS yarns have high wet pick-ups, the slasher may have to be slowed down to eliminate dry can sticking.

Selection of a sizing machine depends on several factors including

- Warp specifications,
- Weaving requirements and production volume.

The output of the sizing machine is determined by the size of the dryer. In so-called "walk-through head-end" beam winder, the beam support/drive unit is independent of the delivery/comb unit. There is a "walk-through" platform in between, which allows better access to the comb, delivery roll and beam. Automatic hydraulic beam loading and unloading, independent hydraulically lifted delivery nip rolls, pneumatically operated expansion, contracting and shifting of the comb are some of the other features of this new system. This concept was developed for large warp beams. The quality of woven fabrics depends to the quality of woven fabrics depends to a great extent on the quality of warp preparation. Therefore, sizing machines are usually incorporated in weaving room control and monitoring systems as shown in Figure 16.2. For trouble-free weaving, a well-slashed warp is a must. Poor slashing may increase loom stops, which in return increases the cost of weaving.

Control system in sizing

The size box is equipped with three rollers with two squeezing nips. The size solution is applied on the warp sheet by a spraying system to provide enough time for solution application; then the size is squeezed by the first nip, and the warp sheet is immersed in the solution and squeezed again by the second nip prior to exiting the size box to enter the drying zone. The size box is smaller than traditional boxes, and the amount of size solution in the box is small. A short pass of the warp sheet provides improved guidance, reduced waste, improved control of size solution temperature and reduced heating energy. The size feed inlet provides flow of the solution in one direction to continuously bring fresh material to the size box. The excess solution overflows into a container connected to the size box and the size solution storage tank for recycling after filtration. The temperature of the solution is controlled by two systems. When the temperature has reached the required level, one heating system may turn-off while the other keeps heating the solution to allow less fluctuation of the desired temperature.

16.3.1 Pre-wet sizing

Pre-wet sizing technology is another recent development, in sizing technology; the yarns are wetted and washed with hot water prior to entering the size box. It is claimed that by doing this, the size add-on can be reduced by 20%–40%, size adhesion is improved, abrasion resistance is increased and hairiness is reduced. Figure 16.3 shows the schematic of the pre-wetting process.

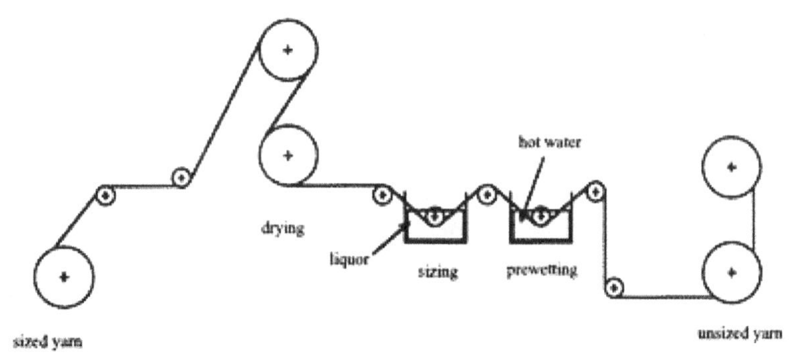

Figure 16.3 Pre-wet sizing process

The improvement in weaving performance is attributed to better encapsulation of the yarn by the sizing agent and better adhesion of the sizing agent to the yarn. The advantages of this system are an increase in tensile strength and abrasion resistance, reduction in hairiness, clinging tendency reduction and reduction in lint formation during weaving.

16.3.2 Sectional warping machines

Typical features of a modern sectional warper are:

- feeler roller to apply material specific pressure to obtain exact cylindrical warp build-up;
- lease and sizing band magazines;
- constant warp tension over the full warp width;
- automatic section positioning with photo-optical section width measurement;
- pneumatic stop brakes;
- warp tension regulation for uniform build-up; and
- automatic warp beam loading, doffing and chucking.

Today headstocks are equipped with advanced design features such as precision direct drive, advanced electronics, smooth doffing and programmable breaking. Automatic hydraulic doffing is accomplished with the operation of one button. Programmable pneumatic braking provides a constant stopping distance regardless of the operating speed or beam diameter. The length of the yarn wound on the beam is controlled with a measuring roller and counter device. The density of the yarn can be controlled by tension, pressure or both. Frictional drive usually results in higher yarn density. In spindle drive, yarn tension and a hydraulically activated pressure roller are used to control density. Some headstocks are designed to run more than one beam width. Manual cutting and knotting takes an average of 8 seconds. For a 640-package creel, it takes 85 minutes for one person to complete the whole creel.

In modern machines, yarn cutting and knotting are done automatically. In an eight-tier creel, automatic knotting and cutting device requires an average 2 seconds per package which totals 21 minutes for the whole creel. The automatic knotting and cutting devices are mounted on rails that are integrated in the creel. There are as many cutting heads as there are tiers. The devices are controlled by a PLC (programmable logical controller). Each package row is approached exactly at traverse/creep speed by means of two proximity initiators. The oscillating suction and gripping tubes offer the yarn ends to the knotting heads, where they are knotted and trimmed. The tails are removed by suction. The current trend in weaving is towards larger warp beam diameters. Today, weaving beams of 1,600 mm diameter are possible extent on the quality of warp preparation. Therefore, sizing machines are usually incorporated in weaving room control and monitoring systems as shown in Figure 16.3. For trouble-free weaving, a well-slashed warp is a must. Poor slashing may increase loom stops, which in return increases the cost of weaving.

16.4 Denting

The drawer takes a bunch of ends which have already been drawn in through the heald eyes, straightens them up, selects them in pair or in any other orders as the case may be, by his left-hand, and finally draws them through the dents of the reed in the proper order by means of the hook with his right-hand. The operation is repeated until all the ends of the bunch have been drawn in. These free ends of the bunch will then be loosely knotted together. The drawing-in of the ends of the remaining bunches will be carried out exactly in the above manner. Sometimes the denting is performed mechanically. An efficient drawer can draw from 1,500 to 2,000 warp ends in plain order per hour. One frame can supply 60–100 looms depending on the total number of ends per beam, the average number of picks per inch in the cloth woven, the average loom speed, etc.

Denting apparatus

Only one operative is required to perform the operation of denting. No power is required to drive the apparatus, except a downward pull of a cord by the thumb of the operative. It can be used in the drawing in frame as well as on the loom. To start with, the operative holds a bunch of ends by one hand, selects the ends in pair or any other order as required by the pattern and inserts these ends in the slot of the drawing in hook. As soon as this is done, the cord, which is held by the thumb of hand holding the bunch of ends is pulled downwards. This will result in the drawing in hook along with the ends hooked in it, being inserted in to the dent of the reed, which has been already kept open by the drawing in blade for that purpose.

The operative now selects the next pair of ends for insertion in the hook when the later returns to its receding position (Figure 16.4). The drawing-in blade mechanically selects and opens the dents of the reed in succession for the denting hook to pass through them. The whole cycle is repeated until all the ends from one selvedge to the other have been dented through the reed. The denting blade can be adjusted to suit reeds of different counts without changing any part of the apparatus. The main work of the operative is to select the ends and put them in the slot of the hook. The selection of dents and insertion of the ends through them involve no eye strain. The apparatus is simple in construction and contains only a few parts. Consequently, however, limited cases where the operation of drawing-in is semi-automatic, i.e., drawing-in of the ends through the healds is carried out alone or where there is scarcity of "reachers". The production of the apparatus depends on the speed of the operative.

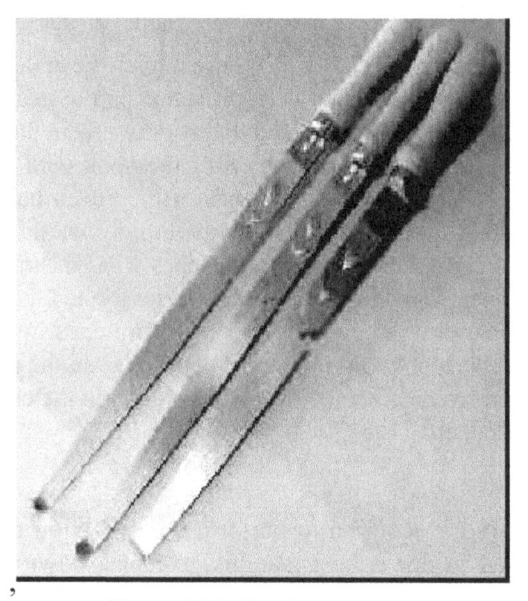

Figure 16.4 Denting apparatus

16.4.1 Semi-automatic reaching in machines

Here, the reaching-in machine is employed for the purpose of end-finding from the weaver's beam, thereby reducing the labour requirements from two workers per set to one worker per set. The worker has only to draw the ends as given by the reaching-in machine through appropriate heald eyes and dents in the reed. It is quicker than the manual drawing-in, the quality of work is better, the cost of drawing-in is less, and the floor space requirement for a given size of loom shed is less as compared to manual drawing-in process. However, since the capital investment is more, it is economical only where the volume of work is sufficient to keep the machine fully engaged (Figure 16.5).

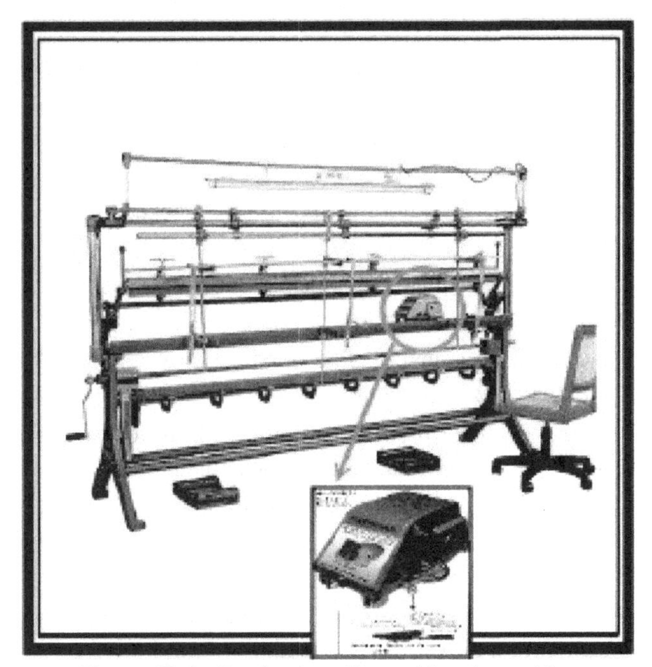

Figure 16.5 Semi-automatic reaching- in machine

16.4.2 Automatic knotting or tying- in

The process is widely used nowadays in mills where the quality of the warp is not often changed and the volume of work is sufficient to keep the tying machine fully engaged. The process can be used only where the new warp is identical to the old warp in respect of total number of ends, counts of the aids and reed and the order in which the ends are to be drawn through the healds and reed. In the case of striped warps, the warp patterns should be identical as well. Both stationary and portable machines are available for carrying out the process. The sequence of operations carried out on these machines is similar to that followed in the case of manual twisting in. The operations of selection and knotting of the ends, cutting the tail ends of the knotted thread and stopping the machine in the event of a thread found missing or broken are performed mechanically and automatically.

16.4.3 Automatic tying machine

An automatic tying apparatus is usually provided with the following important parts of mechanism.

a) A carriage on which the tying apparatus is fitted.
b) Selector needles for picking up threads from the old and new warps one by one in the proper order.
c) Thread carrier for carrying the threads picked up by the selector needles for the next operation of knotting.
d) A knotting unit for knotting the threads picked up by the thread carrier.
e) A shearing mechanism for cutting the tail ends of the knotted threads.
f) A traverse motion for effecting advance movement of the carriage carrying the tying apparatus.
g) An electric lamp for providing additional light for watching the performance of the important parts of the apparatus and also the parts of warp situated immediately near the apparatus.
h) A hand wheel for operating the apparatus manually at the start or for inspection purpose.
i) A driving unit consisting of a small electric motor with necessary arrangement for adjusting the speed.

16.4.4 Stationary tying-in machine

The stationary machine works in the preparation room. The old warp with its healds and reed is carried to the preparation room where the new warp would already be position on the stationary adjustments of the machine, the actual tying-in is started. On completion of knotting of all the ends, the knotted ends of the new warp are pulled through the healds of reed. The ends of the old warp threads are then cut-off. Thereafter, the new warp with its healds and reed is brought to the room and gaiting up is carried out in the usual manner. To prevent the stationary tying machine from lying out of production for the preparation of warps before actual tying-on operation is started, usually two trucks are used nowadays, so that while the automatic knotting is progressing on one of the trucks, the other truck is loaded with the harness and the warp for the next tying on operation.

Figure 16.6 Automatic tying-in machine

As its name implies (Figure 16.6), the machine automatically carries all warp ends to the reed hook in proper sequence. The beam is fitted on the stands of the machine provided for that purpose. The warps are fixed between a pair of clamps without disturbing their relative positions. The first warp end is connected with the automatic reaching-in motion and the current switched on. The automatic device will now present to the reed hook all the warp ends one by one on correct rotation. As soon as any warp end is taken-off by the drawer, the next end is ready for him. In order to minimise eye strain of the drawer, the drawing-in threads is electrically illuminated. Further, the machine is provided with; mechanism for separating the warp ends that might be sticking together. This will ensure only one end being conveyed to the reed hook at each operation.

16.5 Advantages of electronic and control systems in textile industry

- The systems for regulating, controlling and the data recording. Productivity, quality, operational profit of the industry.
- The performance of electronic systems in recent years has made it possible to realise solutions, which was previously rejected as complaints for quality and cost reasons.
- Enable us to quickly identify and systematically eliminate weak points in the production process.
- High production rates.
- Consistency in quality due to continuous monitoring of the process and product.
- Reduced maintenance costs and spares.
- Flexibility in assessing the database and reduced manpower.

Electronic controls in let-off and take-up in weaving

The take-up is used to keep the required thread density in the fabric and wind the fabric on the cloth roll and let-off release the required amount of warp. The electronic system provides important time saving the required pick density is electronically set (no pick wheels are required to change pick density). The accuracy of settings makes it easy to adjust the pick density of the fabric for optimum fabric weight and minimum yarn consumption. The mechanical let-off uses combination of clutches and brakes. The wear, play and backlashes of these mechanical parts result in irregularities in density and is unable to meet ever-demanding quality conscious customers' requirement, The take-up and let-off are driven by separate motors. The electronic motor driven let-off supplies the loom with the necessary warp yarn, maintaining the yarn tension constant from full beam to empty beam. Let-off speed is automatically calculated in context with loom speed, weft density, warp beam diameter and close loop tension control. Precise synchronisation with main motor in forward and reverse direction plus special advance features help greatly to reduce the start and stop marks, thus increasing the grade and quality of the fabric. It also makes cramming/density variation design possible (Figure 16.7).

Let-off and take-up motions are identical in construction, each motion utilises a resolver as the measuring system, connected together with the sensor to a control circuit. Electronic fabric take-up and warp let-off is not only controlling and reacting, but also acting with regard to the future. Absolute sensors measure the warp tension – independently of the position of back-rest roller and mechanical element motion – keeping it constant, even when weaving with splitted warp beams. The accuracy of warp beam settings on the display amounts to 1 cN/end with a filling density resolution of 0.01 picks/cm. Exactly reproducible values for filling density, machine speed, warp tension and contraction support start-mark prevention.

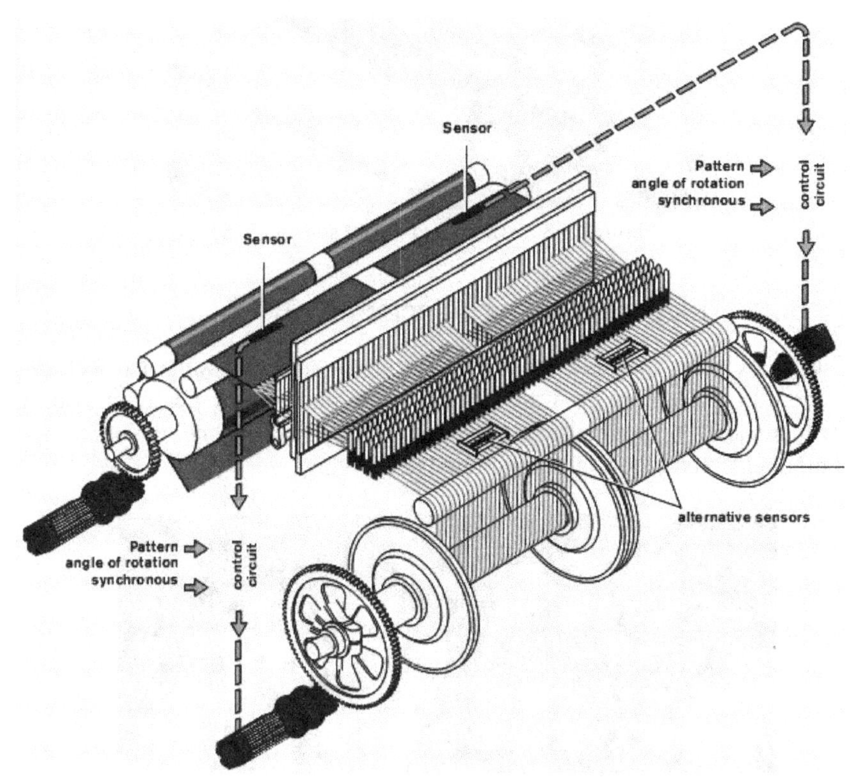

Figure 16.7 Electronic controls in let-off and take-up in weaving machines

16.5.1 Slim through light sensor (STS)

The new, patented filling stop sensor DORNIER STS (Slim through light sensor) shown in the Figure 16.8 is based on the through light principle. It provides highest functional and quality reliability for dark filling colours and finest threads up to 20 den. It can be easily positioned anywhere on the reed with a clip-on attachment according to filling insertion width. The compact design prevents damage to reed teeth.

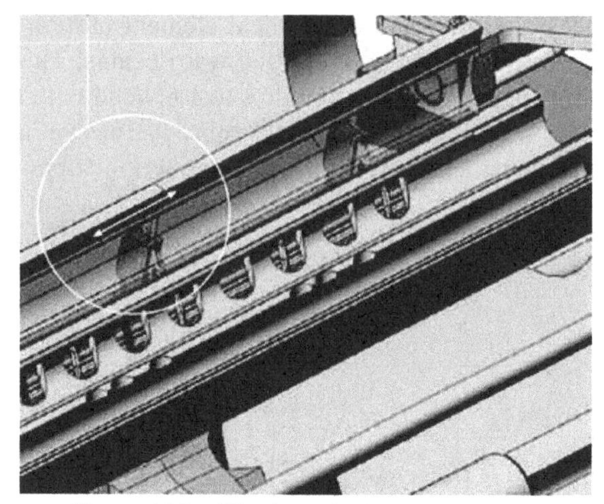

Figure 16.8 Slim through -light sensor (STS)

16.5.2 Electronic colour selector (ECS)

During weaving various colours are used in the fabric in warp and weft direction. In this system the weft colours are changed electronically instead of mechanical slow speed system. The electronic filling tensioner maintains optimum tension on the weft throughout the weaving.

Figure 16.9 Electronic colour selector (ECS)

The electronic colour selector (ECS) (Figure 16.9), and the electronic filling tension (EFT) device with integrated filling stop motion, is based on a new type of stepping motor distinguished by its sturdy construction, extremely small control increment of 0.9 o and high torque setting and function relative to the weaving process are precisely programmable on the machine display. Due to the modular concept individual modules can be added or removed easily. In this way, a single colour machine can be inexpensively upgraded to handle 12 colours. The powerful stepping motors are of a very compact design and permit very small graduated increments which in turn help to perfect sequence in the weaving process. The motors are controlled via a microprocessor. Needle position can be set individually on the colour selector via the display.

16.5.3 Pre-winder switch off (PSO)

GamMax has a Piezo-electric filling detector that stops the machine in case of a filling break. With its pre-winder switch off (PSO) system, the machine continues weaving even if a filling break occurs on the packages or the pre-winder. The pre-winder signals the filling breaks and simply switches to single-channel operation instead of weaving with two channels. PSO is a patented Picanol development.

At a filling break the machine stops and only the harnesses are moved automatically to free the broken pick for removal by the weaver. The automatic pick finder and the slow motion movements are not driven by a separate motor; instead, the pick finding is simply done by the Sumo at slow speed. The required pick finding position is reached with a minimum of reed movements through the beat-up line.

16.5.4 Electronic yarn tension control

Each pre-winder can be equipped with a new type of Programmable Filling Tensioner (PFT). This PFT is microprocessor controlled and ensures optimum yarn tension during the complete insertion cycle. Reducing the basic tension is an important advantage when picking up weak yarns, while adding tension is an advantage at transfer of the yarns and avoids the formation of loops. The tension control enables you to weave strong or weak yarns at even higher speeds. It also drastically reduces the amount of filling stops, and enables you to set an individual waste length per channel and reduce the waste length for some channels. The Yarn Tension Meter is a portable and precise sensor between pre-winder and filling detector to measure and display the filling tension on the microprocessor quickly. Thanks to the immediate feedback, the sensor allows very easy fine-tuning of the yarn tension, which is very helpful when sensitive yarns need to be woven, such as wool. Instead of "trial and error", you can "measure and act". This has a direct impact on the productivity and the quality.

16.5.5 Automatic start-mark prevention (ASP)

ASP: Preventing start-marks at the source. The simple functionality of automatic start-mark prevention saves time and significantly contributes toward quality improvement. All the functions outlined in the illustration can be simply called up on the machine display (Figure 16.10).

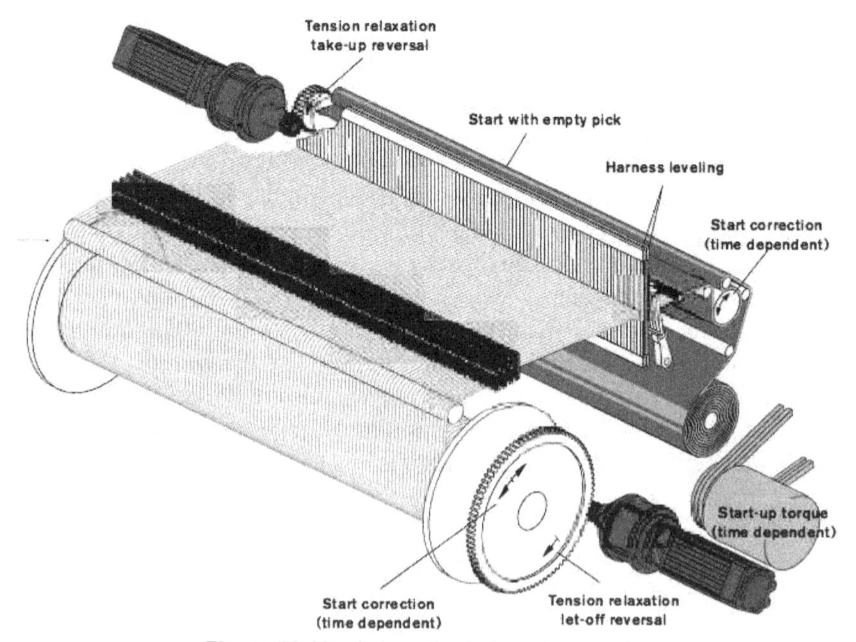

Figure 16.10 Automatic start-mark preventer

Toyoda's electronics technology has created the E-shed, yet another breakthrough in the weaving industry. Its ultimate shedding motion makes weaving easy, even with textile fabrics difficult to weave with conventional shedding motions. Based on the settings from the function panel, the E-shed's main 32-bit CPU controls independent servo-motors that drive individual shedding frames with perfect ease. This flexible system enables not only shedding patterns but also cross-timing and dwell angle to beset from the function panel. All of which leads to greater efficiency, easier operation and higher fabric quality.

16.6 Control system in warp let-off motion in looms

In the modern weaving machines, several weaving machine builders have developed the electronically controlled warp let-off motions. The drive of this system is activated by an independently regulated electric motor, which is controlled by electronic regulator. In Sulzer-Ruti machines P 7100, warp yarn let-off is controlled by a variable speed electric motor. In Picanol looms, the warp yarn tension can be registered by detecting the displacement of the backrest roller by means of and electronic sensor.

Mechanical control system

The mechanical control system for the warp let-off is shown in the Figure 16.11 (A) and (B). In this system the displacement of the back rest roller represents the differences between the warp yarn tensions that is controlled by spring and weight loaded tension. The differences sensed are transferred to the PIV controller to drive the warp beam to a position corresponding to compensate the difference between these two tensions.

The block diagram of a closed loop system and the schematic diagram of the mechanical warp let-off system are shown in the Figure 16.11 (A) and (B).

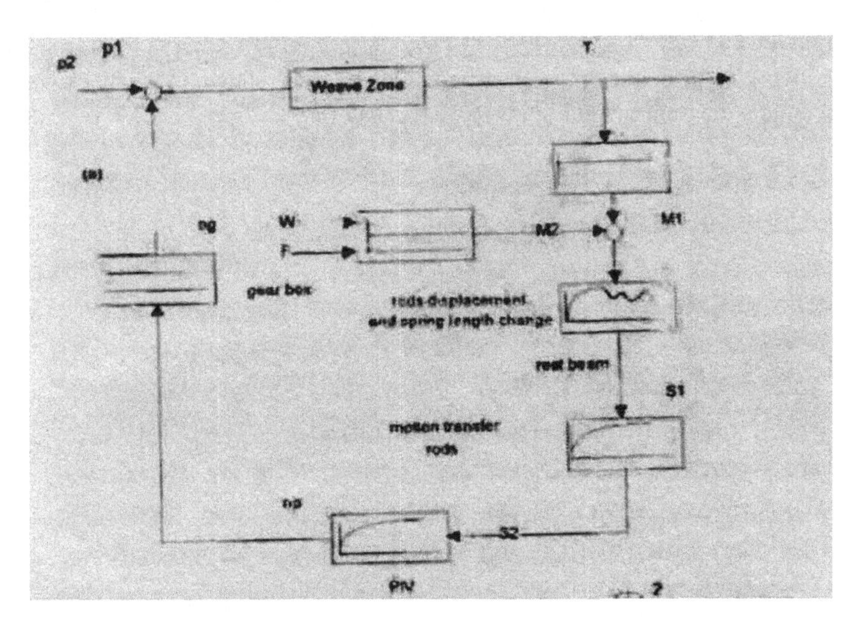

Figure 16.11 (A) Mechanical warp let-off motion Diagram shows the block diagram of mechanical warp let-off

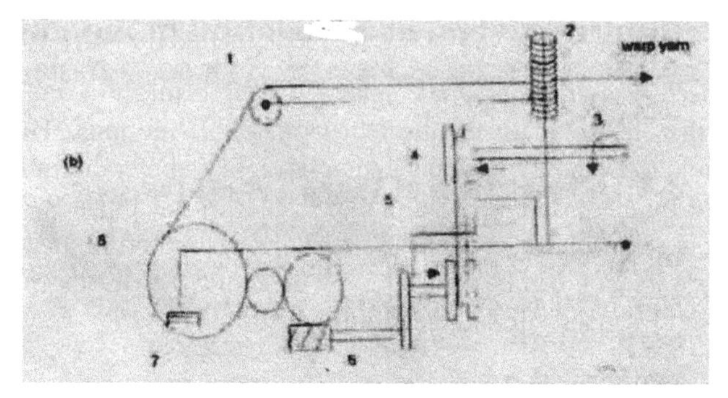

Figure 16.11 (A) Scematic diagram of mechanical warp let-off

1.Back rest roller, 2. Load spring, 3. Drive from loom, 4. PIV drives, 5. Levers for PIV mechanism, 6. Worm gear, 7. Weight, 8. Warp beam.

In the Figure 16.12 (A) and (B), P1 and P2 are the cyclic and varying tensions, respectively. F and W are the spring loaded and weight loaded tensions to displace the back rest roller in M2 value. Warp yarn tension displaces the same roller in M1 value so that the final movement of the back rest is dependent upon the resultant of these two displacements.

16.6.1 Electronic control system

The working principle of the electronic warp let-off is shown in the block and schematic diagram in the Figure 16.12 (A) and (B).

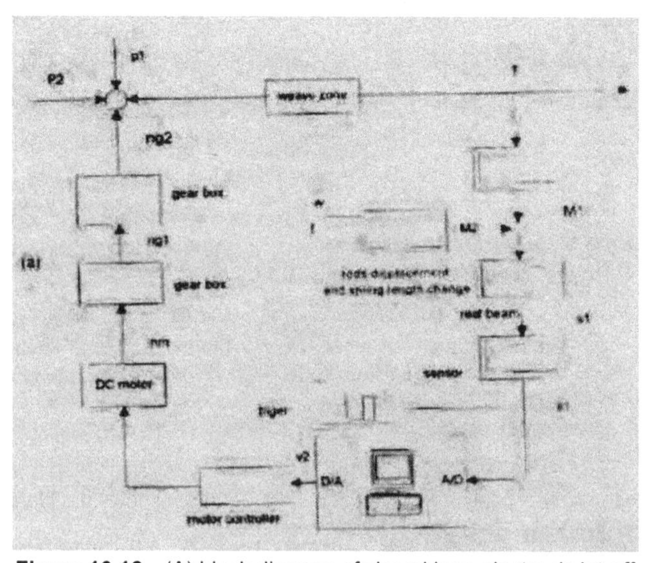

Figure 16.12 (A) block diagram of closed loop electronic let-off

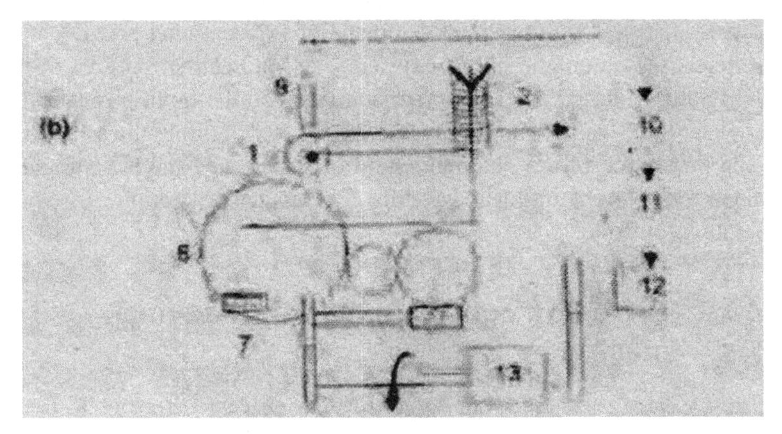

Figure 16.12 (B) Schematic diagram of electronic let-off.

1. Back rest roller, 2. Load spring, 7. Weight, 8. Warp beam, 9. Sensor, 10. Computer, 11. DC motor driver, 12. DC motor, 13. Gear box.

The principle is similar to the mechanical control system and in this system, the movement of the back rest roller displacement is measured continuously by a linear inductive analogue sensor at 75° of running cycle of the main shaft by means of a trigger in an electrical voltage and is transmitted to an A/D card. The measured values are processed by a computer which then the output signal is transmitted to a DC motor via a D/A card, amplifier (Op, Amp) and a motor controller to control the warp let-off units as a compensating function. The block diagram for this system is shown in the Figure 16.13.

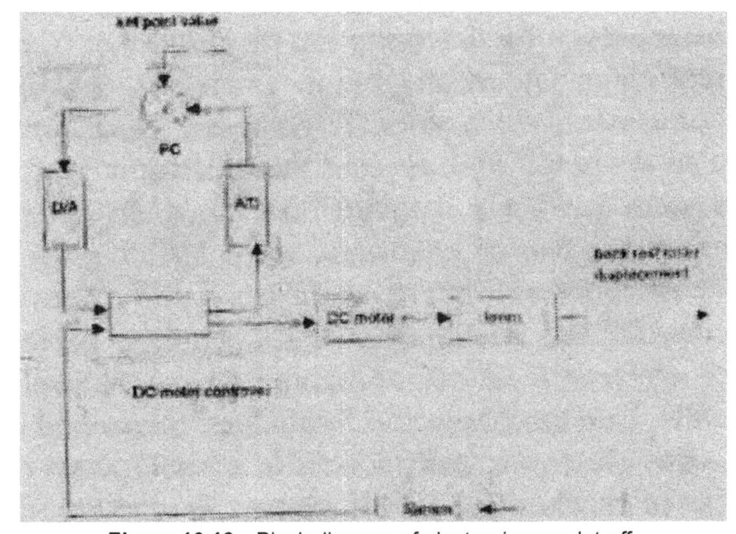

Figure 16.13 Block diagram of electronic warp let-off.

The system is such that the assemblage of PC computer and DC motor controller is the automatic control system that applies the required voltage to the DC motor based on the comparison between the input signal and the required set point value. Hence, it is necessary to determine the relationship between the output of the automatic control system and the actuating error signal. In order to obtain desired results, a computer programme is required to operate a controller.

16.7 Drive and control in modern weaving machines

The latest weaving machines are equipped with microprocessor or PLC units which ensure continuously the control, the drive and the monitoring of the various machine members and of the various functions.

A variety of electronic devices and sensors permits the collection and the processing in real time of the main production and quality parameters. These parameters can also be recorded and transferred through memory cards to other machines or stored for future use (Figure 16.14). The control unit can be connected with outer units (terminals, servers, company managing system) to transmit/receive data concerning both the technical and productive management and the economic-commercial management (Figure 16.15). All this facilitated considerably the weaver's work in respect to machines of previous generation, and enabled to improve the production yield and the product quality.

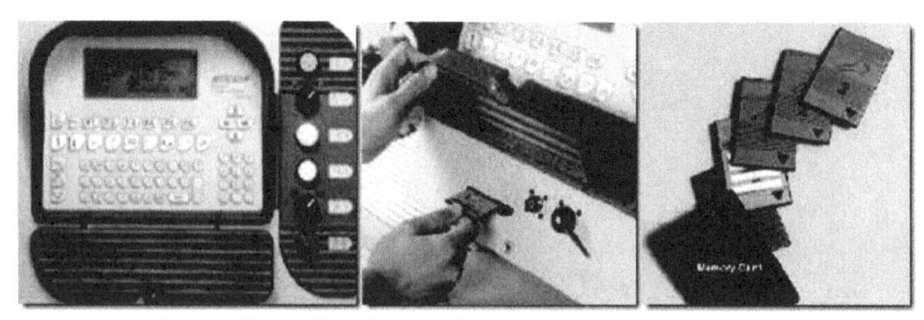

Figure 16.14 Board computer equipped with memory card

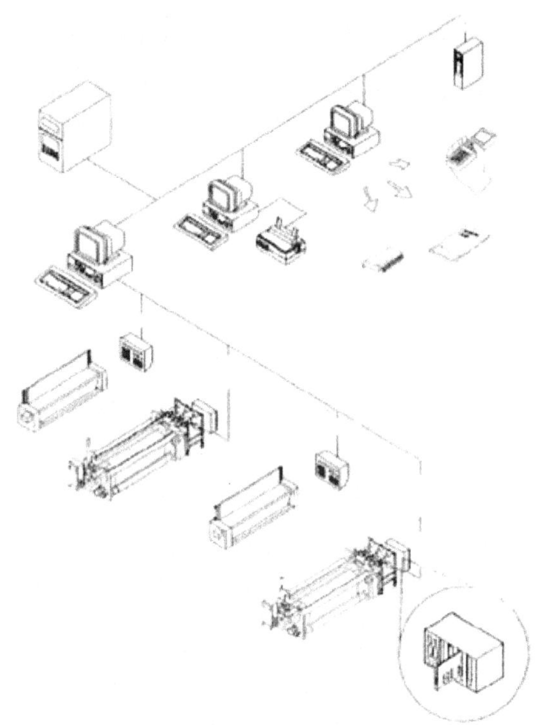

Figure 16.15 Example of a modern monitoring and control network

The main operations which can be carried out by simply keying in the value of the desired parameter on the keyboard of the electronic control unit are:

- Selection and modification of the weft density with running machine, as both the motor driving the take-up roller (sand roll) and the motor driving the warp beam are electronically controlled and synchronised one another. This permits also to combine a programmed automatic pick finding, obtained through correction programs based on the characteristics of the fabric in production, in order to prevent formation of starting marks (after machine stops) electronic selection and control of warp tension through a load cell situated on the back rest roller, which last detects continuously the tension value. This permits the processor to control the movements of the warp beam and of the take-up roller, ensuring a constant tension throughout the weaving operation (Figure 16.16).
- Programming of the electronic dobby and of the electronic weft colours selector.
- Programming and managing of nozzle pressure and blowing time in air jet weaving machines.

- Selection and variation of the working speed, as the machines are provided with a frequency converter (inverter) which permits to modify at will the speed of the driving synchronous motor.
- Statistics.
- Monitoring.
- Managing/programming of all machine functions.

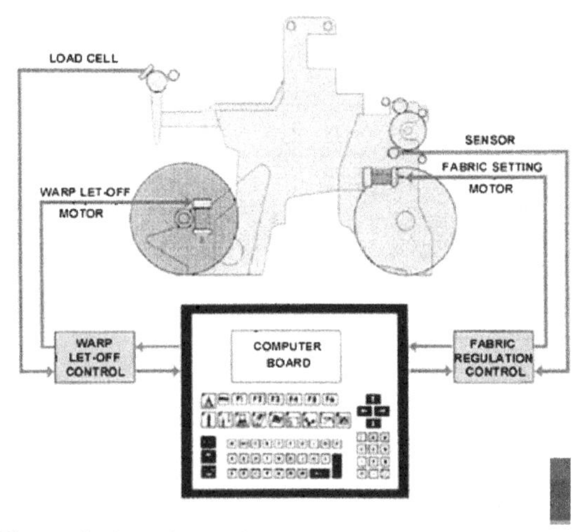

Figure 16.16 Electronic detecting and control system on thread tension. Setting and modification of tension and weft density directly via board computer

16.8 Stäubli Jacquards

Jacquard weaving solutions

The world of Stäubli jacquard machines is a fascinating one, in the entire world; countless weaving mills produce innovative high-quality fabrics with these reliable and multipurpose machines. The variety of ranges from the finest silks to the most complex technical textiles and original decorative fabrics. Stäubli electronic control is high-performance machines, benefiting from many years of technical experience. They are distinguished by their high performance characteristics, the quality of the materials used and their impressive service life. Stäubli Jacquard machines represent a complete range of formats, ranging from 32 to 24576 hooks for producing all types of fabric know-how to all weaving mills, at the international level.

Exacting requirements and high-performance:

Shedding machines guarantee optimal performance and portability. They perfectly fulfil weaving machine and shed forming requirements, geometries, openings and adjustments are almost unlimited. Weave medications are extremely flexible and can be changed rapidly. With a cam motion, a dobby or Jacquard machine, Stäubli systems allow you to get the most out of weaving machines owing to transmission systems, quick frame connects or through a custom made harness or carbon fibre frames. Stäubli attaches major importance to the reliability and user friendliness of its shed forming machines.

16.8.1 Basic construction features of Stäubli Jacquard

Transmission

The DX 100 and DX 110 are delivered with the adaptation elements required for each type of weaving machine pre-installed. The transmission is based on a cardan drive with bevel gears, whose position is determined according to the weaving machine to optimise the cardan shaft angle. For wide microprocessor spaces, the optional addition of a modulator accelerates shed opening. The compact structure of DX 100 and DX 110 machines enables them to be used on standard frames and in low ceiling weaving rooms.

Control box

Each machine has its own electrical power supply and is equipped with an electronic controller and is equipped with a JC6 electronic controller with touch screen. This user-friendly interface between the user and the DX features the latest generation of microprocessor. A flash disk advantageously replaces hard drive technology and patterns can be transferred either via USB key, external drive or the network. The latter offers multiple possibilities and facilitates weaving room management or several remote sites (Figure 16.17).

Figure 16.17 Control box

M 6 Module

The patented M6 module ensures the connection between the lifting mechanism and the harness. Compact, and engineered from composite materials, the M6 module is sealed and wear resistant and consumes very little energy. If required, depending on the loads and speeds in use, M6 module can be equipped with two journal bearings (version A), one journal bearing at the bottom and a ball bearing (C) on the top or with two ball bearings (B). Module replacement is easy and quick. The rollers built into the module make hooking and unhooking the harness cords easy – the system is also available with QUICK LINK (Figure 16.18).

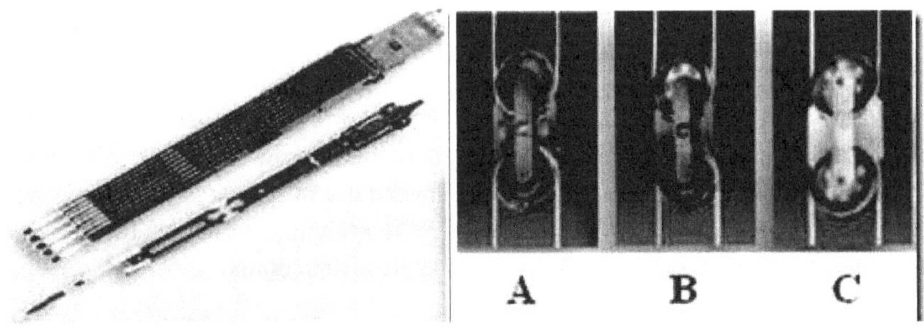

Figure 16.18 Quick link system

Selecting hooks in a module

The double roller enables the upper and lower shed position of the harness cords in double lift.

16.8.2 Low shed position

The electromagnet (f) is activated. The retaining hook (c) does not retain the mobile hook (a) which follows the lowering knife (e) Figure 16.19 (A) and (B). Upper shed position: The electromagnet (f) is not activated. The retaining hook (c) retains the mobile hook (a).

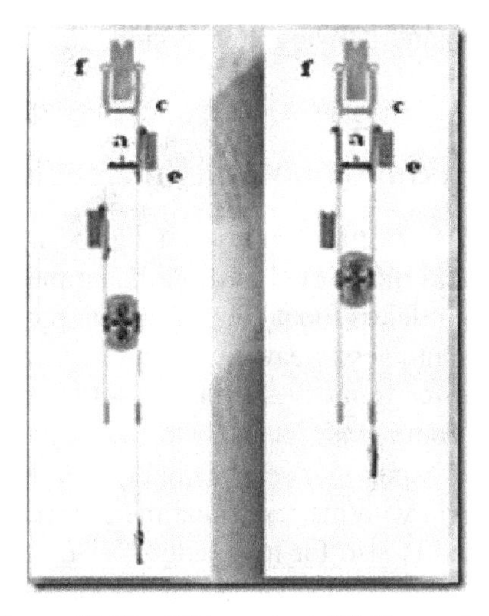

Figure 16.19 (A) Low shed, (B) Upper shed

16.8.3 Quick link

The quick link connection system was developed for speedy hooking and unhooking of the harness cord (Figure 16.20).

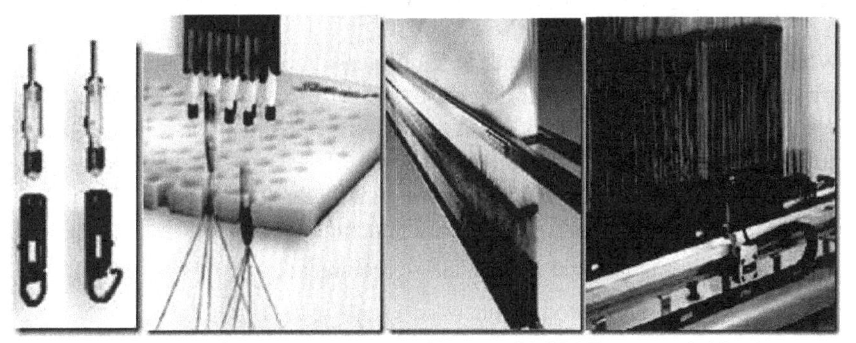

Figure 16.20 Quick link connection Harness BLOBAR automatic blowing

Harnesses

Developed by Stäubli's know-how, harnesses are designed for all applications: labels, silk goods, lining, clothing, upholstery fabrics, table and bed linen, automotive seats, airbags, bedspreads, terry cloth and pile fabric. The links between the harness cords and the heddles, the heddles and the springs are designed to meet all your needs. Blobar an automatic blowing system was designed to facilitate the cleaning of harness springs, at the back of the down pull frame. The pneumatic blowing device limits fly deposit and facilitates manual cleaning.

16.9 Modern developments in weaving machinery

In the last two decades, spectacular progress has been made in the field of weaving technology and the most significant being the replacement of convectional looms by shuttleless looms for increasing productivity and quality of the end product. Shuttleless weaving is making an impact on the textile industry. The changeover from fly shuttle to shuttleless involves both new technology and shift from labour intensive to capital intensive, mode of production. Moreover, for export market, the quality requirements are becoming more and more stringent with the result that the export of the Indian mills is falling. Market demand is also for long lengths of fault-free cloth, which is only possible with shuttleless weaving machines. Increased labour cost without any corresponding increase in productivity is resulting in reduced profit to mill owners. So today, we are in need of shuttleless machines which are weaving from the lightest to the heaviest of fabrics and diversified products using materials like spun, jute, woollen, worsted, metal wire, glass fire, mono and multifilament, etc., with good quality.

16.9.1 Developments in shuttleless machines

When the topic of developments in shuttleless machines is discussed in any group/forum the question that first arises is: “Why Shuttleless Machines?”

In fact the modernisation or automation is not just for reducing manpower. So is the case for installation of shuttleless machines. Our country lags behind in global market competition mainly due to this aspect.

The benefits of installing shuttleless machines are as follows:

- Better and assured quality fabrics produced.
- Higher rate of production.
- Consistency and reliable performance.
- Assured delivery time.
- Flexibility of the machine.
- Scope for manufacturing creative products.
- Better-engineered fabrics, wider fabrics, etc.

Among the shuttleless machines considerable developments are observed in three basic picking principles, i.e., projectile, air-jet and rapier, while the new picking style is multiphase weaving.

Features common in all machines

Several essential features found common with all shuttleless machines are listed below:

- Higher speed.
- Wider width.
- Electronic take-up and let-off.
- Shedding systems -- cam, dobby and jacquard (mechanical and electronic).
- Electronic monitoring of weft yarn flight.
- Electronic warp stop motion.
- Automatic pick finding.
- Quick style change.
- Microprocessor controls with digital display.
- Low noise and vibration.
- Tension free weft supply by weft accumulators.
- Microprocessor control lubrication system.

16.9.2 Developments in projectile machines

1. Colour selection
    - 1 X 1, 2, 4 and 6 colours can be used in weft direction.
    - The system is freely programmable and operated by servo controller.
    - No limitations on feeder position shifting.
2. Electronic weft braker
    - This device keeps a uniform tension on weft.
    - The braking force and the braking duration are programmable.
    - Programme can be given for each pick.
    - The device is driven by stepper motor.
    - Pre-acceleration to weft yarn is given by compressed air, which relieves extra tension in weft while inserting.
    - K3 Synthetic projectile can be used for weaving of delicate yarns.
    - The number of heald shafts operable by cam motion is extended to 14.
    - Speed has been increased up to 1400 mpm (470 rpm) due to improvement in many related mechanisms.

- LED display at signal pole for machine speed, projectile arrival time, angle of machine stop, etc., which helps in monitoring of process.
- Automatic weft brake repair motion enables shifting of feed package to a reserved one in the event of weft break between package and accumulator, no stopping of machine which increases the machine efficiency.

16.9.3 Developments in air-jet weaving

The air-jet weaving machine continues to dominate as the machines of very high speeds. Today, practically (in Indian condition) at 1200 rpm the machine works or wider machine can attain a WIR of 2500 mpm. The system had the disadvantage of higher energy consumption due to the usage of compressed air in picking, which accounts for 60% of total energy consumption.

The machine makers claim a reduction in energy by about 10% (Sulzer, Somet) in their latest models. The developments in picking related systems have helped in expanding the horizon of weft material and count. The yarn colour selection up to 6 or 8 beyond which demand is very rare. That means, the major limitations of the system are being attended and scope for applicability has been increasing.

16.9.4 Modification in weft insertion system

The multi nozzles are divided into two zones and connected directly with separate tanks. The weft yarn requires higher pressure at later part of its flight, and this separation has helped greatly in optimisation of pressure in duration of jet opening. The weft insertion, based on a precise electronic control that includes ATC (Automatic Timing Control), also uses newly developed nozzles, which guarantees optimum weft insertion conditions.

Independent pressure tanks make it possible to set weft insertion pressures at optimal levels; this makes a significant contribution to energy conservation. All settings regarding picking is done by microprocessor keyboard, which reduces machine down time.

Tandem nozzles: In tandem nozzles, the two main nozzles are arranged in series so called tandem nozzles.

Advantages

- It reduces the nozzle pressure.
- Saving in energy.
- Also use of wider weft count range.
- Low pressure weft insertion to occur, making effective for super high-speed operation accommodating yarns with low breaking strength.

Tapered sub-nozzles: It consists of a tapered hole to prevent air dispersion.

Advantages

- It enables stable weft insertion with lower air volumes.
- It stabilises air injection angles during weft insertion.
- The weft insertion is more stable and requires less air.

Tapered tunnel reed: A tapered shape has also been applied to the tunnel selection of reed blade.

Advantages

1. It helps in preventing air dispersion.
2. The weft insertion is more stable and requires less air.

Electronic braking system

One of the serious drawbacks of air-jet picking is the tension peak in weft when brake is applied. The electronic braking system can precisely control braking time and brake stroke, which significantly reduces tension pick, thereby reduction in weft breaks.

Automatic pick controller

For smoother working, all machines have weft arrival time sensing and correction of pressure at nozzles but when package is changed from empty to full package, the arrival time will be delayed and this would be beyond the capacity of such a correction system. With APC (Automatic Pick Controller), this problem is attended.

- It instantly corrects the main nozzle's air pressure for timing control during full cheese changes.
- It automatically adjusts nozzle air-jet pressure, which compensates for variations in the travel timing of weft yarn.

Other developments

- Weft feeder threading is comparatively time consuming and, now the self-threading by pneumatic system is done.
- The weft cutter is electronically controlled and operated by steeper motor. By this, cutter can easily adapt to any cutting time to the accuracy of 1. Style changing time is saved.
- With the help of mechano-pneumatic tucking device can hold the weft at both selvedges firmly during beating and then tuck-in, this

eliminates auxiliary selvedge and weft waste is zero. The system can work upto 850 rpm.

- Almost every machine manufacturer supplies positive easing motion for maintaining constant tension during shedding and beating.
- There is a new shedding concept introduced, in which the heald shaft is directly controlled by Servo Motor. Thus, the total motion of heald shaft can be independently programmed.

16.9.5 Developments in rapier weaving

The rapier machines are emerging as weaving machines of the future. They are not far off from air-jet in production (Speed) rate (up to 1,500 mpm or 600–800 rpm) without scarifying their special status of flexibility. They have been making inroads to heavy fabrics (900 gsm) and also shedding off the known drawback of higher weft waste. The design improvement in rapier gripper permits handling a wide range of yarns without any need for changes.

The machine owes its speed, flexibility and low energy consumption to a combination of high technology and economic design. Style changes can be executed "exceptionally rapidly". Having independent motor drives, this yielded fewer moving parts, fewer gears, fewer oil seals and no timing belts, i.e., there are fewer elements to influence fabric quality, less need for resetting and reduced maintenance. There are no toothed belts, which are prone to wear, and breakage.

Sumo motor

- Saving on energy consumption of more than 10% in comparison with conventional clutch and brake configuration.
- Machine speed setting is done accurately and completely, electronically via the keyboard of microprocessor. This reduces the setting time to zero.
- Speed setting is easy to copy to other machine either with electronic set card or with production computer with bi-directional communication.
- Automatic pick finding becomes faster, which significantly reduces the downtimes for repairing filling and warp breakages.
- The machine can always work at optimum weaving speed in function of quality of the yarn, the number of frames, and fabric construction.

PFL (Programmable Filling Lamellae)

It controls the filling brake ensuring consistent yarn tensions at any time during insertion cycle. The PFL can be installed for each channel between the pre-winder and entry of fixed main nozzle. It has been designed to slow down the filling at the end of insertion. The PFL thus significantly reduces the peak tension of the pick at the end of the insertion and decreases the tendency of pick to bounce back in the shed. As a result of which the filling tip is stretched correctly.

Features

- Lower peak tension in filling yarn.
- Reduced tendency of filling to bounce back.
- Inserted pick can be stretched more easily.
- Adjustments are done by means of machine keyboard and display.
- The settings can be adopted for each filling yarn.

Benefits

- Fewer filling breaks.
- Fewer machine stops.
- Better fabric quality.
- Higher productivity of machine and staff.
- Weaker filling yarn can be used.
- Correct setting of filling waste length and consequently less waste.

ELSY (False Selvedge Device):

The unique ELSY full leno false selvedge motion is electronically driven by individual stepper motor. They are mounted in front of harness so all harness remain available for fabric pattern. This only rapier machine that allows selvedge crossing to be programmed on microprocessor independently of shed crossing even while machine is in operation. So result of resetting can be checked immediately. The easiest position of rethreading can be set by a simple push button. When machine starts, the selvedge system automatically comes to original position.

Electronic take-up and let-off motion

It plays important role. Required pick density can be programmed on microprocessor keyboard. No pick wheel required. The accuracy of settings makes it easy to adjust pick density of fabric with optimum fabric weight and minimum yarn consumptions. By ETU makes it possible to weave fabric having various pick densities.

The electronic link between let-off and take-up is an additional tool to manage the fabric marks. Warp beam driven by electric let off motion through separate drive wheel that stays on loom, ensures trouble-free operation of let-off system and improves fabric quality.

FDEI (Filling Detection at the End of Insertion)

When weaving "lively" yarns, use FDEI system. It checks the presence of filling at the end of insertion. The system detects short picks, rebounding fillings and prevent faults in fabric at right end.

At filling breaks, the machine stops and only the harnesses are moved automatically to free the broken pick for removal of weaver. This is outstanding since automatic pick finder is not driven by separate motor but monitored by hydraulic system. In this way a two speed slow motion becomes a standard luxury to the weaver. The transfer position of filling yarn in centre of fabric is always correct even after changing the cloth for new style.

16.10 Premier loom eye and Uster loom data

Configuration of the loom data

The data collection and processing in loom data takes place through hardware as represented in Figure 16.21. In the loom data configuration, the Universal Stop Box (USB) is fitted on each weaving machine. The USB collects data signals pertaining to various information and transfers the data to the concentrator. The acquisition of the signal may be one of the following three types.

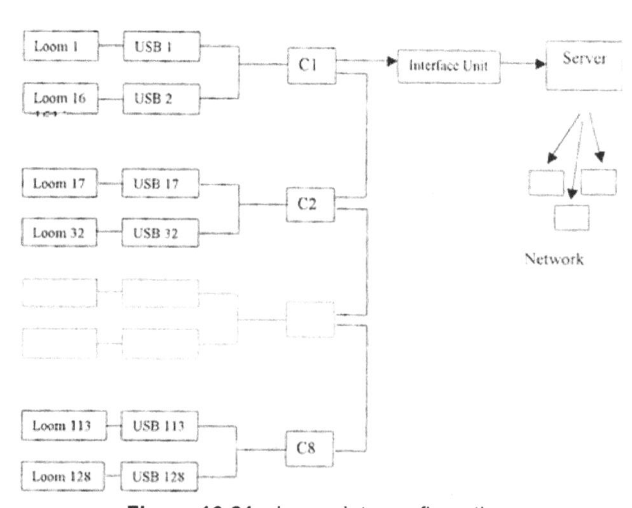

Figure 16.21 Loom data configuration

- Contactor type

  In the contactor type, the machine signals are collected directly from a potential free contactor.
- Voltage type

  In this type, the voltage available at the output of the interface card in the machine is made use of.
- Sensor type.

  In this type, proximity sensors are used to detect machine speed signals by sensing the shaft revolutions.

The machine signals include

- Speed of the machine (picks inserted).
- Stoppages (Warp/Weft breaks).
- Other breaks due to various reasons.

The USB's are connected to the concentrator through a 3-core cable.

The concentrator is used for processing of the machine signals and the information is transferred to a Personal Computer (PC). There is also a provision in the concentrator to enter the reasons for stoppages.

The PC scans the concentrator periodically through an interface unit. The PC does the following functions.

- Sequential scanning of concentrator on polling basis.
- Processing and storage of data.
- Preparation of reports on numerical values and graphical representation of the data.

With this information, the user can communicate with the PC through a user friendly menu driven software.

16.10.1 Information from loom data

Shift and long-term reports

For performance comparisons, all data is collected and stored by individual shift. This allows the user to produce different reports by:

- Machine, machine type or process
- Style
- Operator, Fixer, Supervisor etc.
- Yarn
- Test Group

- Plant or Room
- Time (trend reports)
- Order.

All of these report types allow extensive filtering of the collected data and exception values are highlighted if the values shown are out of limits. For the analysis of longer term data, reports are available by week, month or a user defined time frame.

Material reports

In the material reports, the analysis of the data is related to produce items. This is the warp in weaving preparation and the piece in weaving, finishing and inspection.

All automatic and manually collected data is show in material production maps and numerical reports:

- Stops
- Declared codes (Out of Production-Codes, Call Codes and Activity Codes)
- Quality Codes (Defects)
- Measurements
- Process Data.

Alarm server

The alarm server monitors various criteria like piece or order length reached, machine running without order, planning does not exist, data backup failed, etc. Additional criteria can be added easily according to your needs.

In case of an alarm, various possibilities exist to get the operator's attention:

- Sending of a message to the PC screen of the user.
- Sending an E-mail to the user.
- Sending a SMS to the user (appropriate SMS service is required online).
- Printing a message on a printer.
- Closing of a contact on a machine station.

Production data

On line production monitoring improves the productivity of individual machine by analysis of any back logs. Loom data provides useful information in various departments. The information obtained from the loom data is continuous which enables the weaving manager to detect the shortfalls and corrective actions can be taken for the same. The production data is presented in various forms consolidated machine wise, sort wise, operator wise, group wise according to the requirements. Furthermore, the operator cannot produce low production since each and every minute the production is monitored and also the stoppages including the down times.

Stoppage information

Loom data provides stoppage information in a summarised fashion or as a complete detailed history. A diagrammatic overview of the stoppages also can be obtained. Loom data classifies stoppages in terms of short and long stops each of which can be sub-classified in to several codes representing the actual cause of the stoppage and also the duration.

Instantaneous information

Loom data gives instantaneous status of each machine connected to it which gives the snap efficiency at any instant.

Planning for warp

Loom data gives specific reports on warp planning by giving details such as warp length remaining in the beam, time remaining for the beam to exhaust and other relevant information.

Exception analysis

Loom data provides for easy identification of exception through colour coding in its reports. The limits for the expected values are freely programmable.

16.10.2 Advantages of the loom data

- Loom data gives in formations like production reports, stoppage reports and so on which can be used in variety of ways in weaving preparatory and in loom shed in a weaving mills.
- In weaving preparatory, continuous determination of production facilitate production planning like warp on stock and warp run out production.
- The continuous supervision of production and efficiency of all machines provide for an immediate identification of weak points which can be corrected. A clear picture on down time provides an indication for instituting necessary changes thereby increasing the loom output.
- The availability of the continuous comparison of production data generated for the employees combined with the incentive schemes

results in developing an attitude towards higher productivity among the employees.

- Loom data provides information on fabric inspection giving details on the fault classification pertaining to each roll produced. This is essential to analyse the fault pattern in a roll of fabric for piece roll planning and cutting.

Loom data offers the facility for creating labels containing the details of sort. The label preparation is based on customised layout. Automatic printing of those labels is also possible.

16.11 Online weft yarn tension monitor

The Eltex electronic Tension Monitor (ETM weft) is an on-line device designed for weaving machine applications. The device continuously monitors the weft yarn tension and at the same time acts as a yarn break detector. The ETM weft communicates with the machine computer allowing the option of displaying the weft tension as a graph and/or automatically (Figure 16.22) controls the weft brakes. The yarn tension is measured as a force on the sensing element. The yarn angle across the element is predetermined by two guide eyelets. The necessary deviation is kept as small as possible but still allows the yarn to enter and leave the unit in any direction.

Features

- Monitors the weft yarn tension continuously.
- Generates machine stop at a yarn break.
- Stop motion for double insertion and with Anti-2 function as an option.
- The weft yarn can enter or leave the sensor in either direction.
- Robust design.

Advantages

- Weaving efficiency improvement possible.
- Improves weaving quality.
- Simplifies setting of the yarn brakes.
- Allows repeated style changes in the weft more efficiently and accurately.
- Fast sensor enables monitoring of rapidly changing yarn tension.

- Flash memory technique – future software upgrade possible.
- Overload protection.

Graphical representation of weft yarn tension in weaving

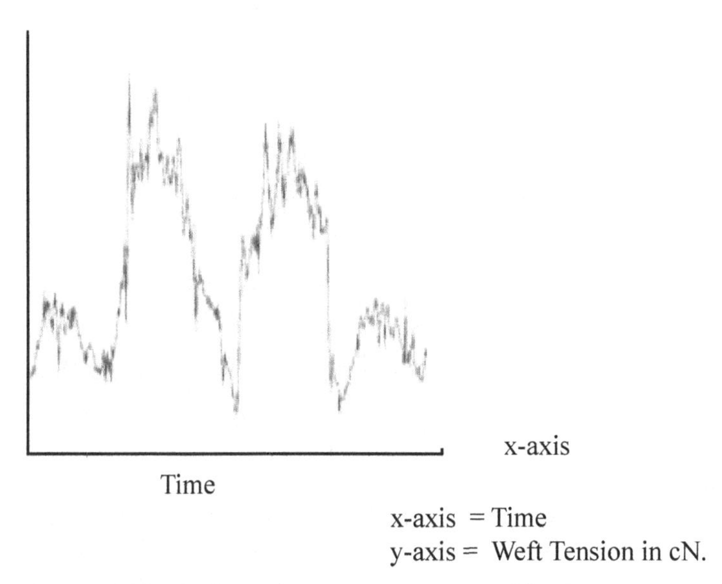

Figure 16.22 On line weft yarn tension monitors (Eltex)

16.11.1 Optical electronic weft feeler

The weft pirn used for this type of feeler is in general covered with a reflective strip which has the property of reflecting a beam of light back to its source. In the Figure 16.23 the light source and the photocell are housed together in the feeler head and both the searching beam and the reflected ray pass through the same optical system.

On the reflective strip of the pirn, incident light ray is directed on the pirn constantly. When there is yarn in the pirn, the light will not be reflected and when the pirn is exhausted, the light ray is reflected back to the feeler head. On reaching the photocell, the reflected light is transformed in to an electrical signal and transmitted to the switch box, which contains the whole electrical supply for the feeler and signals the appropriate selection mechanism to initiate the transfer of pirn. The main benefit of this type of feeler is there is no physical contact between the feeler and the weft yarn but it is quite expensive and nowadays used in filament weaving.

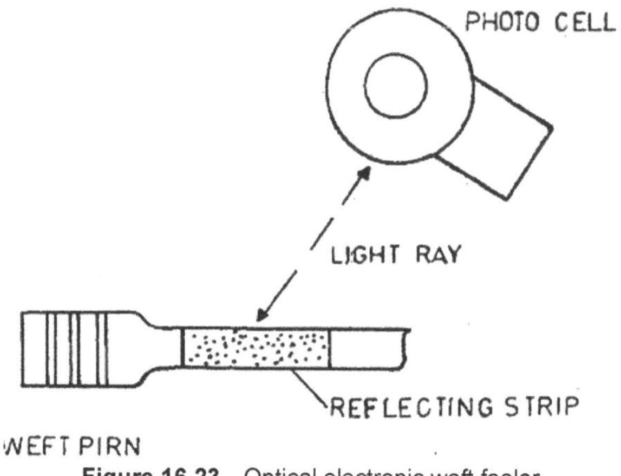

Figure 16.23 Optical electronic weft feeler

16.12 Weaving of yarn-dyed figured cloth

Yarn-dyed figured woven fabrics are generally produced based on small lots, with small size. It requires frequent change in lots causes much loss of time and material with intensive labour work. Murata Machinery Co. has developed a flexible preparatory system for weaving yarn-dyed figured cloth. Its main technological element is a new arranging winder as illustrated in the Figure 16.24. It works according to the program using PC.

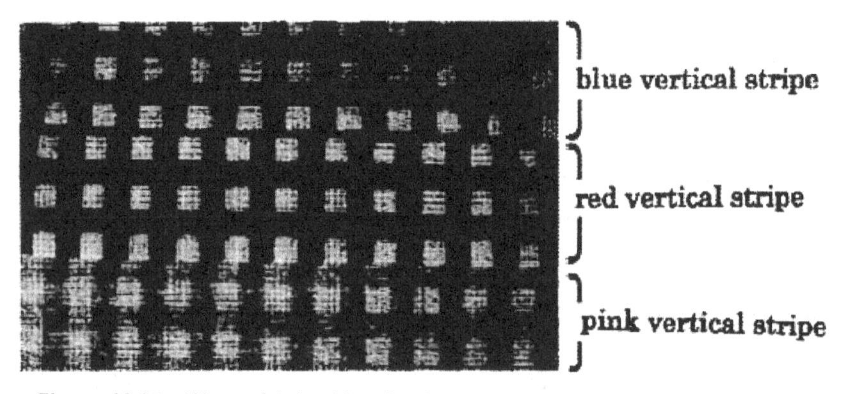

Figure 16.24 Woven fabric with striped pattern with three vertical stripe colours

In the case of cloth having three colour vertical stripes like the fabric shown in the Figure 16.24, a package of warp yarn which sequentially contain blue part, red part and pink part with precisely controlled length for each colour is made by the winder in the order of blue dyed yarn, red dyed yarn and pink dyed yarn as shown in the Figure 16.25. The winder selects a coloured yarn according to the program. After connecting the yarn with the forward yarn, the length of yarn selected is mechanically measured an dthen intermittently wound on a package for warp yarn. Figure 16.26 shows the device for measuring the yarn length.

Packages thus obtained are transferred to the process of sectional warping or beam warping. With this system, the fabrics composed of warp yarn sequentially having several kinds of colours can be smoothly produced with very low cost and less time involvement.

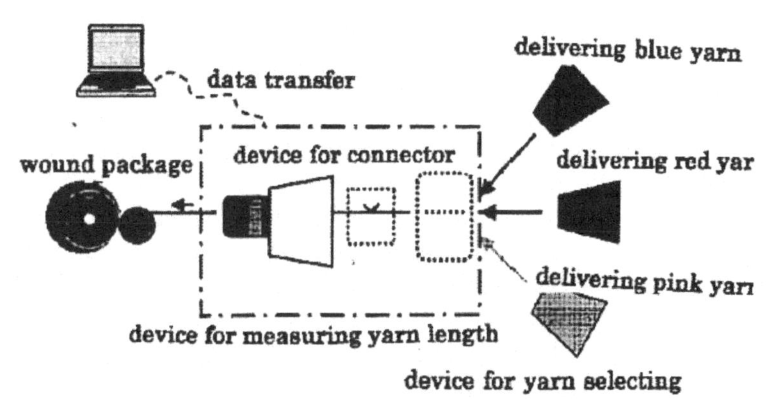

Figure 16.25 Illustration of the arranging winder

Figure 16.26 Device for measuring yarn length

16.13 Automatic fabric inspection

Defects in the fabric is a major concern for the buyers and any quality conscious mills. The faults in the fabric may be due to various reasons and mainly due to inherent defects in the yarn, improper preparation of warp and weft, poor mechanical condition of the machines, bad work practices, material handling and so on. It is impossible to produce a 100% defect-free fabric but it is possible to minimise the defects in the fabrics. Until now, most of the textile factories use the fabric inspection table for the visual examination of the fabrics with illuminated light. Although, human eye has approximately 10,000 × 10,000 = 100 million sensor elements in the eye with a processing capability of human brain equivalent to 50–100 million PCs 25% of the faults in the fabric go undetected during visual examination. This is due to the subjective method of the operator and time consuming, this method fails to produce a standardised fault-free fabric.

To offset the above draw backs, automatic fabric inspection has been developed which eliminates the human intervention. Automatic fabric inspection utilises the image processing technique and Fourier transform has been the fundamental basis for image processing. Software has been developed for storing all the results of inspection which can further be used for analysis an dgiving feed back to back process in very short time. This would help in taking actions at appropriate manufacturing stages and the defect level in the final fabric can be minimised.

One vital aspect of the automatic fabric inspection is getting information back to the weaving department when any off-quality goods are turning up. This is especially good when running defects are detected.

16.14 Applications of computers in the weaving mills

The first digital computer came in to existence in the year 1946 and from their onwards various models and configurations have developed. Earlier, computers came to India for commercial applications. Many commercial and industrial organisations started using computers in various fields of activities. The applications were extended to other areas like planning and control, forecasting, etc. Meanwhile textile industry also started using computers right from the early sixties along with the other industries. In Indian textile industry, there are many potential areas that can be considered for computer application and some of them are listed below:

- Commercial
- Process and product control
- Management information system (MIS)

- Forecasting
- Computer colour matching
- Fabric design (CAD)
- Instrumentation.

Monitoring of loom shed production

In a mill, loom shed is one of the major areas where computers can be effectively used for process control. In many advanced countries, computers are used to monitor the entire loom activity. The loom monitoring system is shown in the Figure 16.27.

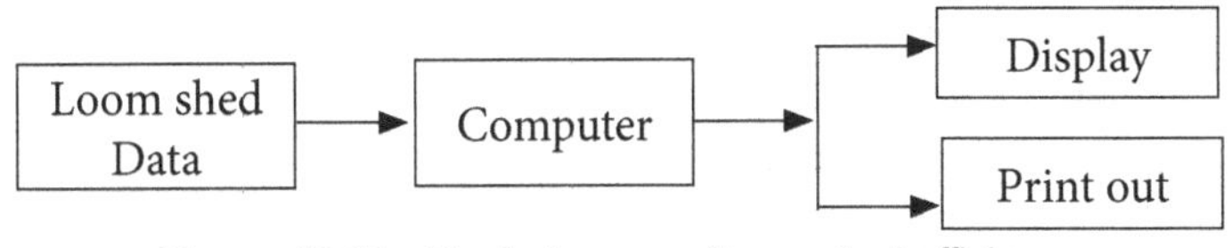

Figure 16.27 Block diagram of loom shed efficiency

In this print outs, the information like stoppages and efficiencies can be obtained for process control analysis. Stoppages show warp break, weft break, beam fall, repairs and maintenance, doffing and others. Efficiencies show loom weaver, jobber, shift, day along with the wages.

In this loom monitoring system, the computers get the information from the loom directly and hence there is minimal or no manual data feeding required. However, such loom monitoring system is required for costly looms like Sulzer or air jet looms. However, such monitoring systems bound to increase the cost and hence it is not followed in most of the mills. Presently, many mills follow various manual systems of process control. At the same time few mills use the computer facility to process their data for process control. In order to have a comparison of one system with the other, it is better to know the pros and cons of each system.

16.14.1 Manual monitoring

Mills are more concentrated to get maximum quantity of quality material from the weaving process. The factors which contribute to the loom efficiency are the type of loom, the sort (material) and the operatives employed.

For control purpose, the performances of the above factors are required to be evaluated. To enable this, the basic data available is the production from each loom obtained either through a pick counter or through the doffed piece length. The same data is required, in addition to the management information system and planning. After necessary calculations, the following control information reports are prepared.

- Shift-wise efficiency of the mills weaving section with details of sheds, jobber sections and individual looms.
- Cause wise stoppages report.
- Sort wise efficiency and production of each sort and number of looms working on each sort.
- Monthly reports of the above.
- Sort wise looms and production
- Average reed and pick.
- Piece work wages and workers.
- Category wise production.
- Balance of cuts on loom at month end.

Computerising the system

The above-mentioned detailed reports can be simplified in a much simpler way with the help of a computer. The data used for the computation is the shift wise loom production. In this format, loom wise, shift wise readings are entered directly from the loom counter. Beam gaiting details are entered from sizer's ticket after getting the new beam on the loom. From the shift register book details of stoppages along with the code number is entered give the results in desired format.

From the input format, the computer reads the day's pick counter (number of picks inserted per loom). It recalls the corresponding readings of the previous day from the transaction file and finds out the differences. It replaces the readings in the transaction file with the latest reading for the calculation of the picks for the next day. The speed of the loom which is available in the loom master file is used to calculate the 100% picks possible without considering the stoppages. The loom efficiency is the ratio of picks calculated from the day's reading and the 100% picks possible in a shift. The same procedure is followed for all the other looms for all shifts. The weaver's efficiency is calculated by grouping the sum of the looms observed by the weaver and sum of the picks inserted per shift with 100% picks possible. In the similar way, jobber efficiency, shed efficiency, machine-make wise efficiency, shift efficiency etc. is worked out. The desired results are printed out daily in addition to keeping them in memory for back-up.

If there is a sort change in a shift on a loom, it is to enter in the computer to replace the sort details in the master file for that loom. This makes the computation of sort efficiency and weaver's wages correct. Furthermore, it is possible to reconcile the meters fed at beam gaiting and the meters produced at the beam fall through computer accounting. This accounting also will give the number of cuts at month end for each loom. Further, it is possible to get a report showing the looms likely to run out in two or three days. By computerising in the present system of going around the department and noting down the "likely beam falls" can be eliminated thereby the manpower used for the purpose can be saved.

16.14.2 Monitoring of loom productivity

The productivity of the loom shed on daily basis can be controlled through a computerised system. To enable this, standard or targeted efficiencies of various sorts running on the loom should be kept on the master file. The speed and sort particulars are already available on the master file. From this, the standard production of each sort can be computed. Thus the efficiency of the loom can be calculated. The machine productivity index, which is defined as a ratio of the standard loom hours required for the given production and the actual loom hours worked can be obtained. By keeping the standard labour requirement also in the master file and recalling the data of standard machine hours required for the given production, the standard labour for the production can be computed. Using this figure and feeding the actual labour employment figure the productivity index can be calculated.

Other potential applications

Another potential area of computer application is value loss computations. In developed countries, computers are used to detect defects from the fabric inspection table and suggest an economic cutting procedure. Instead, it is possible to fix the maximum value of the fabric with a given percent defect by the computer. The computer helps to decide the cutting points on the basis of the defects found on the piece as well as the discounts allowed for various cut lengths. However, this may be economical for all sorts working in the loom shed. In that situation, value loss can be calculated from the packing slips with the help of computer. The advantage of this system is that the required control information can be obtained in addition to meeting the commercial requirements.

16.14.3 Computer application in spinning mills

The efficiency and productivity computation in spinning is another major area of computer application in mills. The production of a ring spinning is recorded after weighment of individual doffs. The manual system followed in spinning mills is the weighment of the individual doffs, is entered in the doff slip and in doff book for further calculations. In computerised system, the doff slip itself can be put in the input format. The stoppages can be fed through a separate format and can be filled in from the log book. The remaining books can be eliminated. The daily outputs from the computer are:

Count wise production/ spindle and efficiency

- Category wise production and
- Cause wise stoppages.

These details can be stored in the memory for cumulative and monthly reports and records. In spinning preparatory, production is measured from the hank meter readings. This work can also be computerised as in the case of loom shed.

16.15 Applications of microprocessors for measurement and control in weaving

Loom monitoring system designed and developed by Physical Research Laboratory, Ahmedabad is an economic micro-computer based system which facilitates decision making at the managerial level for efficiently monitoring production, quality control and preventive maintenance. It consists of three systems namely:

- Multiplexer system
- Central processing unit (CPU) and
- Central computer.

The multiplexer system is designed in three layers to optimise interconnection wiring cost. The system has a maximum monitoring capacity of 256 looms. A Loom Scanner Card connected to each loom has eight discrete and one proportional parameters input. Sixteen such Loom Scanner Card form a cluster at a Node Multiplexer, 16 of which are connected to a Field Multiplexer giving connections for a total of 256 looms.

CPU Supervisor's display is composed of three units which draws information from the Field Multiplexer and selectively communicates it to the Supervisor's display and the Central Computer. The Supervisor's Display is capable of monitoring nine parameters with a maximum sampling rate of four scans per second for each parameter. The display is an LED matrix system representing the loom shed.

The central computer residing at the Manager's room performs two important functions. It communicates with the Supervisor's display via the RS-232 link and is capable of producing records for analysis at an hourly, daily or yearly basis.

16.15.1 Interconnecting system of multiplexer

Figure 16.28 (A) and (B) shows the interconnection system of multiplexer at different levels. From the figure it is clear that all the parameters of all the looms are not brought to the central monitoring points. Instead a master synchronisation scheme is utilised at three different speeds so as to channelize the data from the looms to supervisor's computer at gradually increasing data rate. At each discrete input, the data rate is four samples per second, whereas at the first node at loom, the data rate increases to 64 samples per second. However, at the input of the first multiplexer which collects all the node data, the rate is increased to 1024 samples per second.

In the final phase, field multiplexer communicates all 4096 channel data to field processor unit at the rate of 16,384 samples per second. Field processor consists of three central processing units. The first CPU checks the validity of the data and diagnoses the error. It also conditions and reformats the data for further processing in CPU-2 and CPU-3. CPU-2 works at higher level and is used for data compression, computation and communication to central computer system. CPU-3 is confined to local monitoring and logging of all the loom data. It also accesses both the raw data and processed data. All the three CPU's work in synchronisation to share the total processing load. Maximum length of data cycle is 8 hours which is equivalent to one shift time. For further processing of data, a central computer is required which could look after a number of systems simultaneously and is useful for keeping the records for years.

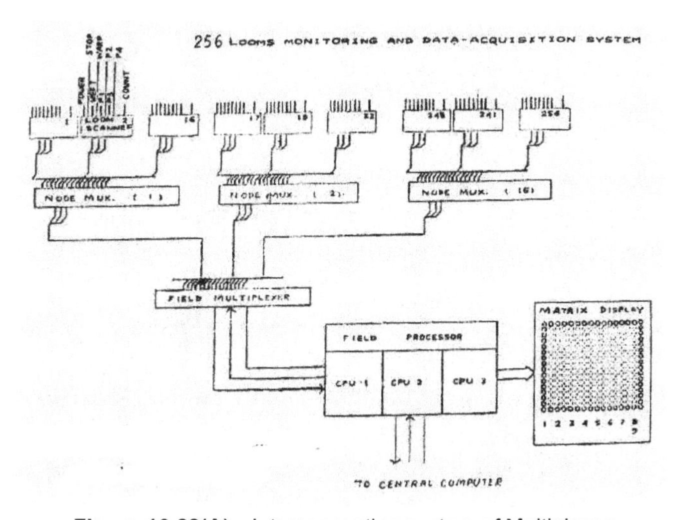

Figure 16.28(A) Interconnection system of Multiplexer

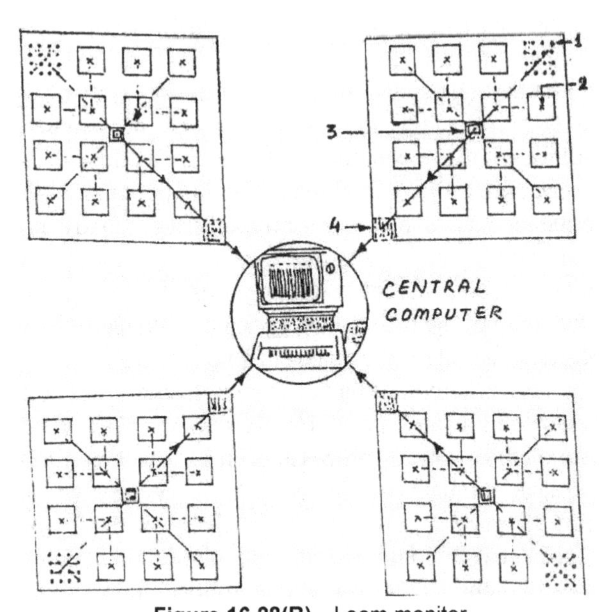

Figure 16.28(B) Loom monitor

1. Loom Scanner, 2. Node Multiplexer, 3. Field Multiplexer, 4. Supervisor's Display.

17

Advancements in on-line monitoring and control of parameters in knitting and sewing processes

17.1 Introduction

Knitting is the fabric manufacturing process which construction elastic and porous fabric. Knitted fabrics can be made much more easily and quickly than woven fabrics at comparatively less cost. Knitted fabrics are generally comfortable in wear even during travel but yet require little care to keep their neat appearance. The tendency of knits to resist wrinkling is another factor to boost-up their popularity. Excellent comfort properties of weft knits have made their entry into formal wears for men and women. But with the technological advancement in manufacturing of cloths and the awareness of consumers to quality, the expectations in knit goods too have gone high. However, knit goods are known for their high structural sensitiveness to deformation during manufacturing process or at their end use. The research work of the past focused on macro level aspects of quality control while the market demand today is on micro level. The quality criteria in the future would be much different than what is being counted today. The improvement of knit structure at micro level calls for better understanding of mechanics of loop formation, fluidity of knit structures and their influence on quality of knit fabrics. The quality of hosiery yarn has to be considered with due weight age to these aspects. If they are not addressed, probably satisfying the customer at global level may become difficult.

17.2 Knitting machines

Knitted fabrics can be classified in to two categories, namely weft knitted and warp knitted fabrics. In the weft knitting, knitted loops made by each weft thread are formed substantially across the width of the fabric. Warp knitted fabric is composed of knitted loops in the warp threads form loops which travel in warp wise direction down the length of the fabric.

Weft knitted fabrics can be conveniently divided in to flat knitted fabric which is made by a machine having straight needle bed, and a circular knitted fabric is made by a machine having the needle set in one or more circular beds.

The introduction of stitching motion and related mechanisms driven by electronic system in these knitting machines have paved way for the production of versatile fabric structures, and in their productivity.

In the knitting industry, important advances were made in order to maximise quality and productivity: the best trade-off between production speed and courses produced during one needle's cylinder rotation were obtained. A set of solutions in the accessories area were also proposed by the manufacturers for production reports, yarn feeding systems and fault detection devices. However, there still are some issues which are not completely solved. In the quality area, fault detection is one of the most important tasks, and the solutions commercially available do not detect all kinds of faults. Moreover, they are incapable of preventing faults. A combination of needle and yarn break detectors with optical sensors is required to fulfil only part of this task.

The solutions fail on some kind of faults, which can seriously impair production and delay the repair of the damaged element. Another important issue is the control of the yarn input tension, where the trip-tape based systems are the most successful between the currently available. The problem occurs when yarn feeding is not continuous: this system is "blind". However, new solutions are now available with feedback for controlling the feeding rate of all feeders. A third issue concerns the production data. The main knitting machine producers have advanced software systems for information purposes, but they are generally very expensive.

In order to detect faults during production, a novel approach was proposed, by monitoring the yarn input tension – YIT. One force sensor located near the knitting zone is enough to detect all the faults that the present solutions are capable (Figure 17.1). This is possible because YIT reflects the knitting process behaviour. If a needle for some reason fails to work, it will be reflected on the YIT and thus quantified. The analysis of the resulting waveform is made by signal processing techniques, namely matching pursuit filters – like AMCD (Average Magnitude Cross- Difference) as shown in Figure 17.2.

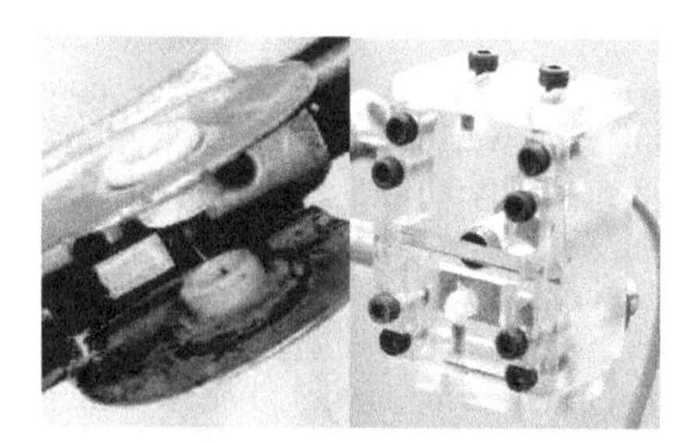

Figure 17.1 Force sensors for fault detection and YIT inspection. Left: sensor based on strain gages. Right: Inductive-based YIT sensor

The signal processing technique produces a signal that clearly marks a fault. It has also been shown that it even allows detecting malfunctions at an early stage, at which defects are not yet being produced in the fabric. At the same time they allow the distinction of false alarms, produced by neps and friction. The cause of the fault is located with an excellent accuracy and precision (within one needle), which saves time and money for repair. It is also possible to identify the kind of fault, by means of a developed pattern recognition system, based on discriminant analysis. The developed system can represent one entire rotation of the needle's cylinder in a single number, thus allowing the use of quality control charts, easier to understand. With this setup, it was observed that the present yarn break detectors, based on gravity principles, introduce some variability on the YIT. Some knitting machine producers, namely those concerned with very high quality products, are replacing these systems by optical devices with no contact with the yarn. An alternative low cost solution was also developed that not only detects the yarn breakage but also presents the magnitude of the YIT which is very useful for tuning purposes (Figure 17.2).

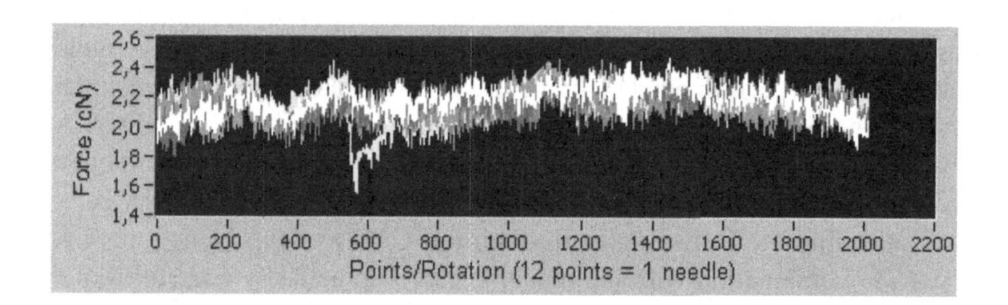

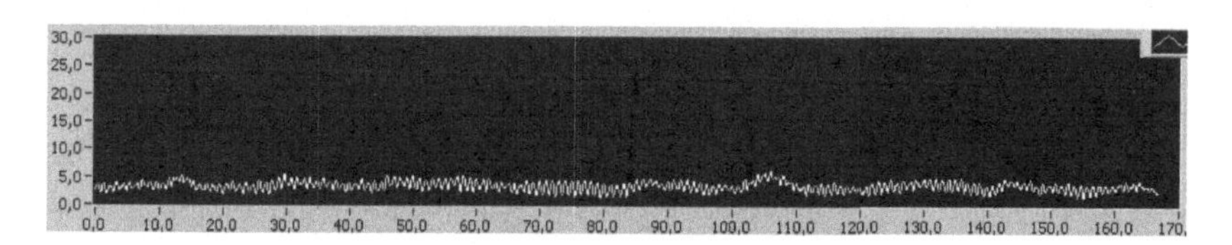

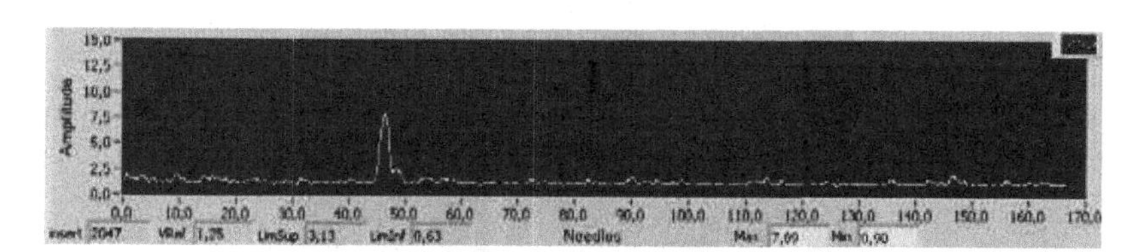

Figure 17.2 Magnitude of the yarn breakage

For the control of the YIT, a novel actuator was proposed and is presently under refinement, which is based on an electromagnetic arm and can be placed in any kind of knitting machine. Finally, a low cost solution was also implemented for acquiring the parameters necessary for determining the production rates and other parameters such as loop length, yarn speed, machine speed, etc. All these proposed solutions are integrated in a single software application called *Monitorknit* and developed with LabVIEW. These sensors communicate with a data acquisition board through a pre-conditioning software programmed board.

17.3 CAD/CAM in knitting

Shimaseike Co. introduced the machine at ITMA 1995, which uses a digital stitch control mechanism, four-bed technology and slide needles instead of latch needles. Four-bed technology ensures higher stitch density. Slide needle which was a newly designed needle for the machine gave rise to higher productivity by its smaller moving distance, and to natural loop configuration by its symmetrical loop formation as shown in the Figure 17.3. Furthermore, the slide needle is capable of forming 12 ways loop forming technique in contrast with six ways technique in latch needle, by which so-called gaugeless knitting can be performed. The knitwear can be made to three dimensionally fit the body and to form silhouette by computer aided designing (CAD). CAD system can produce a visual design in terms of colour/pattern and silhouette. Then the product planner can decide the knit wear to be produced by selecting/confirming the test samples through the CAD system. Hence the CAD system in combination with the machine can be practically a useful tool for mass production.

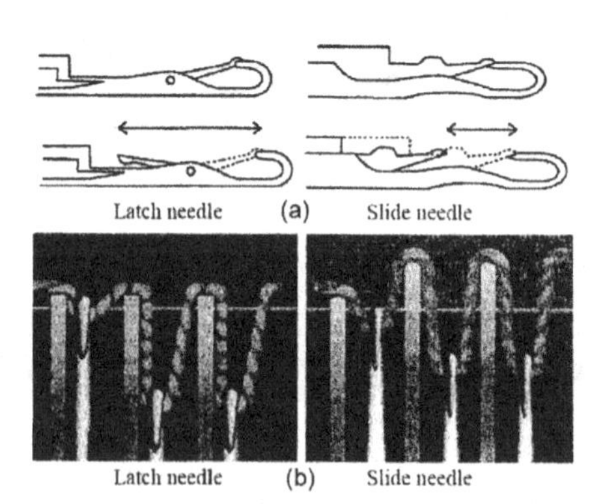

Figure 17.3 Comparison between latch needle and slide needle. (A) Configuration and motion of needles. (B) Loop formation by slide needle.

CAD/CAM is mainly used for jacquard single or double jersey knitting on large diameter circular knitting machines. The content of CAM in piece goods are:

The memory of the controller

The very large memory of the machine built in computer can hold up a few different patterns, which can be called up at any time, or mixing of patterns during knitting. The preparation of jacquard patterns is normally done on CAD units and the final patterns are stored in the data carriers for the knitting machine to read.

Shifting of colours on jacquard fabric

Before knitting, colour yarns are placed on the machine in specified order like

» feeder no 1 2 3 4 5 6 7 8

» yam colour A B C D A B C D

If the same pattern has to be knitted with different combination of colour such as;

» feeder no 1 2 3 4 5 6 7 8 and yarn colour C DJ A B C D or

» feeder no 1 2 3 4 5 6 7 8 and yam colour C D A B C D A B or

» feeder no 1 2 3 4 5 6 7 8 and yam colour D A B C D A B C

Principle

Computer

» The flexibility of making decision, process data (yes or no) makes it ideas for counted loop sequence, branching or jumping operation i.e. electronically controlled knitting pattern preparation and needle selection.

Computer hardware and software

» Hardware: electronic, magnetic and mechanical device.

» Software: programme and documentation for the system.

Main sections of computer

- CPU;
- Memory;
- Peripheral interface

Input and output device

- Input: sensors on knitting machine; keyboards, light-pen; tapes and discs.
- Output: actuators on knitting machine, light; digital and graphic display; tapes and printer.

Knitting machine program and control

Electronic machine eliminates all punch cards, chains, rack wheel, peg drum, element but which are expensive and bulky. Interactive computer graphics enable designer draws pictures and display on the screen. Colour can be added/changed immediately. Pattern can be rotated, stretch, mirror and scale.

Modern V-bed knitting machine control

Machine control

The modern V-bed m/c has many advantages over the pasteboard card control because:

- The electronic parts are cheaper to produce than the mechanical parts;
- Numerical is more accurate and very fast in response;
- Programs are easy to file or store;
- Easy to duplicate the programs.

Two control methods are available in flat V-bed knitting machine. They are

- Numerical
- Computer control

Numerical control

They are the simple control for non-jacquard knitting. It contains a control panel with function keys and control knobs for data input. The major advantages is that no pasteboard card and hole punching is needed.

Computer control

The numerical control is only good for simple knit collars, rib straps and body panels without pattern. Once individual needle selection is involved, computer control must be used.

Modern electronic V-bed m/c has

- Central processing unit: The brain of m/c control with data like pattern, machine control, economiser, pattern development.
- Keyboard: This is used for giving command or makes modification.
- Tape reader: The pattern and data input device.

- Visual display unit (VDU): A monitor to display all information of the knitting machine.

Example: Shima Seika machine manufacturer has control systems incorporated in the flat V-bed knitting machine.

The modern electronic V-bed m/c divided the systems into six sections, namely

- Control tape;
- Pattern tape;
- Pattern development tape;
- Stitch tape;
- Economise tape; and
- Yarn carrier tape.

Control tape

Similar to punch card in mechanical machine, it is a one inch paper tape with eight row holes. The tape is divided into block, one block is the same as one punch tape.

Pattern tape

This sector contains all the pattern details. The design is normally created in a CAD system and is now transferred as a series of holes in the punch tape. The length of tape will depend on the pattern size, the longer the design, the longer will be the tape.

Pattern development tape

This is to control the location of the pattern onto the fabric. It can be mirror the design or repeat the design horizontal/vertically, etc.

Stitch tape

This is used to control the stitch length (loop length) of the fabric. Modern V-bed m/c could have over 30 different setting in the memory for different material and structure design.

Economiser tape

The economiser on the mechanical V-bed is to use the minimum number of pasteboard cards for sweater panel. For the electronic machine with computer control, the system can memorise all the repeat structure and sort out the necessary cards or blocks in the design automatically.

Yarn carrier tape

In cut and sewn panel, the yarn carriers are set at fixed places about 2–4 cm away from fabric selvedge. For fully fashioning panel (or intersia), the knitting width is changing all the time, the yarn carriers must stop in varies position.

The stopping of yarn carriers is controlled by the computer which is either changes the location of the carrier stoppers on the rails or commands the cam carriage to release the working yarn carriers at specific location above the needle bed.

17.4 Industrial sewing machines

The dynamics of the sewing process has been studied by the measurement of the main process variables on lockstitch and over lock machines. Machines have been equipped with sensors and systems allowing the evaluation of thread tensions and consumption, forces developed on the needle-bar, forces developed on the presser-foot bar and presser-foot vertical displacement. The system is composed of the several sensors and their conditioning hardware, a PC with a data acquisition board and software developed in LabView. It allows the integrated evaluation of three of the machine's sub functions: stitch formation, needle penetration and material feeding. The analysis of the stitch formation parameters has revealed very interesting results using both thread tensions as well as consumption measurements. It has been possible to detect several types of localised defects (stitch distortions, skipped stitches and of course thread breakage), using signal processing techniques on thread tension signals. Two other methods have been developed to evaluate correct stitch geometry. The first uses a combination of features extracted from consumption and tension measurements whilst the other compares thread consumption with the theoretically expected values (that can be computed automatically by the software).

Currently, a new thread tensioning device, integrating a compression force sensor and a stepper motor, is in a test stage. This new tensioning device will allow the set-up of an adaptive thread tension control system, using thread consumption as feedback variable and computed theoretical consumption as reference. Thread tensions and the combined tension/consumption parameters will be used in quality monitoring.

The evaluation of quality aspects related to material feeding has shown to be very effective using both presser-foot force as well as vertical displacement measurements. Although the two measurements show complementary information, it has been found that most of the quality problems can be detected using just the displacement variable. These include localised defects (fabric curls or folds) as well as a recurrent feeding efficiency (loss of contact between presser foot and material, resulting in an irregular stitch at high speeds). After this phase of feeding behaviour evaluation, the machine has been equipped with a controllable electromagnetic actuator, which by its own has improved feeding efficiency due to the inexistence of the mechanical spring (Figure 17.4).

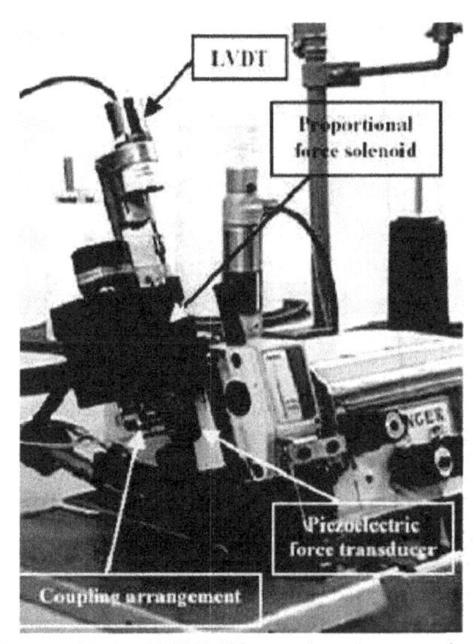

Figure 17.4 LVDT /Piezo electric transducer

Extensive studies have then been conducted to determine force ranges in dependency of material characteristics and sewing speeds. The first step in implementing a control system has been taken by the implementation of a speed-dependant force control. Next, the control loop has been closed using presser-foot displacement as a feedback variable. Several controllers, implemented on the PC using PID and/or fuzzy logic controllers have since then been successfully tested. Current work aims to create automatic teach-in methods to allow the control parameters to be automatically tuned. This will ultimately allow the control system to be fully adaptive. The feeding system will be adjusted to each fabric in a quick and automatic way, by making a simple sewing test before production. The measurement of needle penetration and withdrawal forces is the most difficult one. Due to machine-specific factors, the signals obtained are blurred by parasitic components. Complex filtering techniques have been devised and optimised to attenuate these components and extract the relevant penetration and withdrawal information as accurately as possible. This allows the comparison of different needle sizes, tip shapes or damage, fabric finishing, etc. In Figure

17.4 over lock machine with displacement sensor (LVDT), force sensor and actuator 3 line needle wear detection is possible at constant sewing speed using a force trend analysis. At variable speed, however, the variation of the parasitic components renders the measurement too inaccurate in the current setup.

At present time, a lockstitch machine is being equipped with sensors to measure the same variables. Most of the measurement hardware will be similar, with machine-specific differences. The measurement of bobbin thread consumption is not possible using the same techniques used in the over lock machine due to lack of space for sensors. An innovative contactless method has thus been developed to provide this measurement. On the other hand, this machine allows another setup for the needle penetration force measurement, avoiding parasitic components. The team expects most of the results to be similar on this type of machine, with obvious particularities. The studies are expected to provide information for a generalisation of the developed control and monitoring techniques.

17.5 Electronics in warp knitting

Every section of the textile industry has undergone a formidable change with the introduction of microprocessor electronics and computers. As control mediums in terms of machine functions. And warp knitting is no exception.

Application of electronics has been applied to six areas in the warp knitting sector.

- Jacquard control
- Guide bar control
- Design preparation
- Warp let-off
- Warp take-up and
- Weft insertion.

String jacquards used by Jacquard Raschel warp knitting machines is mainly used to produce ornate patterns of curtains, bed spreads and table cloths. The patterning system utilised is one of the guide deflection by means of pins situated between specially shaped guides. The string jacquard placed above the machine determines whether the pin is raised (guide not deflected) or lowered (guide deflected). The conventional "Verdol" type jacquard uses the conventional "wirework" to raise the harness but it is restricted to 250–300 courses per minute, a restriction imposed by the jacquard.

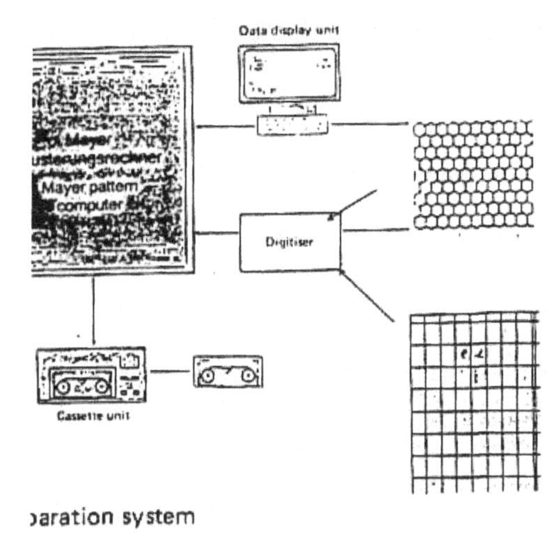

Figure 17.5 Pattern preparation system

With the development and application of computers to jacquard machines (Figure 17.5) the "wirework" was replaced by a series of special levers controlled by electromagnets to form the interface between the computer and the harness cords. The principle is if the magnet is energised, the particular cord is raised, if not it is kept in the low position. With this technique, the production speed is increased by 30%–40% giving a speed of 400 courses per minute. The advent of bubble memory, a small light weight and efficient computer store capable of holding maximum information in a small space and will not lose information in the event of power failure.

Design preparations consist of drafting the design on squared paper in conventional manner, but should be in optically readable colours. The draft is then scanned by the digital reader which feeds the design in to the computer in the form of magnetic impulses. The design may then be viewed in sections and corrected if required by a VDU and keyboard. and when complete is transferred in to a magnetic tape cassette. A portable loading unit consisting of VDU and tape deck transfers the information in to the jacquard computer.

With the development of electronics, the replacement of the pattern chain for guide bar movement solved many problems for the warp knitters as it was costly, time consuming in preparation and machine down time. A further advantage of electronic controls is the design scope. With normal chain links, the maximum number of bars considered was 42, but electronic controls this has been extended up to 56 and electronic guide bar control is available on multi-bar machines. Moreover, the fall plate guide bars give three dimensional effects.

17.6 Electronic beam control let-off motion

The function of electronic beam control let-off motion built in to the computer system serves three functions.

- Control of the warp sheet
- Control of the fabric take-up rollers
- Collective data system for organisational and statistical calculations.

Working principle

Electronic beam control let-off motion is shown in the Figure 17.6. The electronic warp control section consists of a DC motor driving the axle of the beam. Thus as the warp unwinds and its diameter decreases the motor speed must be increased to compensate for the smaller circumference and maintain the warp feed rate constant.

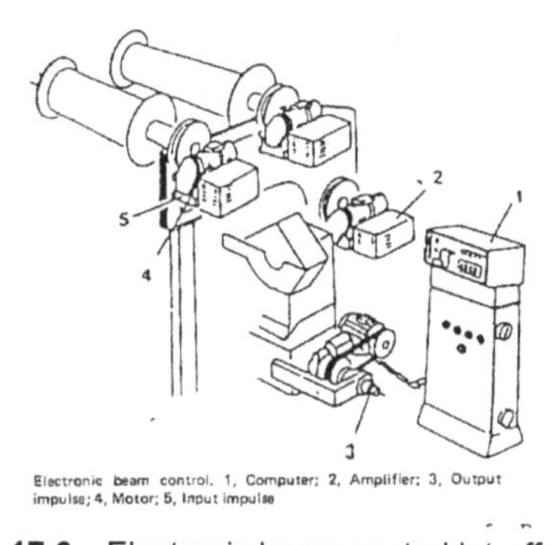

Figure 17.6 Electronic beam control let-off motion

The increase in speed of the motor is controlled by the computer. The amount of compensation required will depend on the rate at which the diameter diminishes which is dependent upon the size of the beam, the run-in and the length of the yarn in the beam.

These factors are used for programming of the computer using four parameters.

- Circumference of the empty beam
- Circumference of the full beam
- The revolutions of the yarn on the beam and
- Run-in required.

Once this information are fed to the computer, the computer accurately controls the beam speed accurately throughout the unwinding of the complete warp from the beginning till the end and need not be adjusted again. This system gives accurate control than the conventional mechanical let-off motion and is also more versatile since it may deliver 20 different warp speeds in any one design as against two for the mechanical let-off motions.

The fabric take-up roller section is also programmed in the similar way, there being 20 different take-up speeds which can be selected at will. The data collection system gives useful information like fabric length, warp length, etc.

17.7 Electronic weft insertion on weft insertion machines

Electronic weft insertion system in raschel warp knitting machine is shown in the Figure 17.7 (A) and (B).

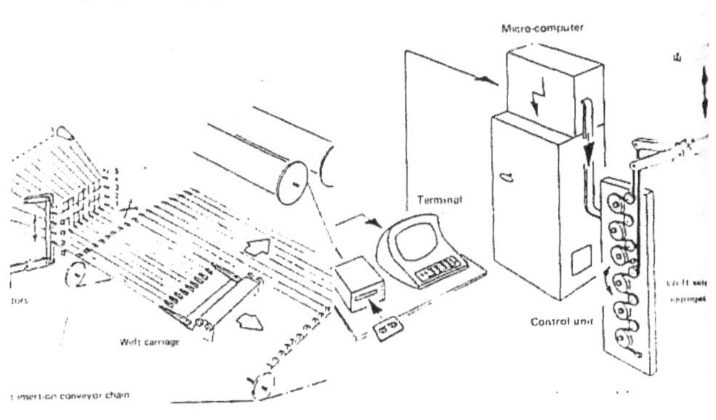

Figure 17.7(A) **Figure 17.7(B)**

Electronic control of the weft selection is another potential area in the warp knitting machine through the use of electronics. The weft is placed across the back of the machine between the rising needles and the warps and the course is knitted, the weft is held between the loops and the under lap in a similar manner to a laid-in thread and the weft traverses across the full width of the machine.

The weft is fed to the knitting elements by means of a two weft conveyor chains projecting out of the back of the machine to feed the weft to a reciprocating weft carriage which supplies the yarns to the conveyor chains projecting from the back of the machine. The weft carriage takes 24 yarns and has the capacity to hold 96 ends, thus providing four yarns for each operating end in the weft carriage. Consequently four ends in a rack are held in the weft station ready to be picked up by the weft carriage.

The system of operation is as follows:

The computer (shown in the Figure 17.7 (B)) selects the required weft (position 1–4) at each of the 24 weft stations situated at the side of the weft conveyor chains. The carriage moves across and picks up the selected wefts, traverses across the machine and places the weft in the right hand side of the conveyor chain. At this point, the wefts are gripped at the right hand conveyor chain and cut between the left conveyor chain and the weft station. The carriage now reciprocates back to the weft station for the next selection of weft.

Since the weft selection is determined by a micro computer, the weft repeat is limited only by the size of the computer memory, which can be equal up to 50,000 courses, giving a repeat length of 50 m of fabric at a standard quality of 10 courses per centimetre.

The computer is programmed by a magnetic tape cassette which is programmed by a VDU and keyboard, in a similar manner to a system used in the multi-bar electronic and electronic jacquard machines. This technique gives the biggest and most versatile weft selection mechanism used on weft insertion machines, and it is ideal for the production of curtain fabrics in which the weft consists of many different types and/or different colours of yarns with long repeats.

18

Online control in wet processing

18.1 Introduction

In today's competitive textile market where the developments and electronic controls are taking place in every machine process right from ginning to weaving, knitting, and automation in wet processing is mandatory and compulsory. It is because in global environment, it is now the survival of the fittest. Hence efforts are required to

- Improving technology
- Enhancing productivity
- Acceptable quality to the customers and
- Involving cost cutting measures on production cost to make the product more competitive.

With few exceptions, the textile industry is mainly involved in the manufacture of coloured products. The major problem experienced, as in any other manufacturing industry is the quality of the product and in the textile industry specifically, the significance of colour quality. Especially in textiles, colour awareness has grown in the retail trade with the pace defined by large chain stores. It is considered not acceptable to display garments in a rack with obvious colour differences between adjacent items. Specifications from the buyers are tightened in such areas as material properties, fading properties, UV deterioration and colour. The colour factor does present its own problems due to the subjectiveness of its nature, yet its control is of great importance as in most cases colour and design are the ultimate deciders in the purchase of the product. For this reason, the dye house is of prime importance in any textile company and for its management, the most important objective is to achieve the optimum unit cost. This is achieved when

- Optimum production is obtained from minimum capital equipment.
- Specific technical requirements are met with a minimum cost in labour, raw materials and energy.

These two basic parameters can only be met when the capital equipment in use of good quality coupled with controlled instrumentation.

Colour is of a subjective nature when visual methods of assessment are used. Exact shade reproducibility is impossible due to the number of variables in the production process, and hence a "commercially acceptable" tolerance is established. Whereas any colourist finds no difficulty in passing or failing a shade to a standard with zero tolerance, the visual assessment becomes subjective when he has to apply pass/fail for shade to standard with an accepted commercial tolerance.

18.2 Wet processing

It involves the processes like desizing, scouring, bleaching, mercerizing, dyeing, printing and finishing. The product after finishing stage is the final product which should have all the necessary characteristics as required by the customer. Hence this is an important and critical stage in textile supply chain and hence modernisation, up-gradation and automation are much needed to meet the requirements. In the last two decades, the application of colour measurement and quantitative colour communication by the use of international systems has gained more importance. With the development in computer technology, it is possible to have ocean of colours and the computerised colour matching systems have become widely accepted in various sectors like paints, ink, paper, textile and in pharmaceutical industries. However, this has only brought some solutions to the problems of quick matching, least metameric, least cost matching, alternating recipe formulation, batch correction, dye quality control, etc. It is customary to exercise a good control in plant colour variation especially within a batch as well as batch to batch. Hence online colour measurement and control system is required to be explored particularly for the following reasons.

The objectives of online colour control systems are as follows:

- Identification of the shade variation as they begin to develop and to rectify and control the same in plant itself.
- Control charts or graphical reports must be available in the shop floor for the personnel working in the plant so that they can compare the shade variation if it exceeds the standard limit.
- Necessary corrective actions can be taken immediately for any abnormalities.
- Redyeing can be avoided and hence reduction in the production cost. Such a system reduces the time needed to correct the shade and hence avoid any colour problem which increases the production throughput and lowers rework cost.
- Higher definition in the print without off shades, light spots and patches, etc.
- Elimination of subjective evaluation by production personnel with difference of opinion about shade uniformity between works and production personnel.

Advantages of online colour control

- Customers and consumer markets are benefited by giving offline continuous colour data which they can check for consistency in shade.
- Colour data file for different shades for different textile products can be maintained for different batches as like Uster data for yarn statistics. This system can obtain a profile of colour consistency from batch to batch or within a batch side to side and end to end.
- It reduces manual sampling and shades sorting as all the shades are controlled to be within tolerance limits by suitable colour control system. This reduces the dye and additive consumption avoiding re-dyeing, etc. and reduces the colour rejects.

Fundamental units of an online colour control system.

Figure 18.1 shows the fundamental units of online colour control system.

This instrument consists of an optical sensor reflectance spectrophotometer with non-contact remote measurement facility on a moving coloured sample. It also scans width-wise to detect side to side shade variation along with the end to end variation. The instrument consists of a microprocessor or PC computer with colour video monitor suitably programmed for analysis and shade variation data, and comparing those shades with set colour tolerances for finding out the error and accordingly amplifies the error signal to take suitable controlling action.

In order to do this, the instrument consists of colour data processing preference or software supporting packages in the microprocessor or PC. Or electronic circuit to send ultimate amplified signal to the controller for activating colour control in the plant itself. Suitable controller either electronic, mechanical or pneumatic type has to be activated by the above error signal for opening more or less dye add valve or to raise or lower the temperature of the dyeing or to control any other variable in the particular dyeing system.

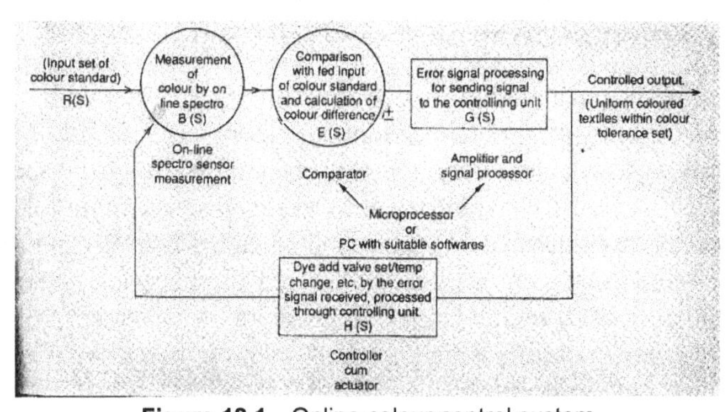

Figure 18.1 Online colour control system

Problems in online remote colour measuring spectrophotometer

Designing a suitable online colour measurement system presents many problems. Many attempts have been made by colour engineers to overcome those difficulties but earned limited success. The following are the reasons:

- Lack of a spectrophotometer suitable for long remote sensing time with non-contact type operation which is capable of rapid (in microseconds) measurements from a moving sample, without interference from ambient light conditions. General integrating sphere (D/0 or D/8 geometry) reflectance spectrophotometer will not solve this problem. However, some specially designed spectrophotometers are now available for remote colour measurements.
- Non-availability of accurate production standards in textile colouration against which tolerance limit can be set or calibration can be done.
- Quantitative colour standards and colour difference standards are not available in the industry for comparison and hence subjective evaluation is still in practice.
- Drawbacks in the conventional colour measurement systems like non-stability, non-reliability and bad performance of the systems designed earlier. The basic units of the whole system are affected by the plant environmental conditions, temperature, humidity, dirt, aerosol, oils, etc.
- Presence of large number of dye variables and again depending on the dye type and the type of fibres, a large number of possible variations in the dyeing process and process parameters would vary. This requires suitable control systems for the different dye variables to be effectively designed out.

18.3 Developments in the new version online colour measuring spectrophotometer with new instrumental control

A new version of technology in spectrophotometry gives an opinion on colour process visibility or to obtain continuous colour variation data (averaging even very short-term colour variation with accuracy of industrial spectrophotometer) for processing with suitable software against set colour tolerance.

Pulsed xenon colour eye-11 spectrophotometer (Ms 4045) and Eagle eye colour surveillance system are two new types of spectrophotometers both having 45/0 optical geometry. Online rapid colour measurement from a moving sample is now possible which is based upon the two new spectrophotometric instrumental concepts.

Concept 1

A pulsed xenon flash tube light source (instead of ordinary quartz tungsten lamp) which produces illumination of very short duration (20–30 microseconds) and high radium intensity (10 watts), high photo electric signal to noise ratio is used. These short duration flashes freeze the motion of the moving sample or substrate and measure the colour within 30 microseconds. Pulsed xenon flash tube of high radiant intensity and 45/0 or special 2/0 geometry instead of ordinary integrating sphere (D/8 or D/0) give a special design to measure at a distance of 1" with larger depth of focus. The high radiant intensity negates interference from ambient light and adequately illuminates dark colour. High photometric signal to noise ratio allows very small total energy delivered to the sample and thus negligible sample heating can occur. The details of such a spectrophotometer head is shown in the Figure 18.2 which is connected to a microprocessor console for colour data processing and tolerance.

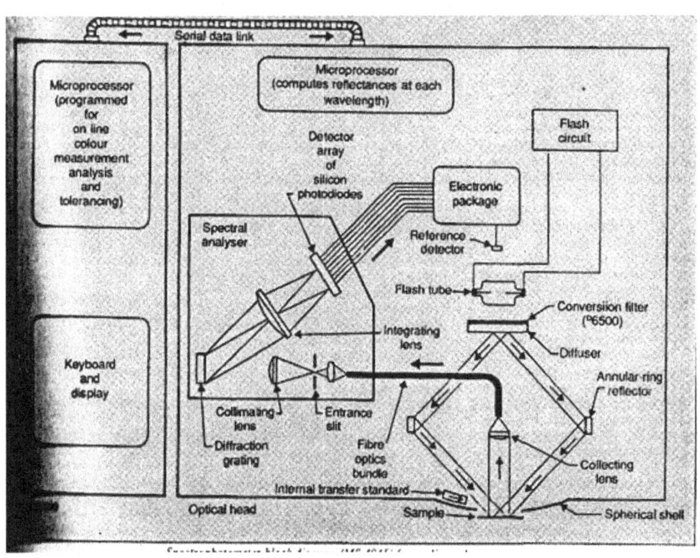

Figure 18.2 Spectrophotometer (MS-4045) online colour measurement

Although, it is possible with this development, in plant environmental conditions like temperature, humidity, oils, aerosols, acid fumes can affect the measurement performance and reliability of the instrument and hence protective measures have to be taken care for efficient working of this online measurement spectrophotometer. A special fitting arrangement is also necessary for scanning side by side colour variation which is not a difficult one.

Concept II

A diffraction grating with detector array of parallel wavelength sensing (instead of ordinary photographic grating) is used in this concept. A grating spectrophotometer generates a complete spectrum in parallel. The use of silicon photo diode detector array local plane enables the simultaneous measurement of all wavelength channels sensing parallely. Thus, a complete spectrophotometer online measurement by this new type of detector array is made consistent with a single pulse of illumination from a moving coloured textile fabric.

A comparison of the configuration is shown in the Figure 18.3 A, B, C and D.

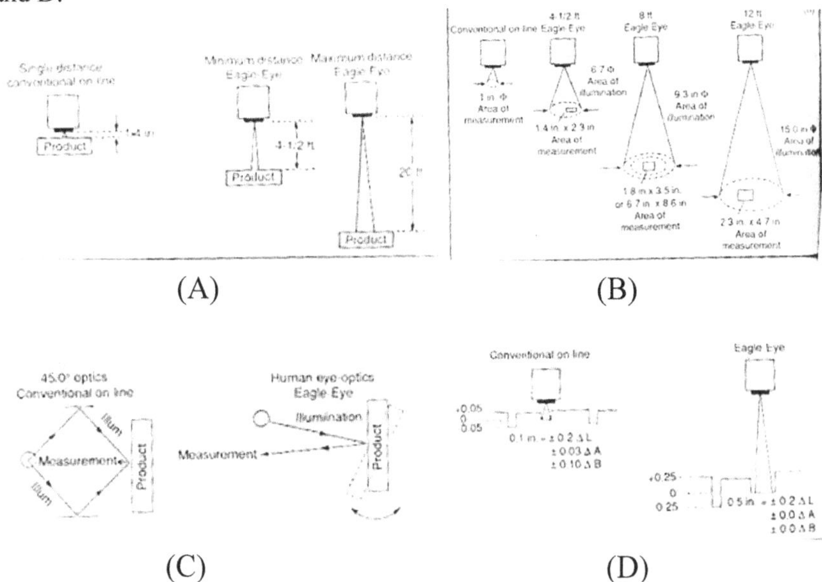

Figure 18.3 Comparison of conventional and remote sensing spectrophotometer

The above-mentioned spectro sensor can be installed either for fixed point measurement or can be suspended by an I-beam above the moving dyed textile fabric, scanning the product width wise for side centre side colour variation measurement. Therefore the online remote colour problem, which a few years back was spectro researcher's subject for brain storming discussion is now solved with the advanced instrumental concept of modern technology through further development on environmental protection, reliability and inter instrument agreement are to be worked out.

From the online remote measurement of colour from moving sample, these data are now required to be compared with the standard tolerance and are to be processed by a processing unit pressed by a control start, etc., and the deviation measured is to be transferred to a suitable signal processor and control unit and actuator for automatic colour control within the plant by controlling a number of dyeing variables is indeed a difficult problem.

18.3.1 Multi variables in dyeing process and colour controllers

The most common problem is to ultimately control the colour either through the dye and valve aperture control or steam flow control thereby controlling dyeing temperature or control of addition of certain dyeing auxiliaries having direct effect on colour uptake. The success of process control in dyeing depends on the type of dye, type of fibre, method of dyeing chosen, the kinetics of dyeing, etc. As the type of control which needs to be exercised varies in different dyeing processes, it is very difficult to control the rate of dyeing by controlling any one dyeing parameter which may not be linearly variable with dye uptake.

In general, all dyeing operations have many variables that must be considered as factors in designing the colour control system. It is because every processing unit have some standard shade. This standard shade has its own L, A, B and E colour difference values obtained by measurement and calculation which act as identifying finger prints for tolerancing the unit's own standards. The processing units are already experiencing working with different variables in dyeing to obtain a standard shade. As the ordered sample with new colour standard of the mill is unknown in nature, it is difficult to simulate and control effectively, which needs a lot of practical consideration, new tolerances and high measurement and control system accuracy.

Case study

A case study has been conducted in thermosol dyeing range. The system consisted of optical spectro sensor, scanner mechanism and edge detector computer and information processing software, mass data storage and supporting colour software. The experiment was chosen to have 30 measurements in 1 minute at 2 second intervals and to repeat the sequence every 10 minutes. The colour data were then sent to the microprocessor console through a serial data link for averaging and statistical data processing. Thus very short-term colour variations were considered.

The plant constantly watched fabric in process on continuous dyeing range at intervals of 500, 1000 yards. The unit initially controlled the manual dye add if end to end variation give lighter and if darker, the temperature is reduced and adjusted the nip roll pressure for correcting the side to side variation. The optical sensor was housed in a cover with a window for the measurement and placing the spectro at a distance from the moving dyed fabric sample on the dyeing range and thus allowing protection of the sensor element from plant environmental conditions. The sensor was mounted under the housing on scanning frame at the exit end of the range, making the measurement possible at both the sides and at centres and compared these to the programmed standard for tolerance. The colour data were then simultaneously transmitted to a monitor located off line for looking at the process visibility at any time by the mill manager. It signalled the variation by changing the colour of the screen or the colour of the graphical representation such as blue for acceptable value, yellow to alert the manager and red indicate that the shade has exceeded the specifications.

18.3.2 Colour definitions

Colour measuring systems work by defining a colour in numerical terms. These figures can then be used to define the position of the colour in "colour space". By comparing the numbers for the standard and the trial, we can define the colour difference numerically. Colour space is built up in three dimensions with black at the bottom, white on top, grey in the middle, with pure colours on the extreme edges and various degrees of strength from each edge to the neutral centre. This type of classification is the fundamental of Munsell system and we can either obtain Munsell Notation, popular in the US, or the more familiar, L.A.B or L.C.H. values.

L.A.B. values are derived from the CIELAB equation (or Opponent Tye system). Certain colours are opposite each other in colour space i.e., black and white, red and green, blue and yellow. A colour cannot be red and green or blue and yellow at the same time but it can be blue and green. Therefore, a CIELAB type colour space can be made up with white and black at the top and bottom, red/green and blue/yellow at the extreme axes. In this way, we have "L" for lightness from black (0) to white (100), "A", (red +/green -) and "B" (yellow + / blue -). This provides an understandable system, but suffers from disadvantage of non-uniformity.

The difference between standard and trial can be shown in the form D (or Delta) L, Da, Db. An additional calculation is done to give DE. Or overall colour difference. Apart from CIELAB, the other equation now in common use is known as CMC equation. This is designed as an equal tolerance equation and overcomes most of the non-uniformity disadvantage of CIELAB. CMC defines colour in terms of L (lightness), C (Chroma) and H (hue). Chroma is the strength of the colour from pure to dull grey. Hue is the degree of colour.

Colour difference is what the observer sees as the difference between two objects. The "observer" can either be a human or a spectrophotometer. Ideally, when checking colour difference, both human and instrumental observers should work together and agree. In order to estimate colour difference, we must have an "observer", two coloured objects, and a good standard light source. Light sources can cause much confusion. Diffused mid-day light in July will be very different to light in January. Normally, all serious colour difference work will be done using one of the four types of light. They are as follows:

- D 65 (artificial day light)
- Tungsten (household bulb)
- Fluorescent (store lighting) and
- Ultra violet.

All of these types of illuminant are available with colour systems. The human eye is very accurate at gauging colour difference and at seeing that an object is red or blue than another. However, it cannot reliably estimate the size of the difference.

Colour systems will measure the two objects, compute their respective positions in colour space and calculate the colour difference. From these delta figures, it can be seen if the sample is lighter/darker, redder/greener or bluer/yellower than the standard and by what degree. A decision now has to be made about acceptability or pass/fail. If records are kept for each batch that is finally passed, a pattern emerges showing the limits of acceptability. In this way, tolerances for DL, Da and Db can be set. Spectrophotometers measure colour by collecting reflectance values at intervals across the visible spectrum from between 400 and 700 nanometres. The present day spectrophotometers measure reflectance every 20 nanometres, so there are 16 reflectance values from which a spectral graph can be plotted. This is a fingerprint of the colour of an object. If the graph is reproduced exactly, then the colour must be the same.

18.3.3 Metamerism

Metamerism means that two objects may match in daylight but not in artificial light. For example, white light strikes a green object. Green light reflects back while red, yellow, purple wavelengths are absorbed. If we view this green object under daylight, some of the light will be reflected back as green. If we view the same object under a fluorescent source rich in red and blue but deficient in green and yellow wavelengths, less light is reflected back. Therefore, the object will look a different colour. For colour matching, there is serious problem because while they may match in daylight, they may not both change colours under artificial light by the same degree, as some mixtures of different dyes and pigments have different absorption characteristics. This is known as illuminant metamerism. There are other two types – Observer metamerism where two independent observers do not see a colour in the same way and Geometric metamerism when a colour appears different depending on the viewing angle.

18.3.4 Non-contact colour measurement

Online non-contact colour measurement and control has been successfully introduced in a number of industries over the last few years. These systems are used on continuous production lines to monitor colour continuously, thus providing data that allows operators to compensate for variations before they exceed specifications. The systems are used in paper, textile, plastics, metal processing and in other industries.

This concept has enabled users to achieve significant production efficiencies and auxillary benefits. The system reduces the time needed to identify variations and enables operators to react quickly to potential colour problems without disrupting manufacturing operations. Online control systems were originally designed for continuous dyeing ranges. They monitor colour continuously on a dyeing range and automatically provide data that permits operators to compensate for shade variations before they exceed colour tolerances. Such installations have reduced time need to identify shade variations. Production personnel can react quickly to potential colour problems without disrupting operations.

A typical system consists of an optical sensor, scanner mechanism, edge detector, information processor, video monitor, mass data storage and supporting software package. The sensor is mounted on a scanning frame at the exit end of the range, just past the last drying cans.

18.4 Computer colour matching system

Computerised colour matching system performs two functions. The first one is selection. It means determinations of one or more practical combinations of compatible dyestuffs which are expected to match with the given shade satisfactorily at reasonable cost. The second thing is that the colour formulations are determined as to how much amount of each colourant should be used to produce a satisfactory match. It means that there are usually more than the combination of dyes which will match the required shade. The computer tries to match the standard with the all possible combinations of dyes and lists the resulting formulas in best match and cost consideration. The colourist can then choose a matching formula from the list considering the closeness of the match, cost and other relevant factors. A computerised colour matching system is shown in the block diagram (Figure 18.4).

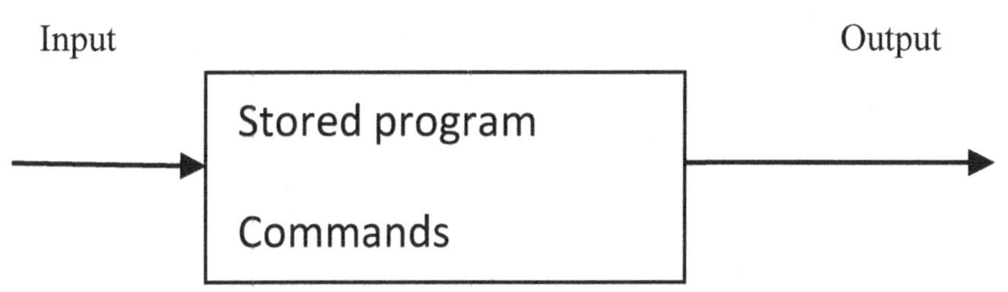

Figure 18.4 Block diagram of computerised colour matching

Output

a) Drive filters
b) Display colour coordinates.

Stored program

a) Input/output.
b) Command recogniser.
c) Data acquisition and storage.
d) Computations display.
e) Self diagnosis and calibration.

18.4.1 Implementation of computerised colour matching

The implementation of computerised colour matching involves many steps to get started. The first and most important one is to make laboratory dyeing which represent multiple concentrations of the primary dyes characterising the affinity of each dye to the fibre. These dyeing are then read individually by spectrophotometer to obtain the percentage of reflectance curves which represent the resultant colour. This data can be directly keyed in to the computer through a direct computer spectrophotometer interface. Graphical plots of the data may be obtained from the system to verify the accuracy, identify errors, analyse the dyes, compare to other dyes and so on.

Working principle

The working principle of computerised colour matching system is shown in the Figure 18.5.

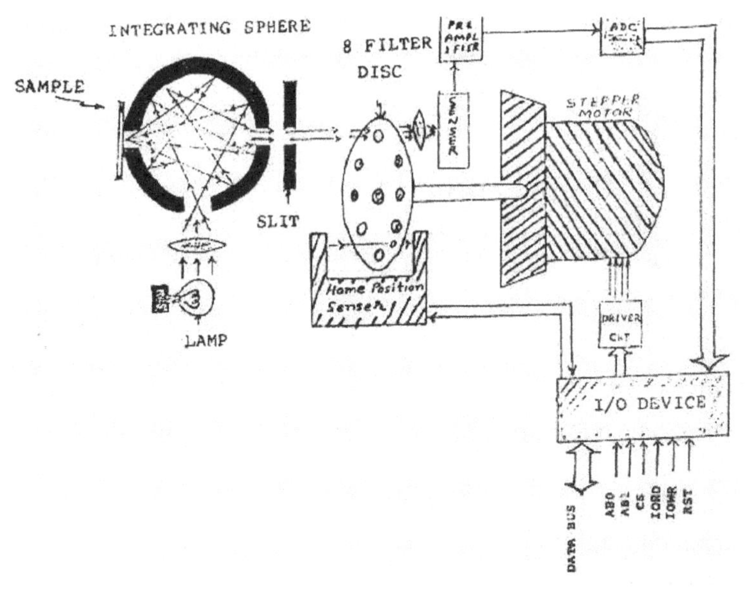

Figure 18.5 Computerised colour matching system

It consists of an integrating sphere with non-polarising optical design to virtually eliminate orientation measurement errors of textured sample. The system uses a single beam technique. Five inch diameter sphere is used for diffuse illumination. The reflected light passes through a rotating filter disc and directed on to a solid state silicon photo detector. The analogue signal is converted in to digital form by DVM interface and the computer converts the data to different point wavelength information.

The microprocessor controls all the functions regarding the operation of the colorimeter. The rotating filter disk is controlled by stepper motor. A total of eight data points representing the visible spectrum from 400 nm to 700 nm are taken for the recommended single filter disc rotation.

18.5 Computer colour control in batch dyeing operations

Process control is one of the most critical aspects of quality assurance in textile operations, such as batch dyeing; there are many variables which are under the dyer's control, but many more which are not. Typically these controllable and uncontrollable factors interact in a very complex way. The dye house of the future must feature even better process controls. Traditional manual control methods in textile processes have been automated using microprocessor systems, with corresponding improvement in process repeatability. However, this mode of control utilises only a minuscule fraction of the total capabilities of modern microprocessor hardware.

In an attempt to improve microprocessor utilisation, there are two avenues to pursue. The most celebrated of these is the macro scale global linking of information referred to as Computer Integrated Manufacturing (CIM), to which much attention has recently been devoted. In typical CIM implementations, sophisticated user interfaces, attractive graphics and computer networking capabilities are employed to make information from machines and machine groups available to managers for real time and post process analysis. The focus is on the use of state of the computing hardware and sensors to acquire data from processes, set up common database formats, link islands of automation, and provide management information in a timely manner in a macro or global sense.

In CIM, individual workstations are connected through networking paths, and data are stored in a standardised central database, accessible to all workstations. The general structure of such a system is shown in Figure 18.6.

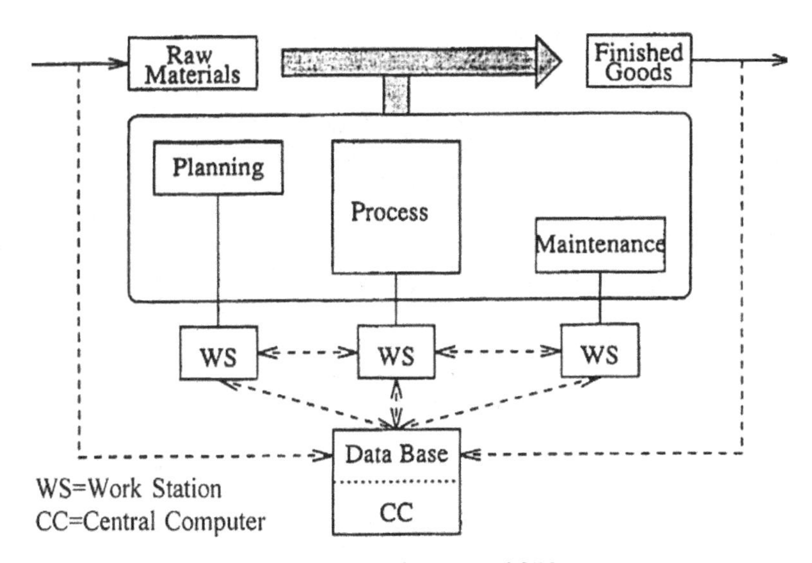

Figure 18.6 Structure of CIM

One of the most important feature of CIM is that it provides data to managers for strategic decision-making. It has made avenues for enhanced microprocessor utilisation is the development of novel control strategies which fully utilise data processing capabilities at the micro level, which use improved control models and also employ improved theoretical and empirical process models, At this time, the basic concept of using microprocessors for control in textile wet processing operations is to automate manual procedures and control methods of the past.

Control hardware has made great strides in recent years. The emergence of low cost, versatile, high speed digital microprocessors has facilitated all manner of textile wet processing applications.

Microprocessor hardware has developed to a high state of reliability and performance, and typically includes multichannel input/output (I/O) capability as well as interfacing to a multitude of peripheral devices, including networking, clustering, and supervisory computers.

Typical control systems include microprocessors with

- control algorithms and interfaces
- multichannel I/O with two way digital/analogue conversion
- synchronous and asynchronous communications abilities
- monitoring and sensors with interfaces
- process devices (e.g. valves, pumps)
- networking, upload/download capabilities.

18.5.1 Structure of a dyeing process control system

Traditional dye process control methods attempt to conform as closely as possible to a specific pre-determined process profile (e.g. time, temperature) to achieve correct results. Discipline is emphasised. Uncontrollable variances are accepted and, in some cases, remedied after the fact, for example by shade sorting or dye adds.

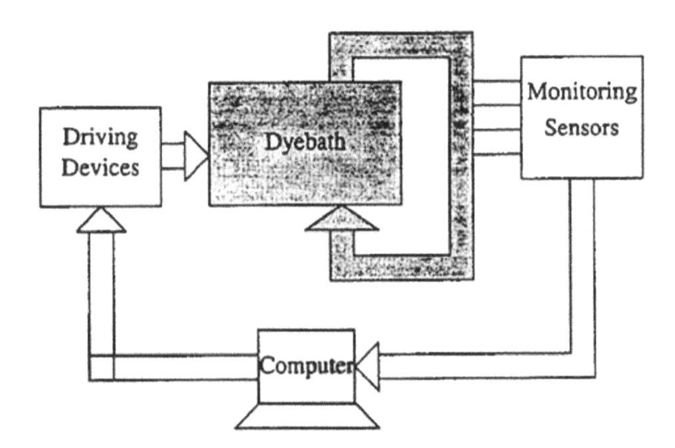

Figure 18.7 Structure of dyeing process control

The approach is to control the ultimate product property of interest (in this case, dye shade) by adjusting controllable process parameters in such a way as to arrive at the desired end result. In this approach, the process may or may not be the same each time it is run, but the goal is only to arrive at the correct result. The process may be varied each time it is run to compensate for non-controllable factors, such as variances in water quality, substrate preparation, and raw materials (Figure 18.7).

The first and most fundamental departure from traditional control concept is the use of predictive result-oriented strategies, as opposed to process conformance strategies. Sophisticated and theoretically sound dyeing models are combined with extensive real time data acquisition to assess the state of the system and predict process outcome (i.e. final dye shade) about every 2 minutes during the dyeing. These predictions are the basis for real time process modifications and departures from nominal process specification. Controllable process parameters are used to offset uncontrollable variances. Another departure from traditional methods is the use of multi/multi control strategies as opposed to the traditional one-to-one approach. For example, in a traditional control algorithm, a standard temperature of 200°F may be the process specification. If temperature deviates from the process specification, the controller will open steam valves to correct. This one to one control strategy senses temperature and controls steam. The novel approach does not control temperature for its own sake, but rather predicts the effect of a temperature variation. If no undesirable effect arises from that situation, the controller does nothing. However, if a problem such as an unacceptable dye shade is predicted then the controller takes action, but not necessarily by opening a steam valve to correct temperature to a nominal value. Rather, action is taken by whatever means will correct the predicted out come to the desired result (shade) a minimum cost and production time. The best action may be, for example, to add salt or change the pH. Of course, there are constraints built in to prevent the controller from taking absurd actions. For example, rate of temperature change or permissible pH or temperature values may be limited. Central to this concept is the ability to accurately predict a result from the present state of the system.

18.6 Control system for the dye house

Varieties of microprocessors are available in the market which differs in operating convenience, basic lay out and other features. Then-Data Comp is one such control system developed in Germany has vast experience in electric relay control and analogue control and more familiar with electronic digital technology. "Data Comp" stands for total control from simple temperature time control through fully automatic operation to a central control unit with data acquisition, and the control systems are all expansible from the basic to a central control unit.

Figure 18.8 shows the overall layout of the Data Comp system. Each dyeing machine is connected to the mains switch gear with colour shop control. This facility enables the machine to be operated by hand or semi-automatically whenever required.

The automatic valves are actuated via push buttons in the luminous circuit diagram. The individual controller DCEB is a completely automatic, autonomous, microprocessor controlled system with facilities for programming and program storage.

Finally, several individual controllers can be operated from the Data Comp DC control and monitoring system. Programming and data logging take place conveniently at the terminal of the central control unit. Data acquisition, data sorting, commercial data processing, technical data processing take place here also. Further, communication is possible with a host computer or with external minicomputers in the colour shop or colorimetry and recipe formulation areas.

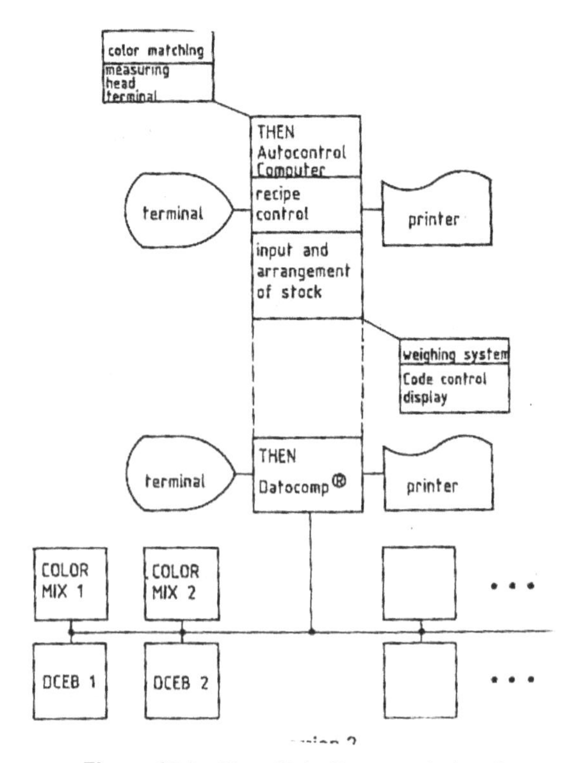

Figure 18.8 Then-Data Comp control system

18.6.1 Individual control

Individual control refers to a control system which is directly allocated to an individual dyeing machine of the dyeing unit. Then-Data Comp DCEB individual control system is supplied in two versions. The DCEB version is a simple temperature controller offering the advantages of VDU and analogue oriented microprocessor control with storage facilities. The functions like "temperature", "gradient", "retention time" and signal can be programmed. With this, the controller decides heating or cooling to take place to control the current temperature of the dye liquor in order to achieve the target temperature. The three way solenoid valves also called as pilot valves are directly controlled by the power outputs of the DCEB and in their operate the pneumatic diaphragm valves for heating and cooling. By installing additional plugs in the PC cards, the control system can be conveniently operated as fully automatic system at any time.

In the fully automatic DCEB system, the CPU is a 8-bit microprocessor which controls input and output units. Now concerning the software, the user can carry out an interactive software program in conventional mode and neither requires programming language nor coding lists. The consistent utilisation of the menu technology simplifies operation as well as insertion, administration and the sequence of the dyeing programs and permits optimum adaptation of the control system to suit the machine.

From the Figure 18.9, the interrelationship between the individual selectable menus on the video screen. By operating the corresponding function key, the control system branches either in to a further select menu or in to a program routine. It helps to differentiate between the basic menus, program sequence, program processing and service routine. Everything is accessible by entering a password. Hence unauthorised altering or correcting the dyeing programs or machine settings is prevented. While writing the program, all controllable functions appear one after the other in a logical sequence in clear text on the screen. For example:

"Fill the dye vessel with water type 1"

"Fill the dye vessel with water type 2"

"Fill auxiliary tank with water type 3".

With the function key the operator with authorised access can configure the necessary program steps. Any illogical steps are rejected by the automatic system and the faults can be corrected immediately. The DCEB numbers the program in steps of ten and the maximum permitted number being 230 steps. Listing of the completed or half finished dyeing program appears on the screen or on a connected printer. This fast and simple means of programming permits individual adjustment to dyeing process, substrate and depth of colour.

After a correction the program can be stored on a digital cassette by means of a four digit program number. The sub-menu "Cassette Routines" permits loading, erasing and storing of dyeing programs and is also used to format and copy the cassettes and produce a list of contents. Before the start of the dyeing program, the required program on the cassette is loaded in to the program memory, checked automatically and then displayed on the screen. When the dyeing program is started, the cassette is not required and the cassette can be used for the other dyeing machines to start up. The digital cassette records 50 programs on each side with maximum 230 steps. In the modern version DCEB system the maximum storage capacity has been increased to 460 steps with the digital cassette recording 25 programs on each side. In order to adapt all the internal parameters correctly like delay times, control data, machine configuration and the time for operation of the dyeing machine, the software package "Service routine" is available. Password is required here also which prevents unauthorised entry.

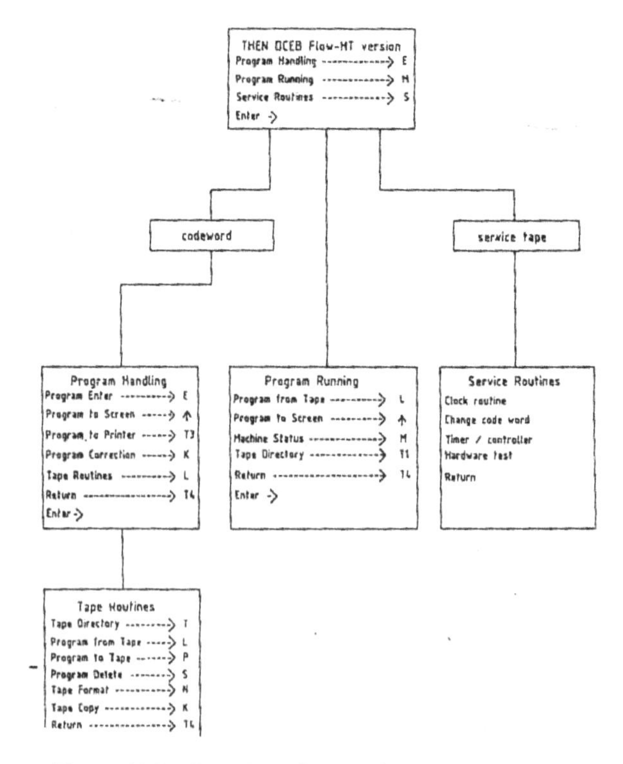

Figure 18.9 Then-Data Comp DCEB individual control

18.6.2 DC control and monitoring

In the wet processing with several individual controllers, the installation of a central control system is recommended. Different machine systems are pooled when programming takes place. The program can be written on a 12" screen with a large alpha-numeric keyboard is much more convenient than with the individual control. The system has all the facilities for complete data collecting and processing. The hardware components of Then-Data Comp DC central control system has a personal computer and the CPU is a 16-bit microprocessor, for programs 320 kbyte RAM capacity is available for the memory. The VDU screen has 80 characters in each of 24 lines. The keyboard is divided in to alpha-numeric keyboard, ten key panels for numerals input and ten function keys.

Figure 18.10 shows the basic menu of Then-Data Comp DC central control system.

The top line shows the date and the time. On the subsequent lines the hours are shown. Below this all current dye batches with preset starting and finishing lines can be seen according to a machine and number. The time elapsed is shown as a bar graph. For example, the responsible person in the dye house can immediately recognise that machines 1 and 5 are not in production, whereas machines 2 and 3 are in the middle of the dyeing program. On machine 4, the dyeing is almost complete, on number 5 the preset time has been exceeded and on machine 7, a new batch has been started.

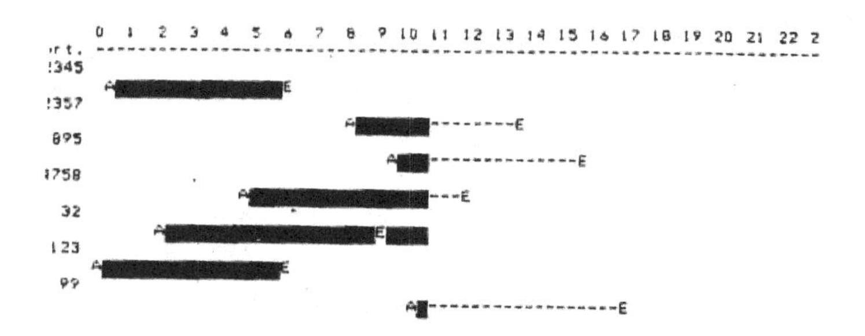

Figure 18.10 Basic menu of Then-Data Comp system

The bottom line shows which function keys have been operated. This is where branching in to the main menu starts i.e., dyeing program processing, evaluation, via the menu "dyeing" the prepared batches are assigned to the dyeing machines. All entered batches are displayed with clear batch number next to the machine number. The order which these are given can be changed and if necessary also the machine allocation.

Activated batches are indicated by a bold display and following completion of the previous dyeing process the dyeing program is transferred in to the next.

18.6.3 Dye house automation

Apart from the use of microelectronics, there are of course other applications of computers and control systems in different areas in the dye house. As shown in the Figure 8.11, right at the top of the list is colorimetry with match prediction and recipe formulation and colour shop with dye and chemical stores. The ultimate goal should be an integrated network in the final phase, so that all applied data are accessible for subsequent use and can be evaluated for commercial, statistical and technical discussions. The auto control system comes in two versions. The first enables existing hardware and software elements such as colorimetry, inventory management, recipe formulation, order processing and weighing-out systems in the colour shop. All these are linked to the Then-Data Comp DC central control system via the auto control computer and ensures a constant data exchange. The second version shown in the figure operates on the assumption that colour measuring and weighing systems are from different suppliers. The associated and missing software packages such as recipe formulation, dye house management and order processing are in this case are separate which are tailored to the auto control system to the requisite hardware configuration.

Assessment criteria include the following:

- Specialist knowledge of the process and the software technology for pre-treatment like bleaching and dyeing equipment is a basic requirement for the suppliers of the control system. It means the dyeing machine; software and the control system should be from the same source.
- Regular update of the software should be guaranteed.
- Dialogue-oriented menu driven technology with professional basic software precludes operation errors and permits operation by personnel who has not been trained in the art of electronic data processing.
- Dyeing processes programming should be simple and quick and later corrections, if any possible.

The most important point is to clarify the reaction of control system in the event of power failure or any defect.

18.7 Yarn package dyeing machines

The first yarn dyeing machine was built in 1882 by Bermaler who utilised an open dye bath and tried with the fluid flow from inside to the outside of the package. Further modification in the machinery was made by the reversal of the pump to change the direction of the fluid flow and the dye bath was enclosed. The change in the fluid flow direction as well as uniform circulation of the liquor and the rate of fluid are the main factors responsible for the levelness of the material obtained. Fluid circulation is obtained by forcing the liquor through a perforated spindle and the package. This two-way direction of the fluid flow is essential to increase the levelness of the dyeing. Yarn package dyeing is shown in the Figure 18.11.

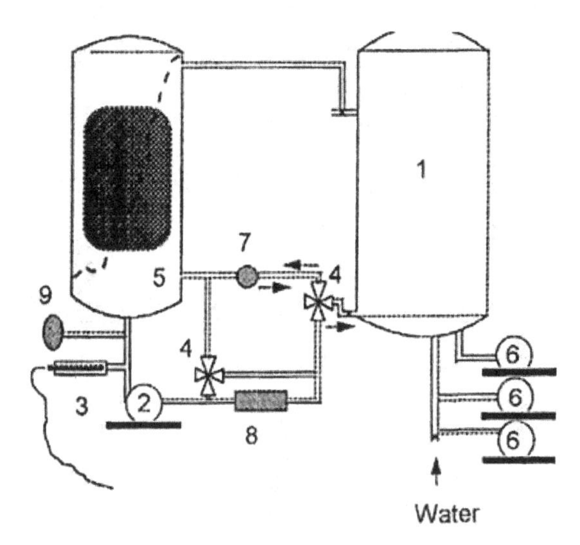

Figure 18.11 Yarn package dyeing machine
Mixing tank, 2. Circulation pump, 3. pH metre, 4. Reversing valve, 5. Dyeing autoclave, 6.Dosing pumps, 7. Colorimeter, 8. Flow metre, 9. Pressure metre.

The important parts in the yarn dyeing machine are summarised as follows:

- Dyeing vessel
- Package carrier
- Circulating pumps for primary and secondary
- Heating devices
- Liquid flow and flow reversal devices

There are some factors influencing the outcome of the dyeing process. The density of the package and its uniformity are, therefore, crucial factors in the outcome of the dyeing process. It means the yarn needs to be wound in to a suitable form such as a cheese, cake or cone. Both metal and plastic formers are used in the dyeing industry. Levelness is very important aspect of the dyeing system. During the dye up-take, it occurs at surfaces which are most available to the flowing liquor, localised over dyeing may occur and it is time consuming to correct it.

The degree of levelness in dyeing for and given dye-fibre combination should have

- The rate of dye liquor circulation and its reversal.
- The package density and the shape of the package.
- The rate of dye up-take in relation to, time, pH and other factors.

The reversal of the fluid direction has been made possible by using either axial flow pumps in which the entire system of pumping is reversed or by using centrifugal pumps in conjunction with four-way valve system. The fluid flow through a package is shown in the Figure 18.12.

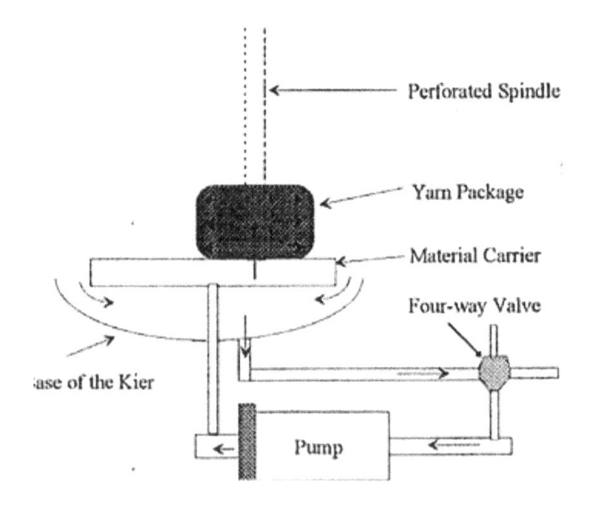

Figure 18.12 Fluid flows in a package

Because of lower concentration of the dye in the liquor which is near to the surface of the fibre or which is trapped in the spaces of the yarn compared to the external dye bath, the rate of dyeing is dependent upon the efficiency with which the dye is transported to the material being dyed, either by the liquor circulating the material or by the materials movement with respect to the dye liquor. The mass transport process in relation to dyeing is defined by two boundary layer.

- The hydrodynamic boundary layer was defined as a layer within which the velocity of liquor rises from zero to 99% of the main stream velocity.
- The diffusional boundary layer was defined as the layer within the concentration of the dye rises from that at the fibre surface to 99% of the main stream.

The latter was regarded as a layer of liquid which hindered the passage of dyeing from the bulk of the dye bath to the fibre surface. As packages are subjected to various mechanical, hydrostatic and hydraulic forces during dyeing, the density of the package should be as uniform as possible as it has a direct bearing on the quality of the dyeing. The density of the package must be optimum so as to obtain an appropriate flow of liquor and prevent the effect of "channelling".

18.7.1 Aero-flow piece dyeing machines

In this type of dyeing machines, the fabric transport takes place by means of a separate gas circuit through humid air or an air-stream mixture without using any liquid. Dyestuffs, chemicals and auxiliaries are dissolved in the processing liquor and injected directly in to an air stream. In this way, the liquor is atomised and evenly distributed on the surface of the fabric. Hence an optimal penetration of the liquor is carried out within the textile material. The liquor ratio is not dependent on the material loading quantities. Figure 18.13 shows the main part of THEN-AIR FLOW machine.

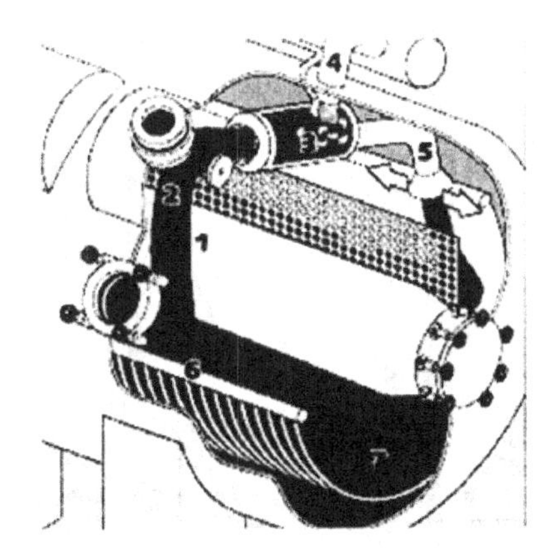

Figure 18.13 Aero piece dyeing machine

After dyeing, rinsing is carried out by spraying fresh water at the section (3). The fabric slides on the rods (6) covered by PTFE and a small amount of excessive liquor and water falls down to the bottom of the vessel. This method can save the consumptions of 50% water, 40% chemicals, 40% energy and 40% operational costs.

18.7.2 Fluid flow in the cylindrical and conical packages

When fluid flows in a packed bed, two flow phenomena come in to play, namely *bulk flow* and *dispersive flow*. These two types of flow create rate-determining steps in the sequence that determines the overall rate of transfer of dye to a given point on the fibre surface at a particular time. There are two types of packages generally used in practice in textile industry like cylindrical and conical packages. Fluid flow across conical and cylindrical packages is shown in the Figure 18.14. When comparing the flow characteristics on these two packages, cylindrical and conical type, the cylindrical shaped packages have better flow characteristics than the conical shaped packages. This is a result of the shape of the conical shaped packages which is subjected to both axial and radial forces. When packages are wound on to compressible springs, the density variation can be minimised. However, it is important to note that even small local variations in the package will directly influence the final levelness achieved in any dyeing process.

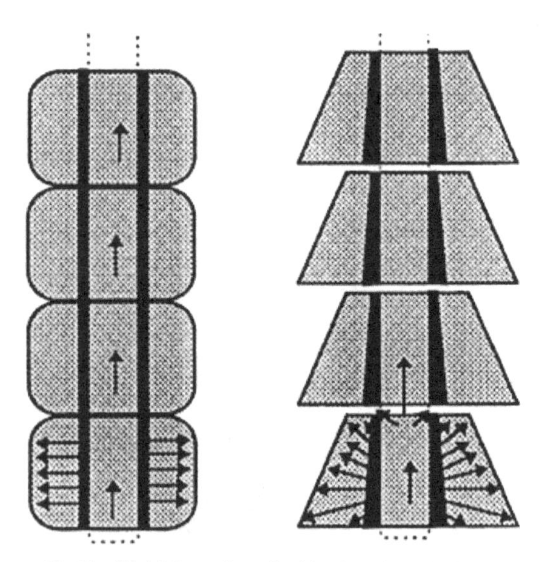

Figure 18.14 Fluid flows in cylindrical and conical packages

18.8 Digital printing

The process of reproducing digital images on physical surfaces is called digital printing. The printing process consists of design development, colour separation, tracking production, screen engraving, sampling, paste preparation and it takes long time to complete the batch. With the modern technology, this time frame has been reduced to within a week. CAD application generally starts with one overall coloured design which is scanned in to the computer and can be modified and coloured. Various colour combinations can be made, stored and a hard copy taken out on a paper or fabric. Ink-jet printers available now can directly print on the fabric using reactive dyes. The physical surfaces can be in any form like a common paper, cloth, plastic, a photographic paper, film, etc. The specialty of digital printing is that, unlike ordinary printing, the ink is not absorbed into the paper. Thousands of impression of the same object can be drawn from one set of plates. Instead of absorption, the ink forms a layer on the surface.

Digital printing refers to methods of printing from a digital based image directly to a variety of media. A dye-sublimation printer is a computer printer which employs a printing process that uses heat to transfer dye onto medium materials such as a plastic card, paper, or fabric.

Digital textile printing is the technology that consists of printing the designs on fabric, directly from the computer, with no other additional step. This means that after creating the required designs, and has them in repeat; it can be printed on the fabric just like printing on paper. It is considered to be the next generation printing and is different from traditional textile printing. The only requirement of the digital textile printing is that the fabrics which have to be used must be pre-treated to confirm about their ink holding capacity. Apart from this, a wide range of colours can also be obtained through the pre-treatment process. Comparison of digital and traditional printing methods is shown in the Table 18.1 and the dyes used in digital printing for different types of fibres are shown in Table 18.2.

Table 18.1 Comparison between digital textile printing and traditional printing methods

Digital textile printing methods	**Traditional printing methods**
Has greater flexibility	Less flexibility
Easy applications	Less convenient in applying
Can be used for versatile purposes	Can be used for limited purposes
Takes less time	Take more time
Having more colours along with photo-realistic effects	Comparatively less colours and absent of photo-realistic effects
Present a variety of exclusive textile designs	Having less number of varieties regarding designs

Table 18.2 Dyes used for digital printing inks and related fibres

Name of digital printing inks	Types of fibres used
Acid inks	Silks and nylons
Disperse inks	Polyesters
Dye based inks	Not used for any fabrics, used for photography only
Reactive inks	All cellulose based fibres: cotton, linen and rayon

18.8.1 Selection of right ink and fabric combination for digital printing

Digital textile printing is fast gaining ground and seeing market acceptance in printing on a variety of fabrics. Diverse markets such as apparel, home furnishings, tradeshow graphics, flag and banner, swimwear, and point of purchase displays are all showing rapid growth.

Compared to analogue techniques, the cycle time from design conception to actual production is significantly reduced by using digital printing. Creating the best image quality for digital textile printing is obtained with proper consideration of several key areas:

- Fabric and ink selection
- Pre- and post-processing of textiles
- Colour software
- Printing equipment

18.8.2 Digital textile printing process

Digital printing of textiles with dye based inks require pre-treatment and post-treatment of the printed fabric for full colour and durability. The chemicals in the fabric coating permit reaction of the dye with the fibre during post processing. After final drying additional treatments can be given to the printed fabric like waterproofing, flame retardant, and so forth.

Ink-jet printing machines have been developed based on ink-jet printers for sheets like paper and film. The inks are jetted on to the fabric through a large number of nozzles such as around 360 by a piezoelectricity. The ink should have lower viscosity with higher surface tension than that of the ink used in screen printing. Such type of machines can be used for dyeing with 16 colours and 600 dpi for 3200 mm width by the speed of 80 m^2/h. Reactive dyeing, acid dyeing, disperse dyeing and pigment dyeing have become applicable to the printing. Images generated by CAD can be, ost precisely and directly transferred to images printed on the fabric. Hence ink-jet printing has many advantages than the conventional printing in terms of flexibility and to quick response.

Fabric selection

Textile fabrics can be broken down into three basic types.

Woven fabrics: These have two sets of yarns running parallel and perpendicular to the selvage called warp and weft. These fabrics are dimensionally very stable.

Knitted fabrics: Formed by interloping of yarns to form loops. Knitted loops are referred to as stitches. Vertical stitches are called wales and horizontal stitches are called courses. Knitted fabrics are dimensionally less stable than the woven fabrics.

Non-woven fabrics: These fabrics have intermingled fibres that are held together by adhesive, thermal bonding, spun bonding, and so forth.

Once the ink set for printing is selected, the selection of fabric needs careful attention to obtain the best image quality.

Surface defects on the fabric can lead to fabric striking the print head resulting in print head or image quality damage. Check for these factors:

- Fabric surface should be free of any lint.
- Fabric surface should be free of any broken fibres or filaments.
- Fabric roll should be wound with uniform tension across the width and length of the roll.
- There should be no wrinkles or creases on the fabric.

Dimension defect is another consideration. Most textile fabrics have a tendency to shrink. Care should be taken to pre-shrink the fabrics and make them dimensionally stable before starting to print, otherwise the printed fabric can change in dimension during post-processing and lead to image distortion. While processing knitted fabrics, care should be taken to prevent excessive stretch during the coating step and also during printing step to prevent image defects due to dimensional stability.

18.8.3 Pre-treatment/coating

In the simplest terms, pre-treatment is an application of a solution to coat a fabric to achieve the following main objectives:

- Better image quality in digital printing by preventing lateral bleed. This is achieved by the use of natural or synthetic thickeners commonly used in the conventional textile printing paste.

- Application of necessary chemicals that are needed for the dye to exhaust onto the fabric and react with the fibres on a molecular level during the heat fixation process. Such chemicals need to be applied to the fabric prior to printing. They cannot be incorporated in the digital printing ink due to the need for stringent purity and low viscosity required of an ink to sustain good quality printing.

Since a universal pre-treatment for all the ink types does not exist, each ink type requires its own unique pre-treatment solution. However, irrespective of what chemistry of pre-treatment is used, the method of application is the same. Fabric is saturated in the pre-treatment solution by dipping the fabric in an open width form, squeezing the fabric out by application of pressure, and drying the fabric in hot air. Most commonly pre-treatments are done using textile equipment called a stenter/tenter frame. It is not essential to invest in pre-treatment equipment or possess detailed knowledge of the know-how of the pre-treatment chemistry to print digitally. Commercially available pre-treated fabrics are becoming more widely available as the digital textile market expands.

Fixation/heat treatment

Once the pre-treated/coated fabric is printed with digital inks it needs to go through a fixation process. This process enables exhaustion of dye from the fibre surface in the printed area into the molecular structure of the fibre. It also enables absorption of colorant on the fibre:

- Ionic forces of attraction in the case of acid inks on polyamide fibres.
- Dissolved in the fibre as in the case of disperse inks for polyester fibres.
- Reacts with the fibre and forms a covalent bond as in the case of reactive inks on cellulosic fibres.
- Binder self crosslinks forming a network in the case of pigment inks.

Wash-off

This final step in the digital printing process is essential for direct printing with all the dye based inks. Wash-off serves the following important functions for the dye based inks.

Removal of excess unreacted dye: In textile coloration with dyes, the adsorption of dyes onto textile fibre is never 100% complete and there is always some loose dye left on the fibre/fabric surface. This dye must be removed to realise the full end use potential of the printed product.

Removal of pre-treatment or coating solution: Coating solution applied on the fabric during the pre-treatment step imparts stiff handle/feel to the fabric. Wash-off process removes the coating solution and makes the fabric feel soft.

Wash-off also makes the white unprinted area come out looking bright and clean. Wash-off is not needed if fabrics are printed with pigment inks. This makes pigment inks particularly attractive since they do not need wet processing.

18.8.4 Benefits of digital textile printing

- The concept of digital printing on fabrics has opened new opportunities for designers, merchandisers and salespersons.
- The process is time effective and also cost effective.
- Assists the customers to control the design and printing process from remote locations.
- Eliminates colour registration of plates or screens.
- Makes Just In Time (JIT) delivery and Quick Response (QR) possible.
- Greatly reduces the need for inventory.

Useful guidelines for digital textile printing

- Choose a piezoelectric printer for printing, it has ink delivery system.
- Instead of heating to force out the ink or dye, one should use electro-potential charge (it moves out the ink onto the surface to be printed).
- Any dye can be used out of the dyestuff for cotton.
- Pigment dyes have become very popular for digital textile printing. These new pigment dyes will not form a stiff film on the fabric that has to be printed.
- Benefits of using pigment dyes are the users don't have to pre-treat, steam and after wash these, as these things are occurred with other kind of dyes like reactive, acid and disperse dyes.
- Wide range of colours. Can be used due to the development of pigment dyes.

18.9 Textile finishing machinery

Conventional concepts in finishing machinery has been modifies and replaced by PLCs, microprocessors, servomotors, individual drivers and many others. Process control and automation are at an advanced stage in the new range of stenters by Babcock, Brueckner and Monforts. Advantages like maintenance free chains, automatic control of fabric width, overfeed temperature controller, moisture and dwell time with high energy efficiency and reduced exhaust pollution, etc. Other drying machines like shrink dryers, tumble dryers by Krantz not only dries, but also controls the fabric shrinkage as shown in the Figure 18.15.

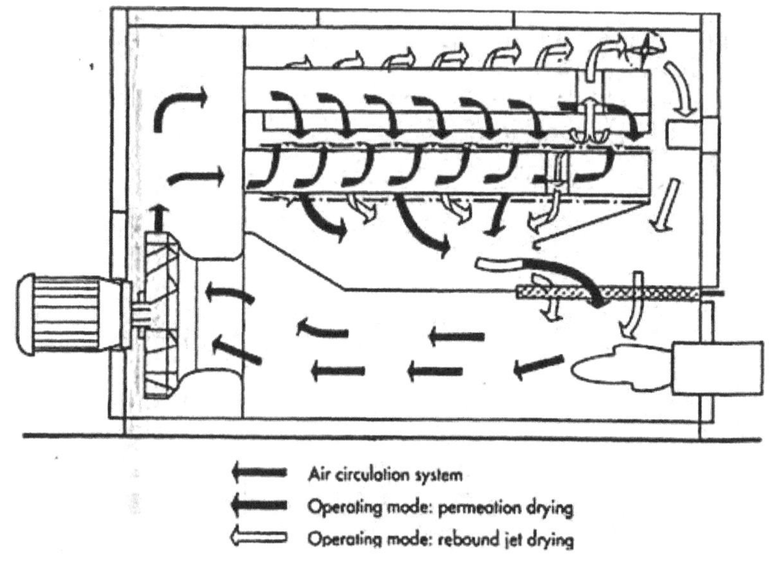

Figure 18.15 Textile finishing machinery

Shrink dryer develops volume and improves its feel in its tumbling zone. Fabric is fed at low tension on to a screen belt with an adjustable overfeed of up to 200%. Drying is carried out by hot air in either of two modes – rebound jet or permeation drying. This semi-dried, pre-heated fabric is then dynamically moved by a special circulating air control systems. The speed of this tumbler process is adjustable according to the finish desired.

18.10 Computer integrated manufacturing in textile factories

In the recent years, the development of online monitoring and control systems to control the process has been in use in the textile industries right from ginning to the finishing process. The development of computer aided design (CAD) and computer aided manufacturing (CAM) have been developed and utilised in their manufacturing operations. Now several industries claim that they have computer integrated manufacturing (CIM) on their plant. In actual practice, this is still in the future but the technology to accomplish this task is the near reality. It requires the overall use of instrumentation, process monitoring and control and how they can be integrated by the use of computers.

Automation in the textile industry is rapidly reaching the critical point where a similar non-proprietary communication standard is needed. The benefits that are expected from such a communication protocol are:

- Ability to integrate multi-vendor control and measurement systems.
- Integrated information flow throughout the plant.
- Reduction in integration costs and associated risks.
- Rapid new process implementation.
- Increased flexibility for network modification and expansion.
- Improvement in the quality of communication service.

Electronic data exchange generally refers to transmitting data between computers, but in textile industry it refers to the sharing of the product information between a vendor and a client. For example, fabric width measurements from a textile company to an apparel manufacturer. Everyone is more familiar with bar coding in all the retail stores. It is the beginning to be used in the textile industry. Bi-directional communication refers to the ability to transmit, as well as receive information to and from production machines. Machines such as looms, can now be adjusted and controlled by a computer in addition to being monitored.

Just In Time (JIT) and Quick Response (QR) are the terms that refer to the manufacturing plant operating theories mainly to reduce inventories, improve flexibility and allow the textile and other related industries to respond quickly to the needs of the customers.

Automation has number of benefits that enhance quick response and facilitates implementation of the JIT management philosophy. Some of the benefits are:

- Reduction of work in-process time.
- Reduction of work-in process inventory.
- Improved flexibility.
- Improved scheduling and production control.

Online quality control is a new approach to controlling quality for the textile industry and this should improve quality by cutting out most of the seconds produced. An automated process must receive materials with consistent quality to efficiently perform a desired task. To achieve desired level of quality, processes must be continuously monitored and automatically controlled at every stage of production, improved quality creates an added benefit by increasing production, efficiency and overall productivity of the process. Continuous monitoring of processes at every stage of manufacturing will be a pre-requisite to maintain consistent quality levels. Examples of such systems developed by many companies are auto levellers in carding and drawing, ring data in spinning, yarn clearers in winding machines. These are all standard features in the modern machines to achieve consistent product quality at all stages.

Machinery manufacturers are seeking the next step beyond continuous monitoring, which is control. A totally automated process must react to the monitoring system in real time to maintain consistent quality. Automation will definitely improve the overall level of quality in manufacturing. A further requirement in automation is the investment in continuous monitoring and control systems. Computer Integrated Manufacturing (CIM) regards manufacturing as a continuous flow process and implies linking together of adjacent operations and overall control systems with central control system. Quite obviously, this is the direction of the development of automation and the use of electronics in textiles. In order to develop JIT and QR, it is better to treat the total textile and related industries manufacturing pipeline as well as individual processes as a manufacturing system that must be integrated by the use of computers. The application of CIM in textile manufacturing will be discussed in detail now.

A diagram of typical CIM system in a textile manufacturing plant is shown in the Figure 18.16.

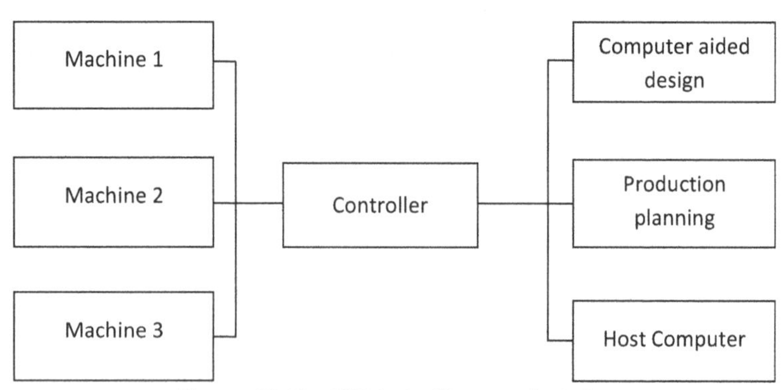

Figure 18.16 CIM in textile manufacturing

From the Figure 18.16 all the manufacturing units are controlled by the computer according to the management decision by input from the CAD system and production planning. With such arrangement, information can be taken from any part of the system at any time without disturbing the operation of the machines. Systems can be designed for one area, such as spinning or weaving or knitting for a vertically integrated operation. Despite the actual installation still consist of monitoring only, the ultimate goal being some form of CIM.

18.10.1 CIM in spinning

Computer integrated manufacturing (CIM) systems are available that control or monitor all the yarn production processes from opening and blending in spinning, winding and twisting operations. Some of the applications include production monitoring, quality monitoring, testing, inventory control, tracking of orders, machinery maintenance, budgeting, mill management and others. Some of the advanced controls are

- Control opening and blending, carding KIT control.
- And individual spindle drives in spinning which offers the flexibility and readily fit in to the CIM concept.
- Sliver weights can be controlled and the levels changed by on machine electronics which can be readily connected to a computer network.
- Online quality control in carding and drawing can perform spectral analysis with the help of spectrogram and also provide a useful tool for analysis of defects.
- Transducers like passive acoustic detection systems which measure the evenness of the slivers that can be easily adapted for all types of cards and draw frames.
- Automatic doffing and automatic piecing in spinning machines performed by robotic mechanisms in rotor spinning help to reduce the number of operatives in the spinning mill.

18.10.2 CIM in weaving

Weaving and knitting machinery manufacturers have been constantly working towards the utilisation of computers in textile manufacturing for many years with their use of CAD, bi-directional communication and artificial intelligence. With the availability of electronic dobby and Jacquard heads, pick finding by automatic and needle selection, etc. Such machines are the most easily integrated in to computer networks of any textile production machines. Figure 18.17 shows the bi-directional communication systems which can be used to control many functions in waving machines. CAD system can be used to produce designs on the fabric and the design can then be transmitted over the network to the weaving machines to produce required designs on the fabric. With these technologies it gives true meaning to the term "quick response" and also the time needed to produce the fabric.

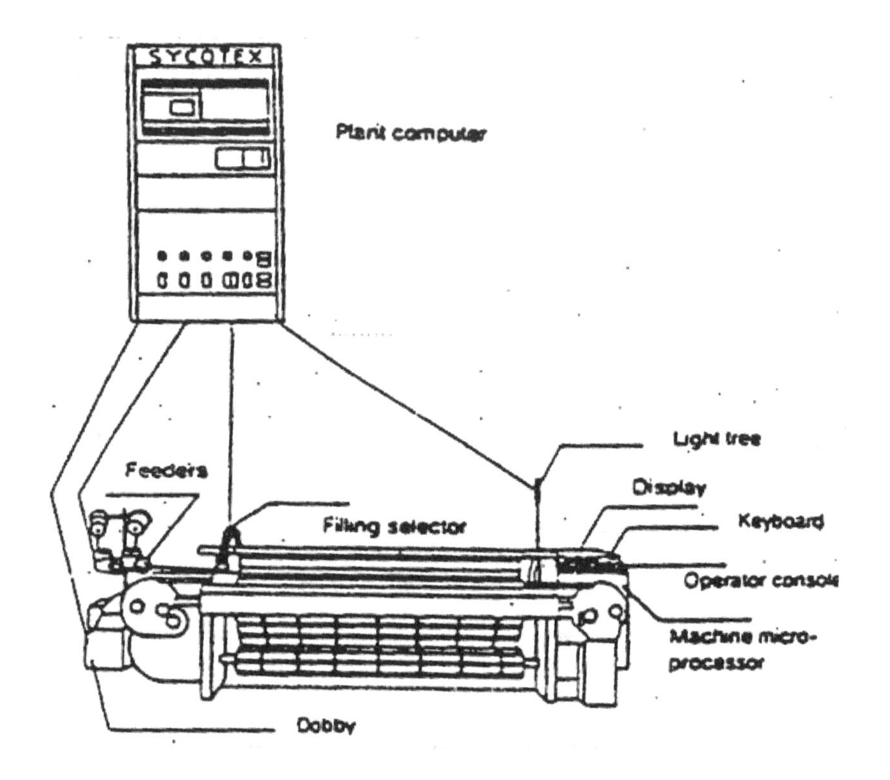

Figure 18.17 CIM in weaving

Combined with these systems, the techniques that are being developed by the loom manufacturers that allow faulty picks to be removed from the fabric ensure a fault free fabric with little labour for mending the defects.

Figure 18.18 shows the artificial intelligence (AI) in weaving which assists in trouble shooting looms. As AI systems are combined with online data collecting systems in a CIM network, the chances for full automation in weaving process are greatly enhanced.

The computer generates sizing machine reports of the sizing machine control systems. This provides a tool for the management to ensure that all the warps are sized identically under normal operating conditions. Obviously, these monitoring systems can be included in a computer network. At the same time, knitting machine manufacturers are making best use of the electronics which provide the machines that are automatic and more versatile and can be conveniently incorporated in to the CIM network.

Similarly, non-wovens are also controlled by electronics and automated material handling techniques which lend themselves in to CIM network.

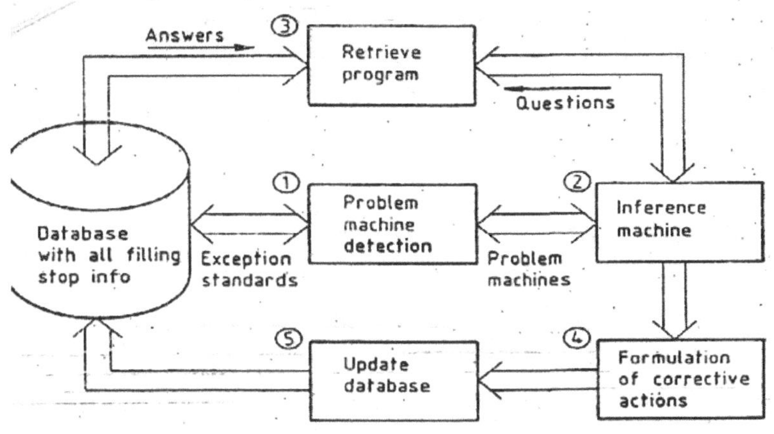

Figure 18.18 AI in weaving

18.11 CIM in dyeing

A typical system of automatic control of dyeing machine is shown in the Figure 18.19. The monitor displays scheduling for any machine and allows the operator to arrange the next lot. Batch weighing updates, inventory of each dye by bulk and container. Any errors in the process can be easily tracked to a particular container if is required.

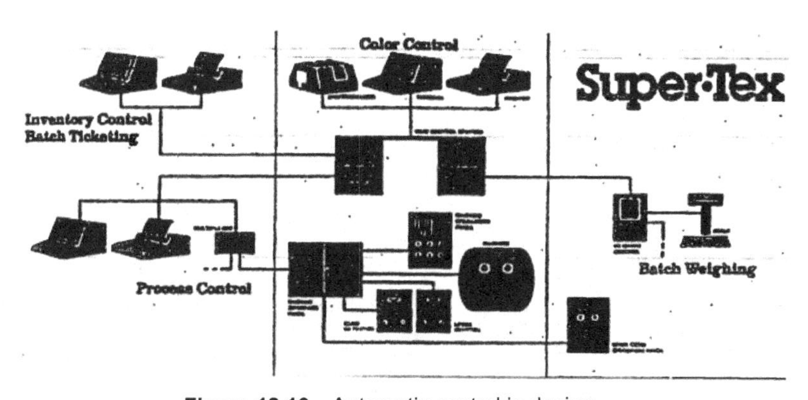

Figure 18.19 Automatic control in dyeing

A "Data Concentrator" accepts inputs and downloads to the central CPU. A parameter table allows each machine to be set-up independently and sets. For example, fill times which depend on the proximity of the machine to header supply line. Exceptions are sent to file, printers and to area exception files. The computer has the latest information about the machine and the reports can be seen within a minute. Automation of the batch dyeing machines and finishing machines using the state of the art electronic technology includes:

- Microprocessor dye machine controllers.
- Management information host systems.
- Chemical room check weighing systems.
- Dye and chemical inventory systems.
- Computer colour matching and control.
- Colour display – colour simulation.
- Bar code based lot tracking systems.
- Energy management systems.
- Real time monitoring of the main frame and dye house host computers for manufacturing information and production planning.
- Automated liquid dispensing system.

Automated colour lab dispensing system helps the amount of colorant automatically adjusted and dispensed for the lab technical personnel.

In the finishing operations, tenter frame automation system with touch screen colour graphics and voice alarms help operators to identify the zones overheat or motors overload as shown in the Figure 18.20. Software control includes:

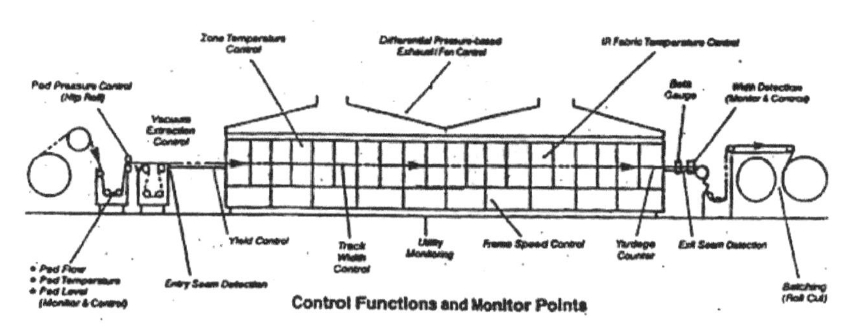

Figure 18.20 Control functions in finishing

- Infra-red fabric temperature, chain speed and zone temperature control.
- Circulation fan control.
- Width monitoring and track control.
- Pad temperature level, flow and pressure control.
- Differential pressure based exhaust control.
- Lot start/stop, lot change over, shut down provisions.

18.11.1 Yarn singeing

The purpose of singeing is to reduce the hairiness of a staple fibre or twisted yarn and at the same time equalise the residual hairiness. The term "gassing" is used often instead of singeing since the process is normally performed with gas burners. The single thread is fed at a high speed through a gas burner which burns away the hairs projecting from the body of the yarn. To do this, breaking strength of the yarn is an important factor which should not be altered significantly. Unsinged yarn (left hand side) and singed yarn (right hand side) is shown in the Figure 18.21.

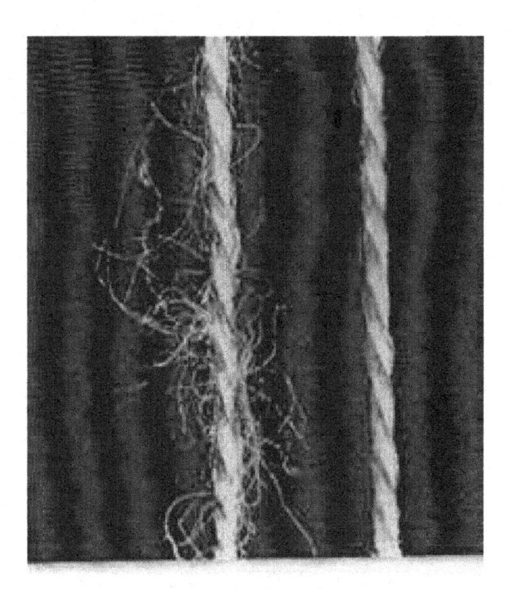

Figure 18.21 Unsinged and singed yarn

The gas burners are designed in such a way that the yarn is exposed to heat uniformly from all sides with environment friendly utilisation of the energy. To assure constant singeing quality it is necessary to produce supply packages of unvarying moisture content. Singeing is performed not only to upgrade cotton sewing yarns but also for many other application areas of staple fibre yarns. These include gassed and mercerised yarns for knitting and embroidering, gassed twisted yarns for poplins or voiles. Blended yarns are singed also because blended untwisted single yarns on a growing scale too recently. Singeing is applied normally after twisting operation and before the next processing stages such as dyeing or mercerising. With the present day singeing machines it is now often possible to omit rewinding after singeing. The savings resulting from this may be considerable making a replacement investment a worthwhile proposition. The technology employed and the pre-requisites for will be discussed in the following sections.

The process stages for the material flow for gas singeing process is shown in the Figure 18.22.

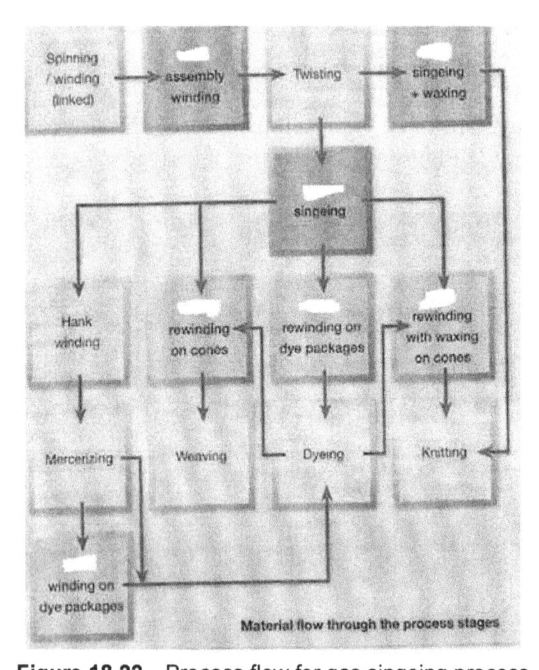

Figure 18.22 Process flow for gas singeing process

18.11.2 High efficiency singeing burners

High efficiency singeing burner is shown in the Figure 18.23.

The burner is made of special steel suitable for cotton, blended yarns and man-made fibres. It is designed as an insert easily fitted in to the basic housing and can be exchanged. Immediately above the burner there is a strong suction which helps to provide effective yarn cleaning. Casings protect against touching the burner, which gets hot during operation. In order to ensure safety, the casings should only be opened when the machine is switched off. The machines may be equipped with optional gas detectors. The purpose is if one of the detectors registers gas contrary to the expectations, all singeing machines in the same room shut down automatically. Suctioning continues on all machines. At the same time the trouble is signalled.

Figure 18.23 High efficiency singeing burner

18.11.3 Unsinged yarn lengths

When a winding position is stopped, its take up spindle together with the spark shield are raised in to an ergonomically ideal position for attendance. The new supply package can be creeled without trouble. With a simple movement of the hand the yarn is drawn in to the guiding elements of the yarn brake. During starting of the winding position, the speed increase of the grooved winding drum is measured electronically. Upon reaching the preselected level the supply package and spark shield are lowered in to the winding position by a pneumatic cylinder. At the same time the yarn is drawn in to the burner. Hence very short unsigned yarn length is assured by this electronically controlled starting sequence.

Figure 18.24 shows the partially singed yarn tail device will allow the unsigned length out of the package.

Figure 18.24 Un singed yarn lengths in gas singeing machine

18.11.4 Waxing during singeing

For certain applications like knitting, yarns must be both singed and waxed. It is indispensable that the waxing is done prior to singeing process as the wax applied evaporates during singeing because of the heat applied. Considerable research has revealed, however, that provided right wax is used applying it immediately before the burner can even yield better results in the form of lower and more uniform friction than has been attained previously. The waxing while singeing process is shown in the Figure 18.25. The waxing device is arranged immediately at the thread brake on the spark shield. The wrap of the thread around the wax roll is adjustable, ensuring uniform wax application unaffected by the size and weight of the roll. With this option, users can do both waxing and singeing at the same time.

Figure 18.25 Waxing during singeing

18.11.5 Dust control

During the process of singeing, more dust will be produced and inevitable. In order to remove the dust from the singed packages, the spinners use additional rewinding for de dusting. This is an extra process which adds to the production cost. Dust removal from the singed packages is essential for downstream operations. Packages with acceptable residual dust content can only be produced by effective suctioning in combination with an air supply above the winding position and a specially adapted yarn path. Such system is shown in the Figure 18.26.

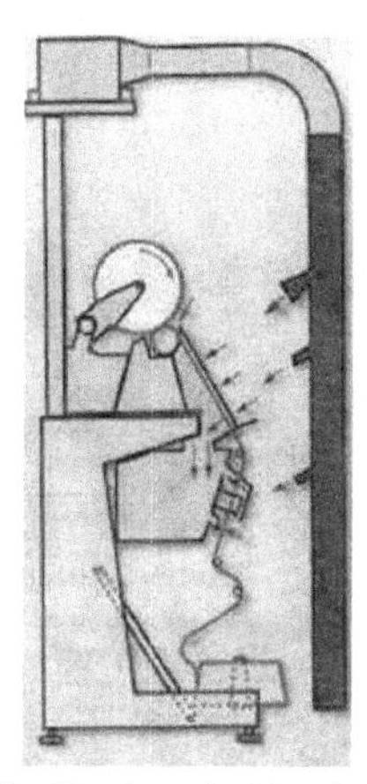

Figure 18.26 Dust control in singeing machine

The newly developed air supply which is an optional one above the machine permits selective delivery of outside air, which depends on climatic conditions, does not need additional heating or cooling. Over the machine, air curtain is formed which is immediately suctioned away by the machine's suctioning. The air supply rate is dimensioned so that about 50% of the machine suctioning rate is covered by the supply above the machine.

19
Pneumatic and hydraulic control valves

19.1 Introduction

In pneumatic control systems, the control valves are required to carry out three types of functions namely direction control, flow control and pressure control. The hydraulic and pneumatic valves are more or less similar. The flapper valves are first developed for low pressure pneumatic systems which are also applied to hydraulic systems also. Hydraulic fluids are non-compressible, pneumatic fluids are compressible. It is because the density of air even at the highest operating pressure is a very small fraction of that for the hydraulic oil. Moreover, the mass involved in pneumatic system is very small. Hence, the fluid forces and Bernoulli's forces which are otherwise large in hydraulic system are negligible for the pneumatic system. The viscosity of air being negligible and the viscous damping with pneumatic valve is negligible.

The lower viscosity causes severe leakage problem in pneumatic system. But, the pressure drop in leakage path is negligible. The flow through the pneumatic system is almost adiabatic.

19.2 Operating procedure of pneumatic valves

Direct operation

The direct operation of the pneumatic valve is shown in the Figure 19.1 (A) and (B), the port can be opened or closed by simply pulling the lever up or down and allowing or stopping the air flow by moving the piston up or down. The necessary force is applied manually by hand.

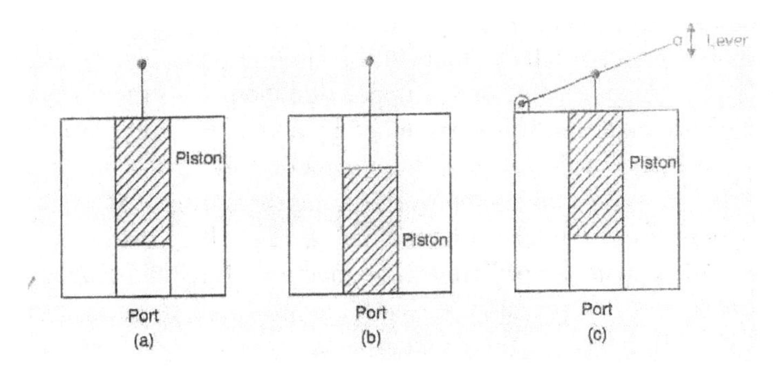

Figure 19.1 Pneumatic valve operation

For ease of operation the valve can be lever operated. The lever is pivoted on left-hand side of the valve. The middle part of the lever is connected to the piston 0rod as shown in the Figure 19.1C. Right-hand side of the lever is moved up or down to move the piston and open or close the valve. Due to the lever action the force required to operate the valve is reduced. The valve can be operated by foot operation.

For applications like spray painting when the hands of the operator is holding the spray gun and the pneumatic valve can be foot operated with the lever moving the piston in or out as shown in the Figure 19.2 (A) and (B).

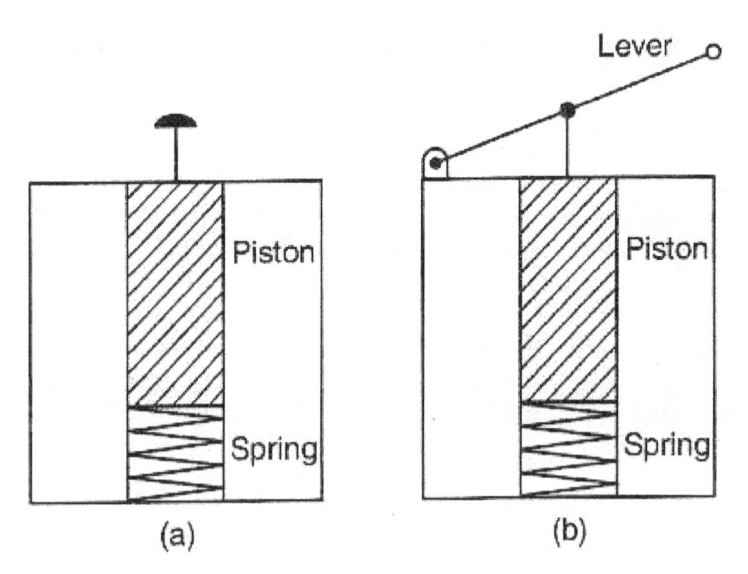

Figure 19.2(A) and (B) Foot operated pneumatic valve

19.3 Solenoid operated valve

In the solenoid operated valve shown in the Figure 19.3, operating force is developed electromagnetically. The lower part of the valve is similar to manually operated valve. The piston rests in a particular position by a spring. To operate the valve on the upper side the part of the piston should be of magnetic material. It is surrounded by a solenoid coil. When the coil is energised by AC or DC supply, the magnetic force pulls in the plunger. The piston rod moves the piston overcoming the spring force. The spring is in compression. In order to change the valve position, the supply to the solenoid coil is switched off. Hence the magnetic force developed is zero. The spring under compression releases, moves the piston up and changes the valve position ON to OFF or vice-versa.

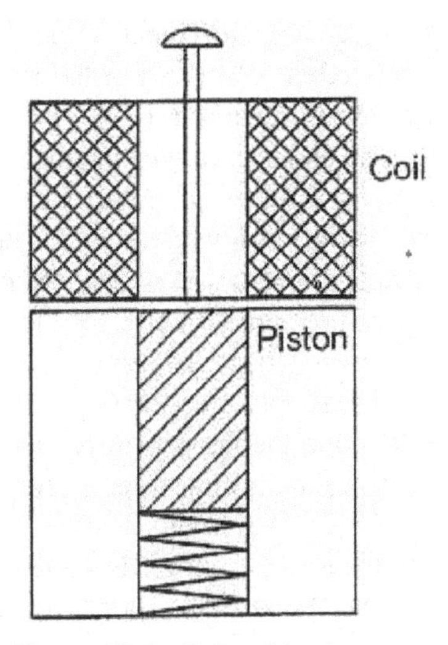

Figure 19.3 Solenoid valve

Instead of spring return both the actions can be achieved by solenoids. In such cases there are two solenoids around the extension of the piston rod. When one coil is energised, the piston rod moves in one direction – the valve opens. In order to close the valve, another coil is energised exerting the force in other direction; the piston rod operates the valve and makes it closed. It will remain closed even if the coil is de energised. There are number of combinations for valve operation. They are double air pressure operated pilot solenoid actuated and air pressure operated, bleed operated, lever actuated leed valves, double bleed operated, etc.

19.3.1 Pneumatic flapper valves

The construction and working of pneumatic flapper valve is similar to that of the hydraulic flapper valves as shown in the Figure 19.4.

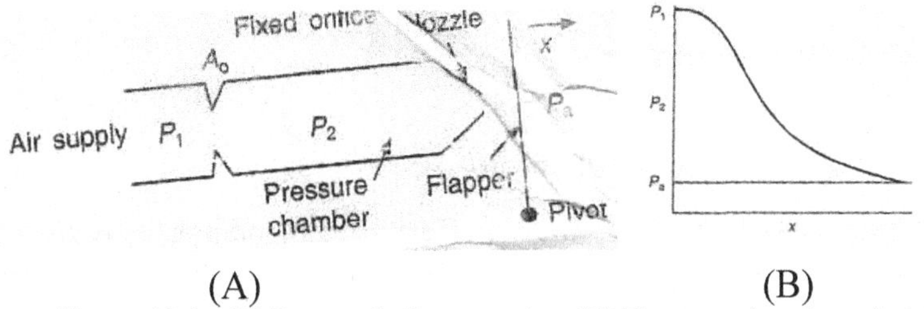

Figure 19.4 (A) Pneumatic flapper valve. (B) Flapper valve characteristics

In this type, there is a pressure chamber in which air pressure is to be maintained. The compressed air comes from the air supply system through a fixed orifice with the cross-sectional area (A_0). The supply pressure is (P_1), the chamber pressure (P_2). The outlet from the chamber is through a nozzle orifice, facing a flapper. Flapper is a flat rectangular plate closing the nozzle orifice opening. The flapper is pivoted at lower end and is kept in vertical position. If the flapper displacement (x) is zero, the nozzle orifice is blocked. There is no outflow. The incoming air pressure (P_1) charges the pressure chamber through the fixed orifice with the area (Ao) to the same pressure (P_1). Hence the chamber pressure (P_2) is equal to supply pressure (P_1). When the displacement (x) is applied in horizontal direction the flapper moves away from the nozzle. From pressure chamber air escapes to the atmosphere where the pressure is (Pa). This creates continuous flow of air. Due to this air flow and continuous obstruction of the fixed orifice with the area (A_0), there is some pressure drop across it and the chamber pressure (P_2) is less than the supply pressure.

When (x) is increased further, after certain displacement the pressure drop is maximum, the chamber pressure is same as atmospheric pressure outside. The flapper valve characteristic is shown in the Figure 19.4 (A). For zero value of (x), the chamber pressure is maximum. It is equal to (P_1), the supply pressure and after certain value of (x), which is generally in the order of millimetre, the chamber pressure is inversely proportional to the displacement. For control application, some linear range of operation is considered where the characteristic is a straight line with a negative slope.

19.4 Types of compressors

Pneumatic system works with air. Compressors are used to compress pneumatic power when the air is compressed at high pressure. The compressors are classified as

- Centrifugal compressors
- Axial flow compressors or positive displacement compressors
- Diaphragm compressors and
- Multi-stage compressors.

19.4.1 Centrifugal compressor

The construction and working principle of centrifugal compressor is shown in the Figure 19.5 (A) and (B) it consists of a circular eye at the centre. In the side view (B), it is connected to the inlet in which the air enters. It has a rotating element called an Impeller. The impeller carries number of curved blades.

The impeller shaft is mechanically coupled to a drive shaft which is given power by an electric motor.

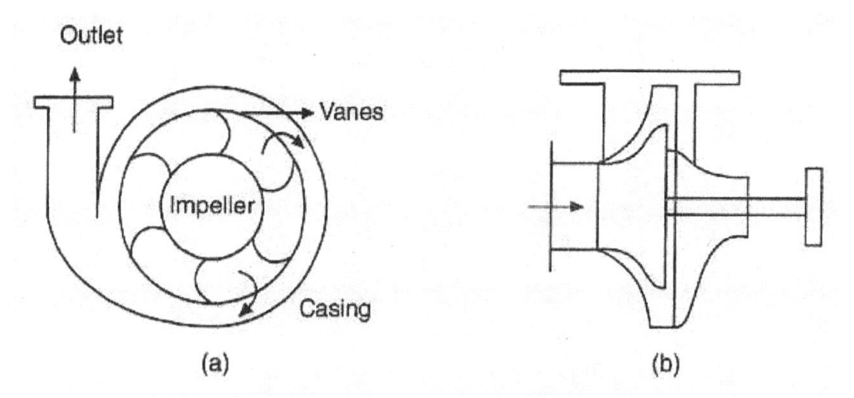

Figure 19.5(A) and (B) Centrifugal compressor

The casing has a circular cross-section. The impeller shaft is eccentric with respect to the casing. Between the periphery and casing the radial distance is increasing from minimum to maximum. The cavity between the impeller and the casing is called volute. When the electric motor drives the impeller shaft the impeller rotates and creates flow of air. The air coming from the eye is thrown outward in the volute due to the kinetic energy. The air is fed out by a diffuser. The kinetic energy of the air is converted in to static pressure as it moves through the volute and diffuser. The pressure developed by this type of compressor is low to the order of 3.5 kg/cm^2.

19.4.2 Axial flow compressor

The construction of an axial flow type compressor is shown in the Figure 19.6.The outer shape of the compressor is cylindrical. The rotor shaft carries a rotating element called runner. The runner is a cylinder with variable radius. The radius is minimum at the left-hand side and gradually increasing to the right-hand side. Hence, the outer surface is tapered. And the unit is called a tapered runner. It carries blades called runner blades. The fixed casing is a hollow cylinder with uniform diameter. It carries fixed guide vanes projecting from inner surface of the cylinder. On the inner side and outlet side the vanes are called inlet guide vanes and exit guide vanes. The inlet opening on left-hand side is larger than the eye of the centrifugal compressor.

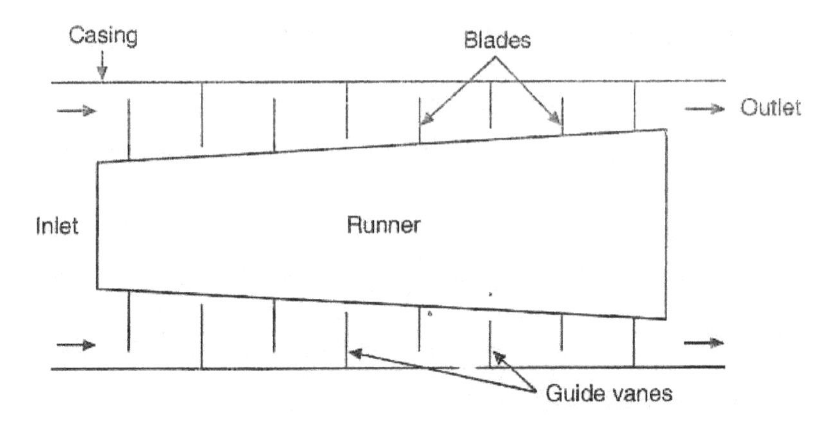

Figure 19.6 Axial flow compressor

The air enters through the inlet guide vanes and when the rotor rotates the air is moved towards to the right. Due to the tapered section of the runner, the available space decreases as the air moves to the right. The air is compressed and the runner blades impart kinetic energy to the fluid. The fixed guide vanes act as diffusers and convert the kinetic energy in to static pressure. The radial height of the blades decreases towards right. The discharge pressure of this type of compressor is of the order of 7 kg/cm^2.

19.4.3 Diaphragm compressor

The construction and working principle of diaphragm compressor is shown in the Figure 19.7 (A) and (B) there is a hollow cylindrical chamber in which there is a flexible diaphragm separating the two sections of the cylinder. The centre of the diaphragm is connected to a diaphragm rod which is connected to the crankshaft of the driving mechanism.

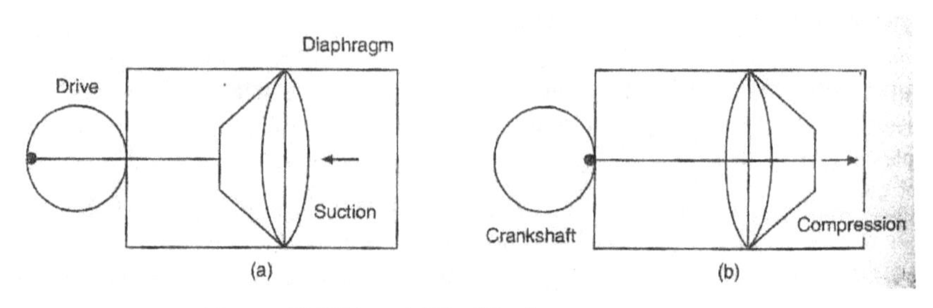

Figure 19.7(A) and (B) Diaphragm compressor

When the crank shaft and diaphragm rod move towards left, the centre of the diaphragm also moves to the left. Due to the curved convex shape of the flexible diaphragm the volume of the left-hand side of the cylinder increases. This creates suction. The atmospheric air enters the chamber. At the end of the stroke when the crank shaft and diaphragm rod give opposite motion to the centre of the diaphragm, the diaphragm deforms in the opposite direction. The volume on the right-hand side of cylinder decreases due to concave shape of the diaphragm and compresses the air and hence the pressure increases. The inlet closes and outlet opens. The capacity of the diaphragm type compressor is limited by convexity and concavity of the diaphragm. It is much smaller than the reciprocating piston compressor.

19.4.4 Multi stage compressor

The construction of a two-stage compressor is shown in the Figure 19.8. For each of the two stages of compression, the compressor has two separate cylinders indicated as cylinder (1) and cylinder (2) in the diagram. The first stage cylinder is larger than the second stage cylinder. Both the cylinders have individual pistons for each stage of compression.

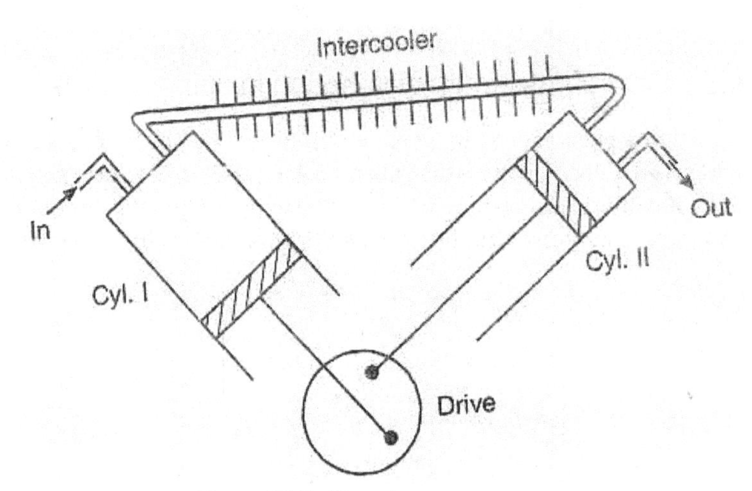

Figure 19.8 Two stage compressor

The two pistons and piston rods are connected to a common crank shaft which couples them to a common motor drive which drives them at the same speed. The larger cylinder is connected to the inlet side which takes the atmospheric air during the suction stage. As the piston moves outward the inlet valve opens and air flows in. After compression, the larger cylinder gives compressed air to the intercooler which connects the output side of the first cylinder to the input stage of the second cylinder. The mechanical arrangement is such that the pistons in the two cylinders work in the opposite direction. When larger cylinder carries out compression and gives air out by opening outlet valve the smaller cylinder starts the suction by opening inlet valve. The inlet air to a larger cylinder is at atmospheric pressure and temperature. At the end of the compression its temperature is increased. The compressed air passes through the intercooler.

The temperature goes back to the atmospheric temperature. This air at higher pressure and atmospheric temperature enters the second cylinder where it is compressed further. After the second stage of compression from the outlet valve the compressed air is available for supply to the pneumatic system. Between the two stages of compression, the air is cooled by the intercooler and hence there is some reduction in pressure. This pressure loss corresponds to the pressure rise due to temperature rise. When the temperature of the air increases, it tries to expand. The intercooler enhances the cooling process which consists of copper tubes with fins to dissipate the heat and maintain temperature.

On the other hand, if simple copper tubes and fins are not sufficient for the cooling purpose, chilled water cooling system is used. It draws the cold water from a water chiller circulates the cold water in the tubing surrounding the airline between the two cylinders. Multi-stage compressors are more efficient.

19.4.5 Compressed air distribution in spinning mills

It is a common phenomenon in a textile mill to find that every care has been taken to select a good compressor to handle adequate volume of air, pressure problems are always encountered. This is due to poor piping distribution system. A typical compressed air distribution system is shown in the Figure 19.9.

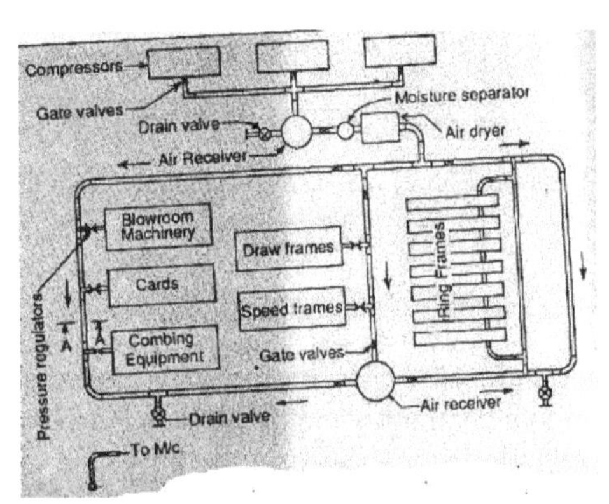

Figure 19.9 Centralised compressed air distribution for textile spinning machinery

Compressor capacity calculations.

Data: Lakshmi Rieter Ring frame model G 5/1.

Volume of air = 1 Nm^3/hr.
Pressure = 6 bar.

Volume given in frr air delivery at 1 ATA, 0 deg C, 0% RH. Hence this volume needs to be corrected to local conditions.

Local data
Altitude of the place = 2,000 m.
Maximum temperature for calculations = 45°C.
V_s = Given volume is 1.00 Nm^3/hr.
T_s = Given temperature 0° = 492 Rankine
T_t = local temperature = 45°C = 573° Rankine

(a) Temperature correction as per perfect gas law:

$V = Vs \times Ts/Tt$

$= 1 \times 573/492$

= 1.164 corm /hr.

(b) Partial vapour pressure

Now
Pa = atmospheric pressure 0.860 kg/cm^3
Ps = Vapour pressure = 0.098 kg/cm^2 at 45°C and 40% RH.
(values from steam tables)

(c) Corrected volume

$= V \times P_b / Pa - (RH\%/100)\, P_2$

Thus, it is seen that under standard conditions 1.00 Nm^3 /hr becomes 1.21 m^3 /hr under local conditions.

The actual compressed air requirement is calculated as

Volume = 1.21 m^3 /hr. Assuming that we have ten ring frames in the spinning department, then
Final volume = 10 × 1.21 = 12 m^3 /hr.
If we give allowance of 65% for usage factor and 15% for air leakage and purge factor then,
Total volume = 12 × 135 / 100 × 100/65
= 25.13 m^3/hour.

= 25.33 × 35.33 /60

= 14.79 CFM or approximately 15 CFM.
= 15 /35 × 1000
= 429 litres/minute.

Useful technical data:

(a) Temperature:

For compressed air calculations, temperature is always indicated in degree Rankine (°R)

1°R = °F + 459.7

1°F = 32 + 1.8°C.

For example, 32 + 1.8 (45) = 113°F.

= 113 + 459.7 = 572.7°R.

19.4.6 Compressor efficiency

Compressors are otherwise called pneumatic pumps. Turbo types and positive displacement piston type compressors are common to control applications. To get advantages of adiabatic and isothermal expansions, multistage compressors are used. Single stage compression which gives compression ratio of 8:10 is not advisable because the presence of lubricating oil during compression will lead to explosion. Hence, multi stage compression with compression ratio of 2:3 per stage avoids this problem and also improves the efficiency.

Adiabatic compression efficiency
= Indicated horse power/Theoretical horse power required for adiabatic compression of air

Isothermal compression efficiency
= Theoretical horse power required for isothermal compression of air/ Indicated horse power

Mechanical efficiency
= Indicated horse power/Brake horse power at compressor shaft
Overall compressor efficiency
= compression efficiency × mechanical efficiency × prime mover efficiency

The overall efficiency of compressor may be between 70% and 85%. One problem is with the isothermal expansion is that even though it is more efficient it is practically impossible to realise it with type of nozzles and nature of loads. Moreover, the time required for isothermal expansion is too high. All these factors limit the efficiency of the pneumatic system from 30% to 50%.

19.4.7 Compressed air in textile industries

Compressed air plays major role in textile machines. It is used in brakes, calendar roller weighting in lap fed scutcher blow room lines, weighting of top rollers in draw frames, speed frames, cone belt shifting in speed frames, auto doffing in ring frames, splicing in auto winding machine, sizing, warping, weft carriers in shuttleless air jet weaving, centralised waste collection plant in air engineering, etc.

Since compressed air is costly, it is essential to calculate the amount of compressed air needed for each application in the textile industry and also the correct size of the compressor required. Moreover, the compressed air should be free from the moisture and dust. The compressed air so generated must be properly distributed throughout the department in a given layout.

Purposes of compression

The principle of any compressor is to deliver air at a pressure higher than that originally existing. Compression is for variety of purposes:

- To transmit power.
- To provide air for combustion.
- To produce conditions more conducive to chemical reaction, and
- To produce and maintain reduced pressure levels for many purposes.

Compressed air in textile machines is mostly used in the spinning preparatory and spinning machinery and also in air jet weaving machines, knitting machinery, etc.

19.4.8 Applications of compressed air in textile industry

Air requirement in the textile industry can be segregated under Low/Medium/Large volume consumers. Typically power looms, process house and ginning mills required low air between 30–500 CFM. Denim spinning, medium PFY industries require medium air between 500 and 1,500 CFM. Large polyester yarn integrated plants, texturing units and integrated cotton textiles may require anywhere between 2,000 and 40,000 CFM. Air quality is one of the major concerns and high quality air requirement is essential in integrated textile units. However, the preparatory segments such as ginning mills, spinning mills and process units can use lubricated compressors with well-designed air filtration system.

The possible saving opportunities in the textile industries are

1. Migration to new technology air compressors in old installations / plants, for example, considering 2-stage rotary or 3-stage centrifugal compressors instead of existing old reciprocating or single-stage screw compressors.
2. Proper evaluation on the performance of air treatment requirement such as dryers, filters and FR units installed near points-of-use.
3. Properly evaluating pressure requirement of different processes and if required segregation of low and high-pressure consumers.
4. System pressure optimisation for unregulated consumers.
5. Evaluating usage of waste heat from compressors (use of waste heat recovery systems for use in boiler make-up or hot water application).
6. Re-look at distribution and storage to minimise the pressure drops.
7. Optimise air consumption and pressure for cleaning air application (this is one of the largest wastage in most of the textile industries).
8. Have effective controls on the operation of the compressors for better optimisation and sequencing.
9. Effective leakage management program and designated team for the same.

Use of compressed air in textile industry

In textile manufacturing units, generally following compressed air powered pneumatic systems are used:

- Spinning machine
- Loom Jet Weaving
- Stacking device
- Printing machines
- Thread detector
- Sewing machine.

Applications

Compressed air is a vital part in textile industry in all phases of spinning, weaving, loom and knitting. The quality of compressed air required to be on various specifications based on the industry.

In spinning

- Pneumatic loading in speed frames and ring frames.

- In Sliver lap/ Ribbon Lap or Unilap.
- In Auto coners.

In Airjet weaving

- Compressed air blown through fine nozzles to transport the weft.

19.4.9 Principles of compressor operation

Every compressor is made up of one or more basic elements. A single element or a group of elements in parallel comprises a single stage compressor. However in many cases due to the excessive compression ratio, temperature may rise or design problems would be encountered. In such cases, elements or group of elements are combined in series to form a multi stage unit. Here there would be two or more steps of compression. Air is frequently cooled between stages to reduce the temperature and volume entering the following stages. Thus, each stage is an individual basic compressor itself.

Dimensions

Before discussing further, we should know the fundamental difference between the mass and weight. Mass is a quantity of matter whereas weight is the force exerted on a given mass by the attraction of gravity and will vary depending on its distance from the centre of the earth.

Terminologies in compressor

(a) Pressure

Pressure is defined as the force per unit area exposed to the pressure. Since weight is really the force of gravity on a mass of material, the weight necessary to balance the pressure force is used as a measure.

For example

Pounds per sq.inch = p.s.i.
Grams / sq.cm = g/cm^2
Kg/sq.cm = kg/cm^2

Measurement of pressure is normally done by a gauge that registers the difference between the pressure in the vessel and the present atmospheric pressure. This is called as *Gauge pressure. (psig)*. This does not indicate the total air pressure. Hence, in order to get the true total air pressure, it is necessary to add the atmospheric or barometric pressure to the gauge pressure. This is called as *Absolute pressure.*

Gauge pressure + Barometric pressure = Absolute pressure

As we know that earth surface atmosphere has a weight. It is the weight of the column of air existing above the earth's surface at 45° latitude and sea level. It is known as International atmospheric and usually expressed as ATA (atmospheric absolute). This is defined as being equal to 14.696 psi A or 1.033 kg/cm^2 at mean sea level.

Some manufacturers indicate pressure in terms of "bar".

1 bar = 10^5 N/m^2 = 10^5 Pa, where N = Newtons and Pa = Pascals.

Converting Newtons in to kg/cm^2, it becomes 1 bar = 1.02 kg/cm^2 or 1.00 kg/cm^2.

If this is a gauge pressure, it is necessary to add atmospheric pressure in order to get the absolute atmospheric pressure. (ATA). The ATA value will vary according to the altitude of the place.

(b) Volume

The second important parameter in the compressor is the quantity of air handled by the compressor and it is indicated in terms of volume. For example

Cu.ft/min = Cubic foot per minute.
m^3/hr = cubic meter / hour.

Volume varies depending on the conditions of pressure, temperature and moisture content. Normally manufacturers indicate the volume at their respective standards. USA standards is 14.696 psiA at 60°F and 0% RH. European standard is 1ATA at 0°C and 0% RH.

(c) Free air delivery (FAD)

FAD is the quantity of air delivered by the compressor when referred in terms of volume that it would occupy at atmospheric pressure, temperature and relative humidity. This is the most commonly used terminology used in the compressor industry since there is always a variation in volume ratio to pressure, temperature and relative humidity depending upon the local atmospheric conditions. It is always better to convert this volume values in to FAD for necessary calculations, the value given in terms of Nm^3/hr can be exactly calculated. Another important parameter to be considered is the state of air. Is it wet or dry? This refers to the vapour content. Water vapour in the air occupies space and the compressor must be able to handle. Water vapour content must be known or otherwise it makes a lot of difference while calculating the capacity of the compressor. Water vapour content, in general is indicated by the RH% which plays a vital role. In general, while calculating quantity of air handled by the compressor, it is necessary to allow for:

- Vapour pressure correction
- Temperature correction
- Pipeline losses
- Air drier or purge losses

Compressor efficiency or usage factor. While we can assume usage factor at 65% it also differs whether it is air cooled or water cooled and also depends on local conditions. It means calculations made at Northern part of India vary with the Southern part of India for obvious reasons.

(d) Relative humidity

Relative humidity is the term frequently used to represent the quantity of moisture or water vapour present in atmosphere.

RH% = Actual partial vapour pressure / Saturated vapour pressure at existing moisture temperature × 100.

(e) Vapour pressure

There are many components in a mixture. Each component exerts its own pressure and it is called as partial vapour pressure. If partial vapour pressure of each component is added it gives the total vapour pressure. It is called as Dalton's Law. Atmospheric air is a mixture of dry air and water vapour. Hence, it is necessary to consider water vapour's partial pressure for all our calculations which can be referred in the steam tables.

19.5 Hydraulic valves

In hydraulic control system, the pressure and flow can be controlled by different types of valves. In hydraulic systems, spool types of valves are more common. It may be of two way, three way or four way spool type valves. The applications of spool valves are relief valves, pressure regulating valves, pressure reducer valves, actuators, pilot valves, etc. Flapper valves are a slightly different category of hydraulic valves.

Spool type valve

Spool is the operating element of the valve. In this type of valve opening or closing of the valve is controlled by a spool. Spool is a short solid cylinder which rests on hydraulic line or near the opening. When the spool moves in the line near the opening it creates the flow opening. The cross-section of the flow path is given by the surface area of the cylindrical or ring shaped gap. In general, it will be πdl where d is the diameter of the pipe, or spool and l is the length of the opening.

Some of the spool type valves are discussed in the following sections:

a) Two-way spool valve

Figure 19.10 (A) and (B) show two types of two-way spool valves. In Figure 19.10 (A), there is a large cylinder in horizontal direction with cross-section (A) and it carries a piston with a piston rod. The piston area on the left-hand side is same as cylinder cross-section. On the right-hand side the effective piston area is the piston area minus cross-section of piston rod. It makes effective piston area on right-hand side as A/2. The cylinder has two openings – one near the left-hand side and another near right-hand side. The piston is able to reciprocate along a horizontal straight line and gives the required output.

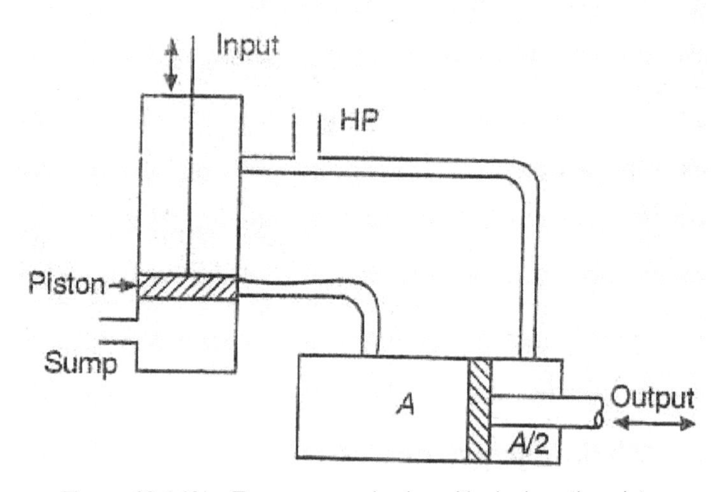

Figure 19.4 (A) Two-way spool valve with single acting piston

There is another small hollow cylinder shown in the vertical direction. It consists of a piston through which the input can be given through the piston rod. This small cylinder has an opening at its lower end which is connected to the sump, another opening slightly upper connecting to left end side of the larger cylinder and near the top it has another opening. The opening on upper side of small vertical cylinder and opening at right-hand end are connected together by a pipe line. From this connecting line one connection goes to the high pressure oil connection.

Initially the piston in smaller cylinder is closing the opening in smaller cylinder for the line leading to larger cylinder. The piston in larger cylinder is in intermediate position between its two openings. The input to the smaller cylinder is a linear displacement of the piston rod. Consider that the piston in smaller cylinder moves down through the opening created, the high pressure oil enters through the smaller cylinder and then to the left side of the larger cylinder. Hence it creates a pressure and force equal to the product of pressure and area (A). At the same time high pressure oil is acting on the right-hand side of the larger cylinder but force is equal to the product of same pressure and area (A/2). Due to the creation of larger force on the left-hand side of the piston, the piston in larger cylinder moves to the right.

If the piston in smaller cylinder is moved up by the input the left-hand side of larger cylinder will not get pressurised oil and hence the force will be zero. But on the right-hand side the high pressure oil will create force that will push the piston in larger cylinder towards left. The oil from that side will go to sump through opening. Thus, the input at the piston rod of smaller cylinder may move the piston down or up. This results in the motion of the piston in the larger cylinder towards right or left. So, the input to piston of smaller cylinder controls the spool valve to control the movement of the piston in the larger cylinder. In this case, the spool valve controls the flow from the high pressure supply to larger cylinder and larger cylinder to sump. Hence it is called two-way spool valve.

In Figure 19.10 (B), the small vertical cylinder has an input which gives vertical movement to the piston rod of smaller vertical cylinder.

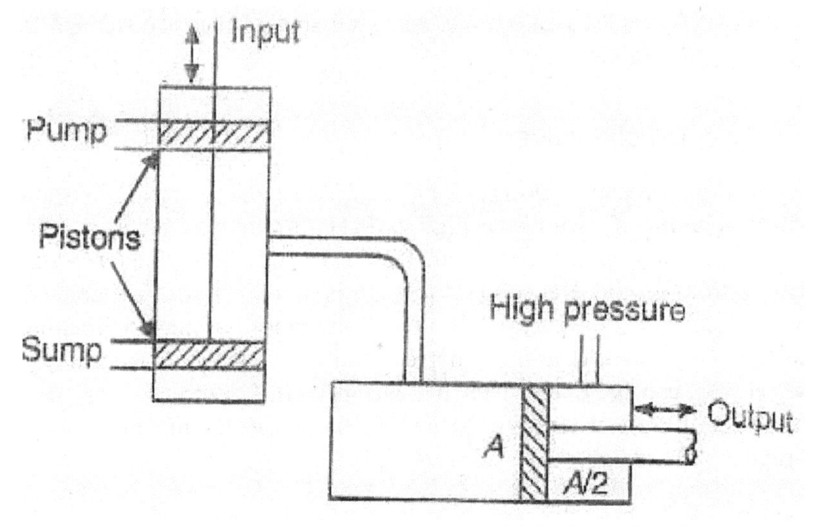

Figure 19.10 (B) Two-way spool valve with double acting piston

It has two small pistons – the upper one initially resting against opening to pump and the lower one initially resting against opening to sump. At the centre the small vertical cylinder has opening with connection to left-hand end of the larger cylinder. Right-hand end of the larger cylinder has connection to high pressure oil. The piston with areas (A) and ($A/2$) on left and right-hand sides is initially at the central position with piston rod connected to the output. When the input is a vertical movement in upward direction the piston in smaller vertical cylinder brings high pressure from pump. Through the central opening it goes to the left-hand side of the larger cylinder and exerts force equal to the product of pressure and area (A), towards right. On the right-hand side also high pressure oil acts and develops force equal to the product of pressure and area ($A/2$) towards left. As a result of this, the piston in larger cylinder moves from left-hand side to the right-hand side due to the larger force.

When the input is a vertical movement in down ward direction the pump is disconnected from the cylinder. Due to high pressure connection on right-hand side of the larger cylinder a force equal to the product of pressure and area ($A/2$) acts on the larger cylinder there is no opposing force. So, the larger piston moves towards left. At this time the oil accumulated earlier on left-hand side of the larger cylinder will be pumped out to the sump.

Thus, linear input to piston of smaller cylinder results in the movement of piston in larger cylinder towards right or left depending on vertical up or down input. The small cylinder represents a spool valve. The valve allows high pressure oil from pump to larger cylinder called actuator in one case and from actuator to sump in the other case so it is called two-way spool valve. This valve is used to control the larger cylinder called actuator and also works like an amplifier.

19.5.1 Nozzle valve

The construction and working principle of nozzle valve used in hydraulic control system is shown in the Figure 19.11. In this type there is a pipeline connection to high pressure oil. The lower end of the pipe has a small opening from nozzle. Hence the flow path is narrowed down. It comes as a jet of oil. Against the jet there is a receiving block having two openings. The receiving block is initially narrow and gradually increasing in size. The jet pipe can be given linear displacement in (X) direction as shown in the Figure 19.11. There is a spring which opposes the movement of the jet pipe.

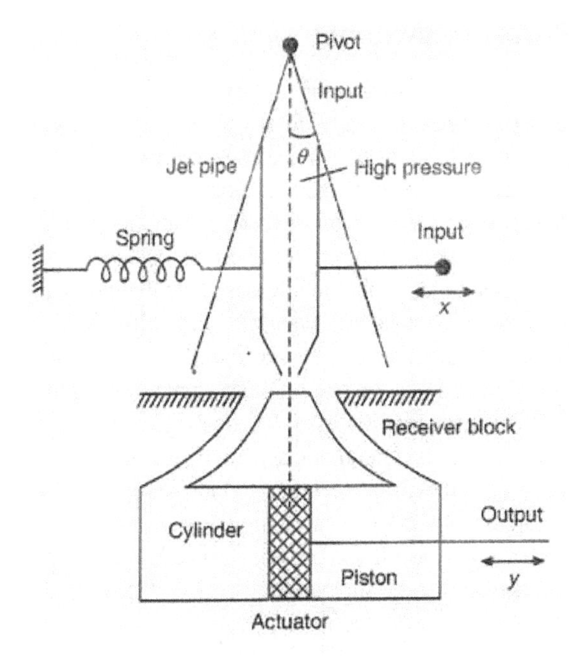

Figure 19.11 Nozzle valve

At the start when the input is zero the spring keeps the end of the jet pipe against the centre of the receiving block. The high pressure oil is equally distributed in the two passages of receiving block. The actuator piston is stationary and the output is zero.

The jet pipe is pivoted at the top in central position. If the input (X) is towards right the jet pipe rotates by an angle (θ) in anti-clockwise direction. The jet moves away from the central position towards right. There is more oil flow from right-hand passage to the right-hand side of the actuator piston. The flow on the left-hand side is less. Due to larger flow from right-hand side the actuator piston moves towards left in horizontal direction. If linear input (X) to the jet pipe moves it towards left, there is more flow of high pressure oil from left-hand side pushing the cylinder piston towards right. The manufacturing of spool type valves requires high tolerances where it is not a problem in nozzle valves. Moreover, the hydraulic resistance and the required input force are also very small. However, it has to overcome the pivot friction and inertia of the jet pipe. The main disadvantage of nozzle valve is high leakage but the frequency response of this device is very good.

19.6 Pressure switches

Some compressors like reciprocating piston give pulsating type of air supply. The pneumatic system may require steady air supply just like DC current. Between compressor and pneumatic supply it is necessary to provide sufficient volumetric capacity just like capacitor in electronic systems. The air coming out from the compressor is first stored in a reservoir or receiver tank and the tank supplies the required compressed air when required. It means when the air is not required, the tank stores the extra compressed air and releases when there is a demand. It works like flywheel in mechanical system or accumulators in hydraulic systems. These receivers's tanks are designed for certain volume of air at certain pressure on the system requirement. In general, the pneumatic system works in the range of 3–15 psi. To account for the pipe line losses and air leakages, the supply air system is designed for 20 psi. When the pneumatic system does not require air or consume less air the compressed air will be accumulated in the receiver tank. During this time, the pressure in the tank goes on increasing. For safety purpose, this pressure should be limited or otherwise if the pressure exceeds the design limit the tank may burst. It is dangerous for the surroundings and also for the operators working around it. Hence all compressors require a pressure operated switch.

Principle

Pressure switch senses the tank pressure continuously and when it crosses the set limit it operates a switch which will cut off the supply to the electrical motor operating the compressor. In the mean time, when the pressure in the tank goes below the set limit due to the usage or leakage, the pressure switch operates the electrical motor to switch on. Like all ON–OFF control systems some dead zone is provided to avoid unnecessary and frequent switching ON–OFF of the electric motor. This is because usually starting current of motors is much more than the full load current of the motor. The construction and working of pressure switch is shown in the Fig.19.5. The pressure switch is mounted on the receiver tank. It is fitted by threaded connection. It consists of a piston which senses the pressure.

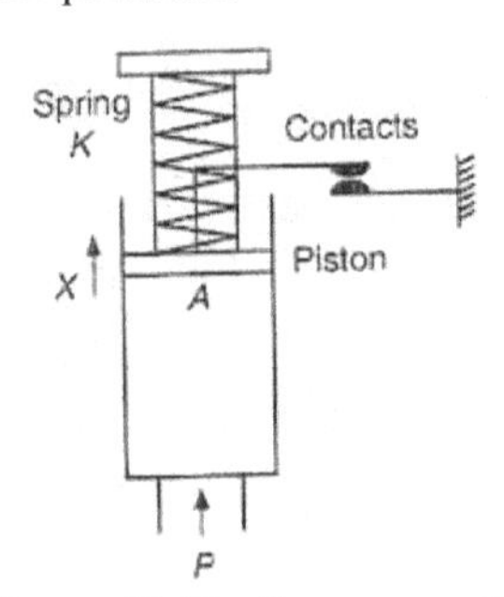

Figure 19.12 Pressure switch

When pressure in the tank increases the piston moves up. On the top side there is a spring. Upper end of the spring is fixed and the lower end is free to move. The upward moving piston has to overcome the spring force which is compressed by the piston force. The pressure (P) acts on the piston area (A), exerts force ($P \times A$) under which spring compression is (X) and (K) is the spring rate.

$$P \times A = K \times X.$$

At the top of the spring there is an adjustment bolt. The spring tension and free length of the spring can be adjusted by tightening the bolt. The purpose is for setting the pressure limit. The free end of the spring is mechanically connected to the electrical contractor. When the pressure is less the spring is free to move. The electrical switch is in closed position. When the pressure increases and the piston reaches certain position it compresses spring and the switch contacts are opened up stopping the compressor motor. When pressure goes down by certain limit the switch again comes to ON position and starting the compressor motor to build up pressure.

19.7 Automation in material handling

Material handling in textiles is the main target for automation. Handling of materials of high component with low value added labour is often a srenous and tedious process which will lead to potential health constraints. Automated material handling eliminates the labour force and indirect benefits like lower health insurance costs and continuous operation.

Some of the automated systems in the textile manufacturing chain are

- Automatic doffing in carding, draw frame, roving, ring spinning and in rotor spinning.
- Automatic splicing and auto doffing in the auto winders.
- Automatic sliver transport system guided by an automatically guided vehicle (AGV) can move sliver cans from carding department to draw frame department. Similar arrangement to move draw frame cans to roving machines or to rotor spinning machines.
- Automatic box loader in rotor spinning, automatic pallet packing and automatic loader for pin trucks.
- Automatic handling of large lap rolls in combers.
- Automatic linking of ring spinning to auto coners. (Link coners)
- Linking of roving frames with automatically doffed roving bobbins via an overhead transport system which feeds them to ring spinning system.

Automation eliminates tedious manual labour, clearly arranged material flow, inventory reduction, shortening of lot run times and production costs.

References

1. A.K. Tayal, *Instrumentation, Mechanical Measuremnets and Control*. Second Edition, Galgotia Publications Pvt.ltd, New Delhi.
2. M.K. Talukdar, P.K. Shriramulu, D.B. Ajgoankar, *Weaving Machines, Management, Mechanisms*. Mahajan Publishers Pvt. Ltd.
3. T. Matsuo, Innovations in Textile machine and Instrument. *IJFTR*, Vol. 33, September 2008.
4. K. Monmoto, *Journal of Textile Machinery Soc*. 61 (2008).
5. Vasu Sarpeshkar, Compressed Air in Textile Industry. *The Indian Textile Journal*, October 1992, 111.
6. J.R. Canada, Evaluation on DS - Computer Integrated Manufacturing Systems, Automation for the Fiber, Textile and Apparel Industries Conference, NCSU, January 13–14, 1987.
7. P.L. Hardy, G.N. Mock, T. Clapp, H. Hamouda. Computer integrated Manufacturing in the Textile Industries. College of Textiles, North Carolina State University.
8. Richard Furter, Uster Technologies, Switzerland. Analysis of yarn faults by a sophisticated classifying system. *International Textile Monitor*. Page 32–38.
9. Yarn clearing – an economical and technical problem, Part II. International Textile Bulletin, 4/1966.
10. The properties and mill operation of electronic clearers, Uster Technologies, Switzerland.
11. Richard Furter, Uster Technologies, Switzerland. Modern sensors and their capabilities to monitor all the bobbins on winding machines. *Asian Textile Journal*. August 2007, 45–48.
12. Peter Gnagi, Modern machines monitor, *Textile Asia*. June 1997.
13. B. Wheatley, Electronics in warp knitting. 30th International Congress of the Knitting Industries, New York.
14. R. B. Kulkarni, CMS Ltd. Mumbai, Fire Control systems in modern Spinning units. *Journal of the Textile Association*. November 1995, 185–186.
15. S.D. Mahajan, Optimising performance of ring data system. *Journal of textile Association*. March–April 2004.
16. Loepfe Yarn Master, Classification of foreign fibers.
17. M.Anbarasan, On-line quality monitoring in roving frames, *Journal of textile Association*. July–Aug 1997.
18. K.V.Joseph, Some applications of computer in Textile Industry, *BTRA Scan*. Vol XVII, No.1. March 1986.

19. Desai, Control system components.

20. J.E. Booth, *Principles of Textile Testing*. Butterworths publication.

21. Preston E. Sasser, *Improving the between laboratory agreement for HVI data.* ITB Yarn Forming, 3/90, p 43–50.

22. Richard Furter, *Physical properties of spun yarns. Uster Application report.* June 2004.

23. Zewelleger Uster, *Foreign Matter Detection and Elimination*. Seminar on 1996.

24. Gabriele Peters, *Taking out the trash. Textile Horizons*. March 2000.

25. Uster Technologies, Uster OM sensor.

26. Christoph Farber, Armin Leder, Trutzschler GmbH & Co, KG, Monchenglabadh/ Germany. "*Modular solution to combat foreign matter problems in ginning and spinning". Melliand International 3/2010.*

27. www.rieter.com, Roving frames.

28. www.ssm.ch. Yarn Singeing.

29. William D. Whipple, Advances in warp preparation machinery. *Warp sizing symposium*. NC State University.

30. Ali A.A., Jeddi et al. Comparative study of electronically and mechanically controlled wap let-off systems. *IJFTR*. Vol 24, December 1999, pp 258–263.

31. BTRA Scan, Instrument transducers: Principles and their applications in textiles.

32. Bhaskar Dutta, Winding: Yesterday and Today. *The Indian Textile Journal*. April 1993, 90.

33. Sanjay Gupta, Recent advances in wt chemical processing machinery. IJFTR. Vol 21, March 1996, pp 57–63.

34. M.R. Shamey, J.H. Nobbs, A review of automation and computer control in dyeing machinery. *Advances in colour Science & Technology*. No.2. July 1998.

35. A.K. Samanta, On line colour control in textile dyeing process. *The Indian Textile Journal*. October 1992, 75.

36. M. Tooley, Spectrum international, UK. Technical aspects of colour measurement. *Textile Month*. September 1988.

37. Zellweger Uster. The properties and mill operation of electronic clearers.

38. Richard Furter, Experience with foreign matter removal systems for cotton. *Uster Application* Report June 2006.

39. S. Donmez Kretzscumar, A. Ellison. Monitoring of the ginning process. Uster Application Report April 2010.

40. P.L. Hardy, G.N. Mock, T. Clapp, H. Hamouda. Computer integrated Manufacturing in the Textile Industries. NC State University.

41. Vivek Plawat. Autolevellers in Draw frames. *ATIRA publications*. 1992.

42. Rieter Instruction manual. Card C4.

43. Rieter Instruction manual. RSB 851 Draw Frame.
44. Rieter Instruction manual. Comber E 62.
45. Rieter Instruction manual. Ring Frame G 5.
46. Rieter Instruction manual. Rotor Spinning R 1.
47. J.E. Booth, Textile Mathematics, Vol II, Butter worths publication.
48. W. Klein, A practical guide to drawing and combing process.
49. Vasu Sarpeshkar. A guide for selection of modern high speed draw frames. *Indian Textile Journal*. June 1993, 62.
50. Neeraj Nijawan, Measurement of Air pressure in the ducts. *Spinning Textiles*. July–Aug 2008.

Index

T

V

W

Y

Z

Printed in the United States
by Baker & Taylor Publisher Services